Hans Jürgen Prömel
Angelika Steger

The Steiner Tree Problem

W0042848

Advanced Lectures
in Mathematics

vieweg

Hans Jürgen Prömel
Angelika Steger

The Steiner Tree Problem

**A Tour through Graphs,
Algorithms, and Complexity**

vieweg

Prof. Dr. Hans Jürgen Prömel
Humboldt Universität zu Berlin
Institut für Informatik
Unter den Linden 6
10099 Berlin, Germany

proemel@informatik.hu-berlin.de

Prof. Dr. Angelika Steger
Technische Universität München
Institut für Informatik
80290 München, Germany

steger@in.tum.de

Mathematics Subject Classification
05Cxx   Graph theory
68Qxx   Theory of computing
68Rxx   Discrete mathematics in relation to computer science
68Wxx   Algorithms

Die Deutsche Bibliothek – CIP-Cataloguing-in-Publication-Data
A catalogue record for this publication is available from
Die Deutsche Bibliothek.

First edition, February 2002

Vieweg is a company in the specialist publishing group BertelsmannSpringer.
www.vieweg.de

Cover design: Ulrike Weigel, www.CorporateDesignGroup.de

Printed on acid-free paper

ISBN-13: 978-3-528-06762-5   e-ISBN-13: 978-3-322-80291-0
DOI: 10.1007/978-3-322-80291-0

Preface

"A very simple but instructive problem was treated by Jacob Steiner, the famous representative of geometry at the University of Berlin in the early nineteenth century. Three villages A,B,C are to be joined by a system of roads of minimum length." Due to this remark of Courant and Robbins (1941), a problem received its name that actually reaches two hundred years further back and should more appropriately be attributed to the French mathematician Pierre Fermat. At the end of his famous treatise "*Minima and Maxima*" he raised the question to find for three given points in the plane a fourth one in such a way that the sum of its distances to the given points is minimized – that is, to solve the problem mentioned above in its mathematical abstraction. It is known that Evangelista Torricelli had found a geometrical solution for this problem already before 1640.

During the last centuries this problem was rediscovered and generalized by many mathematicians, including Jacob Steiner. Nowadays the term "*Steiner problem*" refers to a problem where a set of given points $p_1, \ldots, p_n$ have to be connected in such a way that (*i*) any two of the given points are joined and (*ii*) the total length (measured with respect to some predefined cost function) is minimized. The importance of the Steiner problem stems from the fact that it has important applications in as diverse areas as VLSI-layout and the study of phylogenetic trees. While this might be reason enough to write a book about the Steiner problem, our motivation is a different one.

To handle the modern versions of the Steiner problem one has to combine methods, tools, and techniques from many different areas. The resulting progress has uncovered fascinating connections between and within graph theory, the study of algorithms, and complexity theory. This single problem can thus serve perfectly as a motivation and link for an introduction to these three fields. An additional challenging aspect is that these fields form a bridge between mathematics and computer science. We thus also have to combine different views, approaches and notational conventions within these two communities. Accordingly, we tried to present a coherent introduction which exhibits the interplay between these areas. Hopefully, it will stimulate the reader's interest and appetite for a broader perspective and deeper understanding of the subject. In order to at least partially meet this desire, every chapter closes with an "excursion". These sections are designed to reinforce the concepts and methods introduced for the Steiner problem by placing them in a broader context. Topics for these excursions are themes which we feel represent important and fascinating aspects and are likely to provide motivation for further studies for the reader. Each excursion is self-contained and is not presumed in later chapters.

The book is organized as follows. In the first three chapters we provide a short introduction to the three areas graph theory, algorithms, and complexity. Each introduction could easily fill a book by itself. We therefore had to concentrate ourselves on notions and results necessary for a thorough understanding of the themes in this book. The following seven chapters then cover various aspects of the Steiner problem in particular and algorithm design in general. Every chapter closes with bibliographical notes and a list of problems. Due to space restrictions our notes are far from exhaustive: we do attribute major results and proofs, but beyond that we aim at little more than presenting some selective suggestions for further readings. The problems are aimed at testing the understanding of the concepts of the chapter, including the excursion. A star indicates that the problem is more difficult. It usually requires some creativity and a more lengthy argument.

Acknowledgment

This book is based on courses taught by the authors in Bonn, Kiel, München, and Berlin. We would like to express our gratitude to our students and colleagues who contributed in many ways over several years throughout the development of this book. In particular, we would like to thank those who carefully read and re-read several chapters of it and who suggested improvements and corrections: Clemens Gröpl, Alexander Hall, Volker Heun, Stefan Hougardy, Deryk Osthus, Mark Scharbrodt, Thomas Schickinger, Anusch Taraz. Naturally, the authors are responsible for the remaining errors. We are also grateful to Eva Sandig for several of the drawings and Clemens Gröpl for preparing the index of this book. Last but not least we would like to thank our publisher and the editors of the series for their patience with the authors throughout various delays in completing the final version of the manuscript.

Berlin and München
December 2001

Hans Jürgen Prömel
Angelika Steger

Contents

1

Basics I: Graphs

The birth of graph theory dates back to Leonhard Euler (1707–1783), who studied the problem whether one could stroll around Königsberg and thereby crossing each bridge across the Pregel exactly once. Abstracting this problem into a mathematical setting leads to the notion of a *graph*. While in the 18th and 19th century graph theory played only a marginal role within mathematics, in the 20th century the importance of graph theory dramatically increased, last but not least due to the interplay with computer science. In this chapter we can therefore give only a first glance at this important part of modern mathematics.

1.1 Introduction to graph theory

A *graph* G is a pair $(V(G),E(G))$, usually abbreviated as (V,E), such that E is a subset of the set of unordered pairs of V, i.e., $E \subseteq \binom{V}{2}$. The set V is called the set of *vertices* and E the set of *edges*. In this book we will consider only finite graphs. That is, the sets V and E are always finite. We usually denote the cardinality of the vertex set by $|V| = n$, and the cardinality of the edge set by $|E| = m$. An example of a graph on 10 vertices is shown in Figure 1.1. It is named after J. Petersen, one of the first graph theorists.

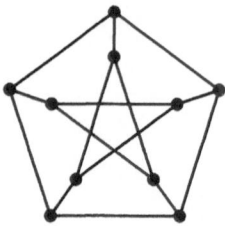

Figure 1.1 The Petersen graph

Two vertices u and v are called *adjacent* in G if $\{u,v\} \in E$. The set of vertices which are adjacent to a vertex v is called the *neighborhood* of v, denoted by $\Gamma(v) = \{u \in V \mid \{u,v\} \in E\}$. The neighborhood of a set $X \subseteq V$ is the set of all vertices which are adjacent to v, it is denoted by $\Gamma(X) = \{v \in V \setminus X \mid \exists x \in X \text{ s.t. } \{x,v\} \in E\}$. The number of neighbors of a vertex v is called the *degree* of v, denoted by $d(v) = |\Gamma(v)|$. A graph is called k-*regular* if all vertices have degree exactly k. The following immediate but nevertheless extremely useful formula relates the number of edges to the sum of the degrees of the vertices:

$$2|E| = \sum_{v \in V} d(v). \tag{1.1}$$

Observe that $\Gamma(v)$ and $d(v)$ are defined with respect to a fixed graph. This graph will usually be clear from the context. In case of ambiguity we will write $\Gamma_G(v)$, $d_G(v)$, etc. to indicate that we mean the neighborhood, degree, etc. within the graph G. An edge $e = \{u,v\}$ is said to be *incident* to the vertices u and v. A graph $G = (V,E)$ is called *complete* if $E = \binom{V}{2}$.

A graph H is a (weak) *subgraph* of a graph G if $V(H) \subseteq V(G)$ and $E(H) \subseteq E(G)$. If, in addition, $E(H) = E(G) \cap \binom{V(H)}{2}$ then H is called an *induced* subgraph of G. In particular, for a subset $X \subseteq V$ of the vertices the graph $(X, E(G) \cap \binom{X}{2})$ is called the subgraph *induced* by the set X. It is denoted by $G[X]$.

A *walk* W in G is an injective sequence of edges $e_1 = \{v_1,w_1\}, \dots, e_l = \{v_l,w_l\}$ such that $w_i = v_{i+1}$ for all $i = 1, \dots, l-1$. The vertices v_1 and w_l are the *end-vertices* and l is the *length* of W. A *path* P is a walk which does not intersect itself, i.e., in which the sequence of vertices $v_1, w_1 \dots, w_l$ is injective as well. A *cycle* (of length l) is a path $e_1, \dots, e_{l-1}$ plus the edge connecting its two end-vertices. An x-y *path* is a path with end-vertices x and y.

A graph G is *connected* if for every pair v,w of vertices there exists a path from v to w. An inclusion maximal connected subgraph is called a *component* of the graph.

A *tree* is a connected graph which is acyclic (contains no cycles). A graph which is acyclic, but not necessarily connected, is called a *forest*. Obviously, all components of a forest are trees. A *leaf* of a tree is a vertex of degree one. The *branching points* of a tree are the vertices of degree at least three. A *spanning tree* of a graph G is a subgraph T of G such that T is a tree and $V(T) = V(G)$.

Lemma 1.1 *Let $G = (V,E)$ be a forest on n vertices with $c \geq 1$ components. Then $|E| = n - c$.*

Proof. We proceed by induction on $m = |E|$. If $m = 0$ there is nothing to show. So assume that $m \geq 1$. Choose an arbitrary edge $e \in E$ and consider what happens if we remove e from G. Clearly, the number of edges decreases by one. On the other hand, the number of components increases by one, as the subtree of G which contains e is split into two smaller trees. □

Corollary 1.2 *Every tree T on n vertices contains exactly $n - 1$ edges.* □

Corollary 1.3 *Every tree T contains at least two leaves.*

Proof. As a tree is connected, we have $d(v) \geq 1$ for all $v \in V$. Let l be the number of leaves in T. Then, using equation (1.1), we have $2(n - 1) = 2|E(T)| = \sum_{v \in V} d(v) \geq l + 2(n - l)$, which is equivalent to $l \geq 2$. □

Lemma 1.4 *A graph $G = (V,E)$ is connected if and only if G contains a spanning tree.*

Proof. Assume that G is connected but not a tree. Then G contains a cycle C. Remove an arbitrary edge e from C. Then the new graph $G' = G - e$ is still connected. We repeat this procedure until the resulting graph is a tree. This shows one direction, the other is immediate, since a tree is connected by definition. □

Theorem 1.5 *Let $G = (V,E)$ be a graph on n vertices. Then the following assertions are equivalent*

(i) *G is a tree.*
(ii) *For every pair $x,y \in V$ of distinct vertices G contains exactly one x-y-path.*
(iii) *G is minimal connected (i.e., G is connected and for all $\{x,y\} \in E$ the graph $G - \{x,y\}$ is disconnected).*
(iv) *G is maximal acyclic (i.e., G is acyclic and for all $\{x,y\} \notin E$ the graph $G + \{x,y\}$ contains a cycle).*
(v) *G is acyclic and $|E| = n - 1$.*
(vi) *G is connected and $|E| = n - 1$.*

Proof.   Assume G is a tree and let $x,y \in V$ be two distinct vertices. As G is connected, G contains at least one x-y path. As G is acyclic, every pair of vertices is connected by at most one path. This shows (i) $\Rightarrow$ (ii). The implications (ii) $\Rightarrow$ (iii) and (iii) $\Rightarrow$ (iv) are straightforward. (iv) $\Rightarrow$ (v) and (v) $\Rightarrow$ (vi) follow immediately from Lemma 1.1. Finally, (vi) $\Rightarrow$ (i) follows from the fact that by Lemma 1.4 every connected graph contains a spanning tree T and that $|E(T)| = n - 1$ by Corollary 1.2.   □

Let us recapitulate our present knowledge by answering the following question. Given a connected graph $G = (V,E)$ how many edges do we need in order to form a connected subgraph which contains all vertices of G? By Lemma 1.4, every such subgraph contains a spanning tree of G. On the other hand, as a spanning tree satisfies all desired conditions already, and by Corollary 1.2 every spanning tree contains exactly $|V| - 1$ edges, this is also the answer to the question.

This was simple. So let us make the problem a bit more interesting. That is, let us now assume that some subset $K \subseteq V$ of the vertices of a connected graph $G = (V,E)$ is given. Again we look for subgraph of G which is connected, contains all vertices from K (but possibly also some additional vertices), and should have as few edges as possible. What can we say in this case? As it turns out, a priori not very much, except that it certainly won't contain more than $|V| - 1$ edges. On the other hand, Lemma 1.4 can again be used to observe that every such subgraph has to be a tree. Furthermore, another simple observation is that all leaves of this tree have to belong to the set K (otherwise we could just leave out the edge connecting the leaf to the remaining vertices). This motivates the following definition. A subgraph T of G is called a *Steiner tree* for K, if T is a tree containing all vertices of K (i.e., $K \subseteq V(T)$) such that all leaves of T are elements of K. The vertices of K are called the *terminals* of T, while the vertices of $V(T) \setminus K$ are called the *Steiner points* of T. A *Steiner minimum tree*[1] for K in G is a Steiner tree T such that the number of edges contained in T is minimum. The problem of finding such a Steiner minimum tree is the central topic of this book. We therefore pause to give the problem a proper name:

MINIMUM STEINER PROBLEM IN GRAPHS:

Given: A connected graph $G = (V,E)$ and a set $K \subseteq V$ of terminals.
Find: A Steiner minimum tree for K in G. That is, a Steiner tree T for K such that $|E(T)| = \min\{|E(T')| \mid T'$ is a Steiner tree for K in $G\}$.

Even though this problem might at first sight look quite simple, we will see in the upcoming chapters that from an algorithmic point of view the problem is highly nontrivial and contains many challenging aspects. From a practical point

---

[1] For historical reasons the term *Steiner minimal tree* is also widely spread throughout the literature. We will, however, use the term *Steiner minimum tree* to emphasize that we mean a shortest tree among all Steiner trees.

of view it is quite often desirable to look at an even more complex version of the problem. This we will define next.

A *weighted graph* or *network* is a triple $N = (V,E,\ell)$ such that $G = (V,E)$ is a graph and $\ell : E(G) \to \mathbb{R}_{\geq 0}$ is a function which assigns to each edge a nonnegative value which can be interpreted as its length, cost, or weight. The *length* of a subgraph H of G is $\ell(H) = \sum_{e \in E(H)} \ell(e)$. If G is an unweighted graph then the length $\ell(H)$ is simply the number of edges of H and is sometimes also called the *size* of H, denoted as $|H|$. For two vertices v and w we denote by $p(v,w)$ the length of a shortest path from v to w. With this notation at hand, we can now define the weighted version of the Steiner tree problem:

MINIMUM STEINER PROBLEM IN NETWORKS:

Given: A connected network $N = (V,E,\ell)$, and a set $K \subseteq V$ of terminals.
Find: A Steiner minimum tree for K in N. That is, a Steiner tree T for K such that $\ell(T) = \min\{\ell(T') \mid T'$ is a Steiner tree for K in $N\}$.

For the time being we postpone the treatment of the Steiner problem to later chapters and continue by introducing some more basic concepts and definitions from graph theory.

A *bipartite* graph is a graph $G = (V,E)$ whose vertex set can be bipartitioned in two sets $V = A \uplus B$ such that every edge $e \in E$ has exactly one of its end-vertices in A and the other one in B. A bipartite graph is also written as $G = (A \uplus B, E)$ in order to specify the bipartition of the vertex set. If $E = A \times B$ then G is called a *complete bipartite* graph. Another neat characterization of bipartite graphs is the following.

Lemma 1.6 *A graph G is bipartite if and only if G contains no cycle of odd length as a subgraph.*

The reader is invited to supply the short proof of this lemma now. We will use it in the next chapter to design an efficient algorithm for checking whether a graph is bipartite.

Another important class of graphs are planar graphs. Loosely speaking, a graph $G = (V,E)$ is called *planar* if G can be drawn in the plane in such a way that no two edges intersect. Or, more rigorously, if there exists a representation of the graph in the plane in such a way that vertices are mapped to distinct points and edges to simple Jordan curves connecting the points of its end-vertices in such a way that any two curves are either disjoint or intersect only in a common endpoint.

The perhaps most famous result in connection with planar graphs is the so-called *Four Color Theorem*: every map can be colored with 4 colors in such a way that any two countries which have a common border are colored differently. To see how the four color theorem relates to planar graphs, observe that every map can be represented by a planar graph and vice versa. The four color theorem is thus equivalent to the statement that the vertices of every planar graph can be colored in such a way that any two adjacent vertices are colored with different colors.

Planar graphs where first considered by Leonhard Euler who found a beautiful relation between the number of vertices, edges, and faces of a planar graph: *Every connected planar graph with n vertices and m edges has $f = m - n + 2$ faces.* (The *faces* of a planar graph are the connected components of the plane which remain after deleting the vertices and edges of the representation of the graph. It can be shown that the number of faces is independent of the chosen planar representation of the graph.) One (of many) consequences of Euler's formula is that planar graphs are relatively sparse (cf. Problems 1.5). In particular, it implies that the two graphs shown in Figure 1.2, namely the complete graph K_5 on 5 vertices and the complete bipartite graph $K_{3,3}$ on two times three vertices, are not planar.

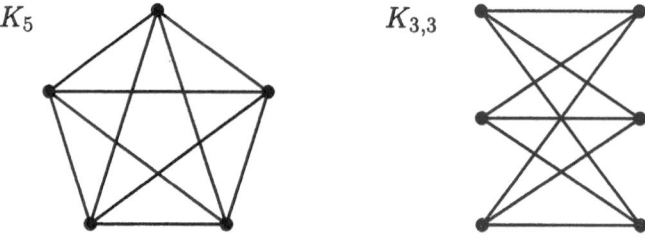

Figure 1.2 Two nonplanar graphs

A famous theorem of Kuratowski states that these two graphs are in fact in some sense the only nonplanar graphs. In order to make this precise we need one more definition.

A graph H is said to be a *subdivision* of a graph G if H is obtained from G by subdividing some of the edges, that is replacing them by paths.

Clearly, if we subdivide an edge of a nonplanar graph the resulting graph will again be nonplanar. That is, all subdivisions of K_5 and $K_{3,3}$ are nonplanar. The Theorem of Kuratowski states that also the converse is true: every nonplanar graph contains a subdivision of K_5 or $K_{3,3}$.

Theorem 1.7 (KURATOWSKI)   *A graph is planar if and only if it does not contain a subdivision of K_5 or $K_{3,3}$ as a (weak) subgraph.*

Example 1.8 The Petersen graph is nonplanar, since it contains a subdivision of $K_{3,3}$, as indicated in the following figure:

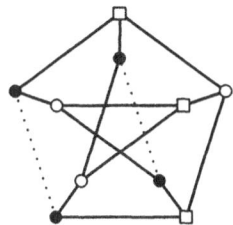

The vertices denoted by a light square resp. circle form the two classes of the $K_{3,3}$, while the solid circles indicate a subdivided edge.

We close this section by considering three different structural properties which a graph $G = (V,E)$ might have. The first one is in some sense related to the notion of a spanning tree. Recall that a spanning tree of a graph is a subgraph which is a tree and contains all vertices of the graph. Similarly, a *Hamilton cycle* of a graph is a subgraph that is a cycle and contains all vertices of the graph. It is named after the Irish mathematician William Hamilton (1805–1865) who first studied such cycles. While we know from Lemma 1.4 that every connected graph contains a spanning tree this is by far not true for Hamilton cycles. Consider for example the Petersen graph in Figure 1.1. Even though it is connected and has minimum degree three the Petersen graph does not contain a Hamilton cycle. For this particular graph, this can be verified by case checking. For general graphs, however, no nice method for proving the non-existence of a Hamilton cycle is known. In fact, deciding whether a graph has a Hamilton cycle is computationally equally intractable as the Steiner tree problem. We will come back to this observation in Problem 3.9 of Chapter 3.

For now we consider a slight variant of the problem. Namely, we do not ask for a cycle which contains every *vertex* exactly once, but for a walk which contains every *edge* of the graph exactly once and that starts and ends in the same vertex. Such a walk is called an *Euler circuit*, named after Leonhard Euler, the Swiss mathematician who considered the problem of finding a closed walk of Königsberg crossing each bridge exactly once. Deciding whether a graph has an Euler circuit is quite easy, as we show in the following theorem.

Theorem 1.9 *A connected graph* $G = (V,E)$ *contains an Euler circuit if and only if the degree of every vertex is even.*

Proof. That the stated degree condition is necessary is obvious. (To see why, just consider walking along the circuit and count how often you enter and leave a vertex.) In order to show that the condition is also sufficient we proceed by contradiction. So assume G is a minimum counterexample, minimum with respect to the number of edges. Observe that in a connected graph every vertex has degree at least one. By the parity condition on the degrees we therefore observe that every vertex in G must have degree at least 2. Formula (1.1) thus implies that G has at least $|V|$ edges. From Theorem 1.5 we deduce in turn that G contains at least one cycle, say C. Remove C from G. Clearly, all components of the remaining graph G' still satisfy the condition that all vertex degrees are even. As G was a minimum counterexample, this implies that all components of G' contain an Euler circuit. Furthermore, due to the connectedness of G, each component of G' must share at least one vertex with C. The Euler circuits of the components of G' can thus be combined with C to an Euler circuit of G in a straightforward way, leading to the desired contradiction. We leave the details of this argument to the reader, c.f. also Problem 2.6 on page 39.   □

Consider the 3-regular graph $K_{3,3}$ on the right hand side in Figure 1.2. Clearly, it contains no Euler circuit, but it does contain a Hamilton cycle. In fact it even contains six different Hamilton cycles. Observe what happens if we remove the edges from one such Hamilton cycle. What is left is a subgraph in which every vertex has degree exactly one. Such a subgraph is called a *perfect matching*. In other words, a perfect matching of a graph $G = (V,E)$ is a subset $M \subseteq E$ of the edges such that every vertex $v \in V$ is incident to exactly one edge in M. In general, a subset $M \subseteq E$ is called a *matching* if every vertex $v \in V$ is incident to at most one edge in M. A *maximum* matching in a graph G is a matching with a maximum number of edges.

We close this section with a beautiful theorem named after Philip Hall, which characterizes those bipartite graphs which contain a perfect matching.

Theorem 1.10 *Let $G = (A \uplus B, E)$ be a bipartite graph. Then G contains a matching M of cardinality $|M| = |A|$ if and only if*

$$|\Gamma(X)| \geq |X| \quad \text{for all } X \subseteq A. \tag{1.2}$$

Proof. The necessity of condition (1.2) is easily seen. Just assume that M is some matching in G of cardinality $|M| = |A|$. Let $G_M = (V, M)$ denote the subgraph consisting of the edges in M. As $M \subseteq E$, this implies

$$|\Gamma_G(X)| \geq |\Gamma_{G_M}(X)| = |X| \quad \text{for all } X \subseteq A.$$

To see that condition (1.2) is also sufficient consider a graph $G = (A \uplus B, E)$ which satisfies this condition. Clearly, we may assume that G is minimal in the sense that removing any edge results in a violation of condition (1.2). Let us first assume that there exists a vertex $b \in B$ which has two neighbors a_1 and a_2 in A. By our assumption on the minimality of G the deletion of either edge $\{a_1,b\}$ and $\{a_2,b\}$ results in a violation of condition (1.2). That is, there have to exist sets A_1 and A_2 such that $a_i \in A_i$, $|\Gamma(A_i)| = |A_i|$, and such that a_i is the only vertex of A_i which is adjacent to b. Note that this implies in particular that neither a_1 nor a_2 is contained in $A_1 \cap A_2$. Combining these observations with condition (1.2) we observe that

$$
\begin{aligned}
|\Gamma(A_1) \cap \Gamma(A_2)| &\geq |\Gamma(A_1 \setminus \{a_1\}) \cap \Gamma(A_2 \setminus \{a_2\})| + 1 \\
&\geq |\Gamma(A_1 \cap A_2)| + 1 \geq |A_1 \cap A_2| + 1.
\end{aligned}
$$

and hence also

$$
\begin{aligned}
|\Gamma(A_1 \cup A_2)| &= |\Gamma(A_1) \cup \Gamma(A_2)| = |\Gamma(A_1)| + |\Gamma(A_2)| - |\Gamma(A_1) \cap \Gamma(A_2)| \\
&\leq |A_1| + |A_2| - |A_1 \cap A_2| - 1 = |A_1 \cup A_2| - 1,
\end{aligned}
$$

which contradicts condition (1.2). Hence, we know that in fact every vertex of B has degree at most one. On the other hand, condition (1.2) clearly implies that every vertex in A has degree at least one. Together, both facts immediately imply that G contains a matching of size $|A|$. □

We note that this is just one of many ways to prove Hall's Theorem. In Problem 2.7 of Chapter 2 the reader will be asked to find a rather different proof, which also leads to an efficient algorithm for finding such a matching.

1.2 Excursion: Random graphs

The aim of this section is to introduce basic facts from probability theory and to show how they can be applied to prove surprisingly strong results in graph theory with little effort. In fact, probabilistic methods will reappear a couple of times throughout this book, reflecting their increasing significance in modern graph theory and algorithm design.

A *probability space* Ω is a set equipped with a probability measure. In this book we will consider only finite sets Ω. The probability measure is then simply an assignment of probabilities p_ω to every $\omega \in \Omega$ so that $0 \leq p_\omega \leq 1$ and $\sum_{\omega \in \Omega} p_\omega = 1$. One of the main aims of probability theory is to study the probability of certain events. An *event* is a subset $A \subseteq \Omega$. Its probability is the sum of the probabilities of the ω's contained in A. It is denoted by

$$
\Pr[A] := \sum_{\omega \in A} p_\omega.
$$

A typical example of a probability space is the set $\mathcal{G}_n$ of all graphs on n vertices, in which each graph is equally likely. What does this "uniformity" condition imply for the probability distribution? To answer this question, we first note that throughout this section we assume that the vertex set of a graph in $\mathcal{G}_n$ is $V = \{1,\ldots,n\} =: [n]$ and that the graphs are considered as labeled graphs. This means that e.g. the two graphs G and H given by $V(G) = \{1,2,3\}$, $E(G) = \{\{1,2\},\{2,3\}\}$ and $V(H) = \{1,2,3\}$, $E(H) = \{\{1,3\},\{2,3\}\}$ are considered as different graphs. Using this assumption it is easily observed that $|\mathcal{G}_n| = 2^{\binom{n}{2}}$. The probability distribution for the probability space of the set $\mathcal{G}_n$ of all graphs on n vertices is thus given by

$$p_G = 2^{-\binom{n}{2}} \qquad \text{for all } G \in \mathcal{G}_n.$$

Given any graph theoretic property, let p_n denote the probability that a graph picked randomly from $\mathcal{G}_n$ has this property. If $\lim_{n\to\infty} p_n = 1$, we say that "almost all" graphs have the considered property, and contrary, if $\lim_{n\to\infty} p_n = 0$, we say that "almost no" graphs have the considered property.

Let us start by considering the graph property of having no isolated vertex. A vertex $v \in V(G)$ is called *isolated* if G contains no edge incident to v, i.e., if $\Gamma(v) = \emptyset$. Surely, if G has an isolated vertex then G is not connected but not necessarily vice versa.

Proposition 1.11 *Almost all graphs have no isolated vertex.*

Proof. There are $2^{\binom{n-1}{2}}$ graphs containing one particular vertex as an isolated vertex. Hence, the number of graphs in $\mathcal{G}_n$ containing at least one isolated vertex can be bounded from above by $n\,2^{\binom{n-1}{2}}$.

In order to prove that almost all graphs do not have any isolated vertices, it thus suffices to show that the quotient $n\,2^{\binom{n-1}{2}} / 2^{\binom{n}{2}} = n\,2^{-n+1}$ tends to 0, which is obviously the case.   □

In fact, also a stronger assertion is true. A typical graph does not only have no isolated vertices but it is even connected.

Theorem 1.12 *Almost all graphs are connected.*

Proof. Every disconnected graph in $\mathcal{G}_n$ contains at least one "bad" subset of at most $\lfloor \frac{n}{2} \rfloor$ vertices, where "bad" means that no vertex of this subset is joined to any vertex outside of the subset. If we fix such a bad subset, say of size i, then there are

$$2^{\binom{i}{2}}\, 2^{\binom{n-i}{2}} = 2^{\binom{n}{2}-i(n-i)}$$

many graphs which are disconnected because of this bad set. Hence, we can bound the number of disconnected graphs in $\mathcal{G}_n$ from above by

$$\sum_{i=1}^{\lfloor \frac{n}{2} \rfloor} \binom{n}{i} 2^{\binom{n}{2} - i(n-i)}.$$

In order to prove that almost all graphs are connected, it suffices to show that the quotient of this number and the number of all graphs on n vertices tends to 0. Obviously, this quotient amounts to

$$\sum_{i=1}^{\lfloor \frac{n}{2} \rfloor} \binom{n}{i} 2^{-i(n-i)} \leq \sum_{i=1}^{\lfloor \frac{n}{2} \rfloor} n^i \, 2^{-i(n-i)}.$$

For n sufficiently large, the largest summand on the right hand side is the first one, i.e., the one for $i = 1$. Therefore, the sum can be bounded from above by $n^2 \, 2^{-n}$ which tends to 0 exponentially fast. $\square$

Theorem 1.12 implies that if we randomly pick a graph from $\mathcal{G}_n$ it will most likely be connected. Next, we will go one step further. Namely, we restrict the length of the connecting path between any two vertices. A graph G is said to have *diameter* k if the maximum over the lengths of the shortest paths between all pairs of vertices $u,w \in V$ is k. Formally, $diam(G) := \max\{p(u,w) \mid u,w \in V\}$, where $p(u,w)$ denotes the length of a shortest u-w path. Our aim is to show that almost all graphs have diameter 2.

While, in principle, this can be done by a counting argument as in the proof of Theorem 1.12, it becomes slightly more complicated (the reader should try!). In order to simplify the computations it is now time to observe that we can describe the underlying probability space of $\mathcal{G}_n$ also in a different way. Namely, given n, we let

$$\Omega = \prod_{e \in \left[\binom{[n]}{2} \right]} \{0,1\},$$

where each term models the presence or absence of a particular edge. We can generate a member of Ω by letting each pair $e = \{i,j\}$ be an edge with probability $1/2$, independently. The graph $G_{n,\frac{1}{2}}$ generated in this way is called a *random graph*. The straightforward observation that

$$\Pr\left[G_{n,\frac{1}{2}} = G \right] = 2^{-\binom{n}{2}}$$

for every fixed graph G shows that this formulation is indeed just another description of the probability space $\mathcal{G}_n$ in which all graphs are equally likely. One advantage of this formulation is that it allows us to use tools from probability theory. In order to exploit this, we introduce a few basic facts from finite probability.

A *random variable* X is a mapping which assigns to each element of Ω a real number. The expectation $\text{Ex}[X]$ of the random variable X is the weighted average over the images of X, i.e.,

$$\text{Ex}[X] := \sum_{\omega \in \Omega} X(\omega) p_\omega . \tag{1.3}$$

As a first example, consider the random variable Y which assigns to each graph in $G_{n,\frac{1}{2}}$ the number of its isolated vertices. Unfortunately, it is a quite tedious task to compute $\text{Ex}[Y]$ by using the formula from equation (1.3). To overcome this difficulty, we use a very important property of the expectation:

Proposition 1.13 (LINEARITY OF EXPECTATION)   *Let $\lambda_1,\dots,\lambda_k$ be arbitrary real numbers and let X and $X_1,\dots,X_k$ be random variables on the same proba- bility space Ω such that $X = \sum_{i=1}^{k} \lambda_i X_i$. Then $\text{Ex}[X] = \sum_{i=1}^{k} \lambda_i \text{Ex}[X_i]$.*

Proof.   Just use the definition of the expectation to obtain

$$\sum_{i=1}^{k} \lambda_i \text{Ex}[X_i] = \sum_{i=1}^{k} \lambda_i \sum_{\omega \in \Omega} X_i(\omega) p_\omega = \sum_{\omega \in \Omega} p_\omega \sum_{i=1}^{k} \lambda_i X_i(\omega) = \sum_{\omega \in \Omega} p_\omega X(w) = \text{Ex}[X].$$

$\square$

For a vertex i, let Y_i be the indicator variable for the event that i is an isolated vertex, i.e., $Y_i(G) = 1$ if i is an isolated vertex in G and $Y_i(G) = 0$ otherwise. Obviously, $\text{Ex}[Y_i] = (1/2)^{n-1}$, since there are $n - 1$ potential edges which have to be absent to ensure that i is an isolated vertex. The random variable Y which assigns to each graph in $G_{n,\frac{1}{2}}$ the number of its isolated vertices can be expressed as

$$Y = \sum_{i=1}^{n} Y_i .$$

Hence, by linearity of expectation

$$\text{Ex}[Y] = \sum_{i=1}^{n} \text{Ex}[Y_i] = n \, 2^{-n+1} .$$

Clearly, this expectation tends to zero when n tends to infinity. Can we use this fact in order to reprove Proposition 1.11? To do so, we introduce a very simple even though immensely useful inequality.

Lemma 1.14 (MARKOV'S INEQUALITY)
Let X be a nonnegative random variable on a finite probability space Ω and $t > 0$. Then

$$\Pr[X \geq t] \leq \frac{1}{t} \text{Ex}[X] .$$

Proof. We again just use the definition of the expectation to obtain

$$\mathrm{Ex}\,[X] = \sum_{\omega \in \Omega} X(\omega) p_\omega \geq \sum_{\substack{\omega \in \Omega \\ X(\omega) \geq t}} X(\omega)\, p_\omega \geq t \cdot \sum_{\substack{\omega \in \Omega \\ X(\omega) \geq t}} p_\omega = t \cdot \mathrm{Pr}\,[X \geq t].$$

□

For our purposes, the following immediate consequence of Markov's inequality is very useful. It is known as the *first moment method*. (This name indicates that there also exists a second moment method. We shall introduce it on page 19.)

Corollary 1.15 *Let $X_1, X_2, \ldots$ be a sequence of nonnegative integer random variables on finite probability spaces Ω_n. Then*

$$\lim_{n \to \infty} \mathrm{Ex}\,[X_n] = 0 \qquad \text{implies that} \qquad \lim_{n \to \infty} \mathrm{Pr}\,[X_n = 0] = 1.$$

Proof. Just apply Markov's inequality for $t = 1$ and observe that $\mathrm{Pr}\,[X_n = 0] = 1 - \mathrm{Pr}\,[X_n \geq 1]$.

□

Surely, the random variable Y which counts the number of isolated vertices is a nonnegative integer-valued function. Hence, combining Corollary 1.15 and the fact that $\mathrm{Ex}\,[Y] = n2^{-n+1}$ we have reproven the result from Proposition 1.11 that almost all graphs have no isolated vertex. While this proof technique might look like an overkill for such a simple result, we emphasize its usefulness by proving the following theorem.

Theorem 1.16 *Almost all graphs have diameter 2.*

Proof. Consider the random variable Z which counts the number of pairs of vertices which have no common neighbor. Clearly, if $Z = 0$ then G has diameter 2. It thus suffices to show that Z is equal to zero with high probability. To see that this is indeed the case we first apply linearity of expectation to calculate $\mathrm{Ex}\,[Z]$. For every pair i,j of distinct vertices let Z_{ij} be an indicator variable for the event that the vertices i and j have no common neighbor, i.e., $Z_{ij} = 1$ if $\Gamma(i) \cap \Gamma(j) = \emptyset$ and $Z_{ij} = 0$ otherwise. Obviously, $\mathrm{Ex}\,[Z_{ij}]$ is just the probability that the vertices i and j have no common neighbor. As there are $n - 2$ potential common neighbors and every one of them is connected to i *and* j with probability $1/4$, we have $\mathrm{Ex}\,[Z_{ij}] = (1 - 1/4)^{n-2}$. As

$$Z = \sum_{1 \leq i < j \leq n} Z_{ij}$$

linearity of expectation implies that

$$\mathrm{Ex}\,[Z] = \binom{n}{2} \left(\tfrac{3}{4}\right)^{n-2},$$

which tends to zero for n tending to infinity. Hence, an application of Corollary 1.15 concludes the proof of the theorem.

□

Next we go slightly further and show that in fact an even stronger property is true for almost all graphs.

Let $G = (V,E)$ be a graph, k be a fixed integer and $\delta = (\delta_1,\dots,\delta_k) \in \{0,1\}^k$ be a 0-1 sequence of length k. A k-tuple $(w_1,\dots,w_k)$ of vertices is called δ-extendable if there exists a vertex $v \in V \setminus \{w_1,\dots,w_k\}$ such that $\{v,w_i\} \in E$ if and only if $\delta_i = 1$. The graph G is called k-extendable if for every sequence $\delta \in \{0,1\}^k$ every k-tuple of different vertices is δ-extendable. Note that having diameter 2 is a consequence of being $(1,1)$-extendable.

Theorem 1.17 *For every k, almost all graphs are k-extendable.*

Proof. Let X be the random variable which assigns to every graph the number of k-tuples of vertices which are not δ-extendable for at least one $\delta \in \{0,1\}^k$. If $X = 0$ then the graph is obviously k-extendable.

For a k-tuple w let $X_w = 0$ if w is δ-extendable for every $\delta \in \{0,1\}^k$, $X_w = 1$ otherwise. Then

$$\mathrm{Ex}[X_w] = \mathrm{Pr}[X_w = 1] \leq 2^k \left(1 - \tfrac{1}{2^k}\right)^{n-k},$$

as for each of the 2^k many δ-vectors the probability that none of the remaining $n - k$ vertices is connected to the vertices of w in the desired way is exactly $(1 - \tfrac{1}{2^k})^{n-k}$. Since $X = \sum_w X_w$, we infer from this equality that

$$\mathrm{Ex}[X] = n(n-1)\cdots(n-k+1) \cdot 2^k \left(1 - \tfrac{1}{2^k}\right)^{n-k} = \mathcal{O}\left(n^k \left(1 - \tfrac{1}{2^k}\right)^n\right),$$

which tends to 0 exponentially fast. Now applying Corollary 1.15 yields that $\lim_{n\to\infty} \mathrm{Pr}[X = 0] = 1$, i.e., almost all graphs are k-extendable. (See Table 1.1 for the definition of the big oh symbol.) □

If G is k-extendable then obviously G contains every graph on $k+1$ vertices as an induced subgraph. Hence, we obtain the following corollary of Theorem 1.17.

Corollary 1.18 *Let H be any fixed graph (e.g., the Petersen graph, a path of length 5, or a complete graph K_r on r vertices). Then almost all graphs contain H as an induced subgraph.* □

Let us pause to reconsider what we have proven so far: almost all graphs are connected, almost all graphs have diameter 2, almost all graphs are k-extendable. Is this really surprising? Actually not very much, as it turns out that almost all graphs on n vertices are rather dense, that is they contain "many" edges. More precisely, as we will show now, almost all graphs on n vertices contain roughly $n^2/4$ edges.

Table 1.1 Asymptotic notation: the Landau symbols

notation	definition	meaning
$f(n) = O(g(n))$	$\limsup\limits_{n \to \infty} \dfrac{\lvert f(n) \rvert}{\lvert g(n) \rvert} < \infty$	$f(n)$ grows at most as fast as $g(n)$
$f(n) = \Omega(g(n))$	$\liminf\limits_{n \to \infty} \dfrac{\lvert f(n) \rvert}{\lvert g(n) \rvert} > 0$	$f(n)$ grows at least as fast as $g(n)$
$f(n) = \Theta(g(n))$	$f(n) = O(g(n))$ and $f(n) = \Omega(g(n))$	$f(n)$ grows as fast as $g(n)$
$f(n) = o(g(n))$	$\lim\limits_{n \to \infty} \dfrac{\lvert f(n) \rvert}{\lvert g(n) \rvert} = 0$	$f(n)$ grows slower than $g(n)$
$f(n) = \omega(g(n))$	$\dfrac{\lvert f(n) \rvert}{\lvert g(n) \rvert} \to \infty$	$f(n)$ grows faster than $g(n)$

Clearly, linearity of expectation implies that the expected number of edges of a random graph $G_{n,\frac{1}{2}}$ is $\frac{1}{2}\binom{n}{2} = \frac{n(n-1)}{4}$. In order to show that actually the number of edges of a graph $G_{n,\frac{1}{2}}$ is with high probability close to its expected number, we need another notion from probability theory.

The *variance* of a random variable X is defined as

$$\operatorname{Var}[X] = \operatorname{Ex}\left[(X - \operatorname{Ex}[X])^2\right]$$

and measures the expected difference of X from its expectation. Note that evaluating the square together with linearity of expectation implies that the variance is also computed by the formula $\operatorname{Var}[X] = \operatorname{Ex}[X^2] - (\operatorname{Ex}[X])^2$. Another basic inequality from probability theory relates the probability that a random variable deviates from its expectation to the variance of the random variable.

Lemma 1.19 (CHEBYSHEV'S INEQUALITY) *Let X be a random variable on a finite probability space and $t > 0$. Then*

$$\Pr\left[\lvert X - \operatorname{Ex}[X] \rvert \geq t\right] \leq \frac{1}{t^2}\operatorname{Var}[X] .$$

Proof. Just apply Markov's inequality to the random variable $Y = (X - \operatorname{Ex}[X])^2$. □

Let us now see how Chebyshev's inequality can be applied to show that a random graph has about $n^2/4$ edges.

Proposition 1.20 *Let E_n be the random variable that assigns to each graph on n vertices the number of its edges and let $\alpha(n)$ be a function which tends to infinity arbitrary slowly. Then*

$$\lim_{n\to\infty} \Pr\left[|E_n - \tfrac{1}{2}\tbinom{n}{2}| \geq \alpha(n)\cdot n\right] = 0.$$

Proof. Let X_{ij} be the indicator variable for the event that $\{i,j\}$ is an edge. Then

$$E_n = \sum_{1\leq i<j\leq n} X_{ij}.$$

Hence, $\mathrm{Ex}\,[E_n] = \tfrac{1}{2}\tbinom{n}{2}$, and by Chebyshev's inequality, it is enough to show that

$$\mathrm{Var}\,[E_n] = o(\alpha(n)^2\cdot n^2).$$

Using the formula $\mathrm{Var}\,[E_n] = \mathrm{Ex}\,[(E_n)^2] - (\mathrm{Ex}\,[E_n])^2$ we see that the only missing term is $\mathrm{Ex}\,[(E_n)^2]$. Using linearity of expectation once more (and the fact that $Z^2 = Z$ for a 0-1 variable Z) we deduce that

$$\mathrm{Ex}\,[(E_n)^2] = \sum_{1\leq i<j\leq n} \mathrm{Ex}\,[(X_{ij})^2] + \sum_{\substack{1\leq i<j\leq n\\ 1\leq k<l\leq n\\ \{i,j\}\neq\{k,l\}}} \mathrm{Ex}\,[X_{ij}\,X_{kl}]$$

$$= \tbinom{n}{2}\cdot\tfrac{1}{2} + \left(\tbinom{n}{2}^2 - \tbinom{n}{2}\right)\cdot\tfrac{1}{4} = (\mathrm{Ex}\,[E_n])^2 + \tfrac{1}{4}\tbinom{n}{2}.$$

Therefore,

$$\mathrm{Var}\,[E_n] = \mathrm{Ex}\,[(E_n)^2] - (\mathrm{Ex}\,[E_n])^2 = \tfrac{1}{4}\tbinom{n}{2} = o(n^2\alpha(n)^2),$$

which completes the proof of Proposition 1.20.   □

Letting $\alpha(n) = \log n$ we observe that, for example, almost every graph on n vertices has at least $\tfrac{1}{2}\tbinom{n}{2} - n\log n$ and at most $\tfrac{1}{2}\tbinom{n}{2} + n\log n$ edges.

To obtain more subtle information about properties of graphs, we refine our probability space. Namely, given n and a value $0 < p < 1$ we again consider

$$\Omega = \prod_{e\in\binom{[n]}{2}} \{0,1\},$$

but now generate a member of Ω by choosing each $e = \{i,j\}$ as an edge independently with probability p (instead of $1/2$ as we did previously). The "random graph" generated in this way is denoted by $G_{n,p}$. In fact, $G_{n,p}$ is a probability space containing all graphs on n vertices where each graph with m edges has probability $p^m(1-p)^{\binom{n}{2}-m}$. Note that the expected number of edges in $G_{n,p}$ is $p\binom{n}{2}$. Similarly, as in the proof of Proposition 1.20 one can also show that the actual number of edges is close to its expected value with high probability, see Problem 1.11.

Let us return to considering the graph property of being connected. Intuitively, almost all graphs are connected because they contain a lot more edges than necessary to be connected. The next goal will be to determine the smallest p so that $\lim_{n\to\infty} \Pr[G_{n,p}$ is connected$] = 1$. As it will turn out, this "smallest" p will not be a constant, but will tend to zero when n tends to infinity. To capture this phenomenon we will therefore henceforth allow that $p = p(n)$ is a function of n. In order to put this problem into a broader perspective, we introduce the notion of a threshold function.

Definition 1.21 *Let Q be a property of graphs. A function $t(n)$ is called threshold function for Q if*

(1) $p(n) \gg t(n)$ *implies* $\lim_{n\to\infty} \Pr[G_{n,p}$ *has property* $Q] = 1$

(2) $p(n) \ll t(n)$ *implies* $\lim_{n\to\infty} \Pr[G_{n,p}$ *has property* $Q] = 0$.

(Here we use the notation $p(n) \gg t(n)$ as short hand for $p(n) = \omega(t(n))$, cf. Table 1.1. Similarly, $p(n) \ll t(n)$ has the same meaning as $p(n) = o(t(n))$.)

At first sight, there is no reason why some graph property should have a threshold function. In fact, not every property does. E.g. the property "*contains an even number of edges*" obviously has no threshold function. One can, however, show that every property which is monotone with respect to adding edges indeed has a threshold function. Note that threshold functions are not unique - multiplying a threshold function by a constant yields again a threshold function. The smallest interesting threshold function is $t(n) = 1/n$. This is, e.g., the threshold of being nonplanar (see Problem 1.12) and of containing a triangle. Let us call a graph *triangle-free* iff it does not contain a triangle, that is, a complete subgraph on 3 vertices, as a subgraph. Then $G_{n,p}$ is almost surely triangle-free if $p(n) \ll 1/n$ and $G_{n,p}$ contains almost surely a triangle (in fact, lots of them!) if $p(n) \gg 1/n$, cf. Problem 1.13.

Let us consider again the graph property of having no isolated vertex. What is the threshold function for this property?

Let Y be the random variable which counts the number of isolated vertices in $G_{n,p}$. Then $\mathrm{Ex}[Y] = n(1-p)^{n-1}$. Let us consider this formula for various values of $p = p(n)$. Clearly, if p tends to zero very quickly (e.g. for $p = \frac{1}{n}$) then $\text{Ex}[Y]$ tends to infinity. On the other hand, if p is sufficiently large (e.g. for $p = \frac{1}{100}$) then $\text{Ex}[Y]$ tends to zero. What happens in between? As $n(1 - p)^{n-1}$ is a continuous function, there must be some function $p = p(n)$, where $\text{Ex}[Y]$ is equal to 1. Such a function seems to be a natural candidate for a threshold function. So let's try to determine such a function p. Assume for a moment that $p \ll 1/\sqrt{n}$. Then, by Taylor's formula

$$(1 - p)^n = e^{n \ln(1-p)} = e^{-n \sum_{i=1}^{\infty} \frac{p^i}{i}} = e^{-np+o(1)}$$

and hence, $\text{Ex}[Y] = (1 + o(1)) \cdot ne^{-np}$. But this implies that for $p = \ln n/n$ we have that $\text{Ex}[Y] = 1 + o(1)$. A good guess for the threshold function therefore seems to be $t(n) = \ln n/n$. The proof that this function satisfies the first criterion of a threshold function follows immediately from Markov's inequality.

Proposition 1.22 *Let $p(n) \gg \frac{\ln n}{n}$. Then*

$$\lim_{n \to \infty} \Pr[G_{n,p} \text{ has no isolated vertex}] = 1.$$

Proof. If $p(n) \gg \ln n/n$ then $\lim_{n \to \infty} \text{Ex}[Y] = 0$. Hence, from Corollary 1.15 it follows that $\lim_{n \to \infty} \Pr[Y = 0] = 1$.   □

Now assume that $p(n) \ll \ln n/n$. Then, obviously, $\text{Ex}[Y] \to \infty$. Unfortunately, we cannot conclude from this without further thoughts that $\Pr[Y = 0] = o(1)$. Intuitively speaking, it could happen that a graph which contains one isolated vertex with high probability contains already many isolated vertices. In other words, few graphs having many isolated vertices could push the average number of isolated vertices towards infinity, even though almost all graphs contain no isolated vertex. (For a formal example consider a random variable X with the property that $\Pr[X = 0] = 1/2$ and $\Pr[X = n] = 1/2$. Then $\text{Ex}[X] = n/2$ tends to infinity, even though X is equal to zero with probability 1/2.)

In order to show that the random variable Y satisfies $\lim_{n \to \infty} \Pr[Y = 0] = 0$, it suffices to assure that its values are reasonably concentrated around $\text{Ex}[Y]$. Here the variance comes in handy. More precisely, we will use the following easy consequence of Chebyshev's inequality.

Corollary 1.23 *Let $X_1, X_2, \ldots$ be a sequence of random variables on finite probability spaces Ω_n. Then*

$$\text{Ex}[X_n] \to \infty \text{ and } \text{Var}[X_n] \ll \text{Ex}[X_n]^2 \text{ imply that } \lim_{n \to \infty} \Pr[X_n = 0] = 0.$$

Proof. Since $\mathrm{Ex}\,[X_n] \to \infty$, we may assume that $\mathrm{Ex}\,[X_n] > 0$. Hence, Chebyshev's inequality yields

$$\Pr\,[X_n = 0] \;\leq\; \Pr\,[|X_n - \mathrm{Ex}\,[X_n]| \geq \mathrm{Ex}\,[X_n]] \;\leq\; \frac{\mathrm{Var}\,[X_n]}{\mathrm{Ex}\,[X_n]^2}\,,$$

from which the corollary follows immediately.   □

An application of Corollary 1.23 is called the *second moment method*. The name comes from the moments of the random variable. In general, the kth-moment of X is the expectation of X^k. As $\mathrm{Var}\,[X] = \mathrm{Ex}\,[X^2] - \mathrm{Ex}\,[X]^2 \geq 0$, the assumption that $\mathrm{Var}\,[X] \ll \mathrm{Ex}\,[X]^2$ implies that the second moment of X has the same growth rate as the square of the first moment.

Now we are in the position to prove that $t(n) = \ln n/n$ satisfies also the second criterion of a threshold function.

Theorem 1.24 *Let* $p(n) \ll \frac{\ln n}{n}$. *Then*

$$\lim_{n\to\infty} \Pr\,[G_{n,p}\ has\ no\ isolated\ vertex] = 0\,.$$

Proof. Let Y be the random variable which counts the number of isolated vertices in $G_{n,p}$. Recall that $\mathrm{Ex}\,[Y] = (1 + o(1)) \cdot ne^{-np}$, which tends to infinity if $p = p(n) \ll \ln n/n$.

In order to be able to apply Corollary 1.23, we only need to show that $\mathrm{Var}\,[Y] \ll \mathrm{Ex}\,[Y]^2$, i.e., that $\mathrm{Ex}\,[Y^2] = (1 + o(1)) \cdot \mathrm{Ex}\,[Y]^2$. Recall that $Y = \sum_{i=1}^{n} Y_i$, where Y_i is the indicator variable of being an isolated vertex. Hence,

$$Y^2 \;=\; \sum_{i=1}^{n} Y_i^2 + \sum_{i\neq j} Y_i Y_j$$

and therefore
$$\mathrm{Ex}\,[Y^2] \;=\; \mathrm{Ex}\,[Y] + n(n - 1)\mathrm{Ex}\,[Y_1 Y_2]\,.$$

Note that $Y_1 Y_2 = 1$ if and only if both vertex 1 and vertex 2 are isolated. This requires the absence of $2n - 3$ edges. Thus,

$$\mathrm{Ex}\,[Y_1 Y_2] \;=\; (1 - p)^{2n-3} \;=\; (1 + o(1)) \cdot e^{-2np}\,.$$

From this we obtain

$$\begin{aligned}\mathrm{Ex}\,[Y^2] \;&=\; \mathrm{Ex}\,[Y] + (1 + o(1)) \cdot n^2\, e^{-2np} \\ &=\; \mathrm{Ex}\,[Y] + (1 + o(1)) \cdot \mathrm{Ex}\,[Y]^2\,.\end{aligned}$$

Since $\mathrm{Ex}\,[Y] \to \infty$ this relation implies that $\mathrm{Ex}\,[Y^2] = (1 + o(1)) \cdot \mathrm{Ex}\,[Y]^2$. Now applying Corollary 1.23 yields $\lim_{n\to\infty} \Pr\,[Y = 0] = 0$, which is the desired result.

□

Corollary 1.25 *The function* $t(n) = \frac{\ln n}{n}$ *is the threshold function for the graph property of having no isolated vertices.*   □

Though there are obviously less graphs in $\mathcal{G}_n$ which are connected than graphs which have no isolated vertex, both properties share the same threshold function.

Theorem 1.26 *The function* $t(n) = \frac{\ln n}{n}$ *is the threshold function for connectedness of graphs.*

Proof. In the light of Theorem 1.24 it suffices to show that for $p = p(n) \gg \ln n/n$ the random graph $G_{n,p}$ is almost surely connected. In fact, from Proposition 1.22 we already know that for $p = p(n) \gg \ln n/n$ the graph $G_{n,p}$ almost surely does not contain an isolated vertex. We now generalize the idea of that proof in order to show the more general result we are aiming at now.

In order to simplify notation, let $\alpha(n) := p(n) \frac{n}{\ln n}$, where $\alpha(n) \to \infty$. Let X_k be the random variable which counts the number of components on k vertices in $G_{n,p}$. Then, for $k \leq n/2$ we have

$$\mathrm{Ex}\,[X_k] \leq \binom{n}{k}(1-p)^{k(n-k)} \leq n^k\, e^{-pk(n-k)} \leq n^{k\left(1-\frac{1}{2}\alpha(n)\right)}.$$

Now let $X = \sum_{k=1}^{\lfloor \frac{n}{2} \rfloor} X_k$ be the random variable which counts the number of components of size less than $\lfloor \frac{n}{2} \rfloor$ in $G_{n,p}$. Surely, if $X = 0$ then $G_{n,p}$ is connected. Observe that

$$\mathrm{Ex}\,[X] = \sum_{k=1}^{\lfloor \frac{n}{2} \rfloor} \mathrm{Ex}\,[X_k] \leq \sum_{k=1}^{\lfloor \frac{n}{2} \rfloor} n^{k\left(1-\frac{1}{2}\alpha(n)\right)} = \mathcal{O}\left(n \cdot n^{1-\frac{1}{2}\alpha(n)}\right).$$

As $\alpha(n) \to \infty$, this implies that $\lim_{n\to\infty} \mathrm{Ex}\,[X] = 0$ and we therefore obtain from Corollary 1.15 that $\lim_{n\to\infty} \mathrm{Pr}\,[X = 0] = 1$.   □

Actually, a careful inspection of the proofs of Proposition 1.22, Theorem 1.24, and Theorem 1.26 shows that we can actually prove something more than we have claimed. Namely, that whenever $p(n) = c\frac{\ln n}{n}$ with $c > 1$ the random graph $G_{n,p}$ does almost surely not contain isolated vertices and is even almost surely connected, whereas for $c < 1$ the graph $G_{n,p}$ is almost surely not connected, it even contains almost surely an isolated vertex. (The reader should compare this result with the definition of a threshold function which only requires the identification of the threshold up to a multiplicative constant.)

The curious reader should be informed, without proof, that if we choose $p(n) = \frac{\ln n}{n} + \frac{c}{n}$ then

$$\lim_{n\to\infty} \mathrm{Pr}\,[G_{n,p} \text{ is connected}] = e^{-e^{-c}}.$$

The function $t(n) = \ln n / n$ is the threshold for the graph property of having no isolated vertices as well as for the seemingly stronger property of connectivity. Even for the property of containing a Hamilton cycle this function serves as a threshold function. What about the property of having diameter 2? Here the fact that this property is stronger than the ones considered above is reflected in an increase of the threshold function.

Theorem 1.27 *The function* $t(n) = \sqrt{\frac{\ln n}{n}}$ *is the threshold function for the property of having diameter 2.*

This theorem can be proven using the techniques of this section. We leave the details as an exercise for the reader.

Problems

1.1 Show that a tree on n vertices has exactly $2 + \sum_{v\,:\,d(v)\geq 3}(d(v) - 2)$ leaves.

1.2 Let $T = (V,E)$ be a tree on at least three vertices in which no vertex has degree two. Show that this implies that there exists a vertex $v_0 \in V$ that is adjacent to at least two leaves.

1.3 Show that $d_1 \leq \cdots \leq d_n$ is the degree sequence of a tree if and only if $d_1 \geq 1$ and $\sum_{i=1}^{n} d_i = 2n - 2$.

1.4 Prove *Euler's formula*.

1.5 Show that $|E| \leq 3 \cdot |V| - 6$ for every planar graph $G = (V,E)$ on at least 3 vertices. Show also that $|E| \leq 2 \cdot |V| - 4$ for every triangle-free planar graph $G = (V,E)$.

1.6 Deduce from the previous problem that neither the K_5 nor the $K_{3,3}$ are planar.

1.7 Show that every planar graph contains a vertex of degree at most five and that, similarly, every triangle-free planar graph contains a vertex of degree at most three.

1.8 A *k-coloring* of a graph $G = (V,E)$ is a mapping $c : V \rightarrow \{1,\ldots,k\}$ such that $c(x) \neq c(y)$ for all $\{x,y\} \in E$. Use the result of Problem 1.7 to give a short proof of the fact that every planar graph is 6-colorable.

1.9 Prove that every k-regular bipartite graph contains a perfect matching.

1.10 Show that every graph $G = (V,E)$ contains a bipartite subgraph with at least $\frac{1}{2}|E|$ edges.

1.11 Let $\epsilon > 0$ be a fixed constant and $0 < p = p(n) < 1$ be such that $pn^2 \to \infty$. Show that the random graph $G_{n,p}$ contains with probability $1 - o(1)$ at least $(1 - \epsilon)p\binom{n}{2}$ and at most $(1 + \epsilon)p\binom{n}{2}$ many edges. Can the assumptions on p be weakened?

1.12 Prove that $p(n) = 1/n$ is a threshold function for the graph property "G contains a triangle".

1.13 Prove that $p(n) = 1/n$ is a threshold function for the graph property "G is nonplanar".

1.14 Show that the random graph $G_{n,\frac{1}{2}}$ contains with probability $1 - o(1)$ at least $n^{1/3}$ many vertices with degree exactly $\lfloor \frac{n}{2} \rfloor$. Can you improve this result? (Hint: Use Stirling's formula $n! = (1 + o(1))\sqrt{2\pi n} \cdot n^n e^{-n}$.)

1.15 Prove Theorem 1.27.

Notes

There are many textbooks on graph theory. Two standard references are the textbooks by Bollobás (1998) and Diestel (1997). In these books the reader will also find more detailed expositions and further references to many topics we could barely touch upon (or not even that) in our short introduction to modern graph theory.

References for various aspects of the Steiner tree problem will be given throughout this book. Here we just mention two. For an easily readable introduction we recommend an article by Bern and Graham (1989). This article also contains some more detailed remarks on the history of this problem. The monograph *The Steiner Tree Problem* by Hwang, Richards, and Winter (1992) on the other hand covers the state of the art (in 1992) and is a rich source for further references.

The four-color theorem was first proven by Appel, Haken, and Koch (1977). Being based on calculations done by a computer program, their proof was heavily disputed for many years. A much shorter proof (which nevertheless still uses computer calculations) was given by Robertson, Sanders, Seymour, and Thomas (1997). The interested reader is referred to http://www.math.gatech.edu/~thomas/FC/fourcolor.html for more details on the history and the latter proof of the four color theorem. For a thorough introduction to matching theory we refer the reader to the monograph by Lovász and Plummer (1993).

The study of random graphs, and probabilistic methods in general, has been a very successful and flourishing area in the last decades, with far reaching consequences in mathematics and computer science. The starting point for this field was an article by Erdős and Rényi (1960) entitled "On the evolution of random graphs". A rich source for all kinds of results in connection with random graphs is the classical monograph of Bollobás (1985). For an excellent introduction which also covers many results and developments of the more recent years the reader is referred to the book of Janson, Łuczak, and Ruciński (2000). See also the notes of Chapter 8 for more references.

2

Basics II: Algorithms

What is an algorithm? Clearly, everybody has seen many kinds of them already: recipes, directions, computer programs are all different variants of algorithms. Just as this abundance of examples facilitates the task of explaining the ideas and aims of the abstract notion of an "algorithm" it makes a formal definition difficult. In this chapter we introduce the concept of a random access machine and use it to define what we will henceforth mean when we speak of an algorithm.

2.1 Introduction to algorithms

The first rigorous definition of computability goes back to Alan Turing (1912-1954) who defined a simple imaginary machine consisting of a semi-infinite tape subdivided into cells containing letters from a finite alphabet, a read/write head and a set of instructions which determine the actions to be performed. These *Turing machines*, as they are now called, permitted the first precise definition of *computability*. By now, many different concepts of computability have been introduced. While these various models are seemingly different, they have all been shown to be essentially equivalent. Even more, a conjecture known as *Church's thesis* states that all "solvable" problems can be solved on a Turing machine and hence on any of these machines. In this book we will base our notion of computability on the concept of a random access machine (RAM), which combines the advantage of incorporating many features of "real world" computers with the fact that it is equivalent to the Turing machine.

Random Access Machines.

A random access machine (RAM) consists of a processor and a memory composed of an (unlimited) number of cells denoted as $M_1, M_2, \ldots$ which can contain any finite (even though arbitrarily large) integer. A RAM is controlled by a sequence of instructions (the program), which is stored in an additional read-only memory. A so-called program counter contains the line number of the program which is currently executed. At the start of the program this counter is set to 1. Compare Figure 2.1 for a schematic representation of a RAM.

The instruction set of a RAM resembles the instruction set of real computers. Just as in the real world there is some freedom in including or not including instructions which can be simulated by others. In this book we will assume that the basic instruction set of a RAM consists of the following instructions:

$M_i \leftarrow 1$ (*"write constant 1 in memory cell M_i"*),

$M_i \leftarrow M_j + M_k$ (*"write the sum of the numbers contained in cells M_j and M_k into cell M_i"*),

$M_i \leftarrow M_j - M_k$ (*"write the difference of the numbers contained in cells M_j and M_k into cell M_i"*),

$M_i \leftarrow M_j * M_k$ (*"write the product of the numbers contained in M_j and M_k into cell M_i"*),

$M_i \leftarrow \lfloor M_j / M_k \rfloor$ (*"divide the number in M_j by the number contained in M_k and write the integer part of the result into cell M_i"*),

$M_i \leftarrow M_{M_j}$ (*"write in M_i the number which is contained in the cell whose index is contained in M_j"*),

$M_{M_i} \leftarrow M_j$ (*"write the number which is contained in M_j in the cell whose index is contained in M_i"*),

go to m **if** $M_i > 0$ (*"if the number contained in M_i is greater than 0 continue with the mth instruction of the program; otherwise continue with the next"*),

halt   (*"stop the execution of the program and return the number contained in memory cell M_1"*).

Strictly speaking, an *algorithm* is nothing else than a sequence of instructions for a RAM. From a rigorous mathematical viewpoint this approach of defining algorithms via a random access machine is quite satisfactory. From a more practical point of view there are, however, some drawbacks. Formulating an algorithm using only the instructions of a RAM makes the algorithm extremely lengthy and complicated, perhaps even incomprehensible. It is like programming a computer in assembler instead of using a modern high level programming language. On the other hand, clearly, every instruction of such a high level programming language can be simulated by a sequence of machine instructions. In this book we will often formulate our algorithms in colloquial English only, sometimes mixed with Pascal-like instructions — with the tacit assumption that there should be

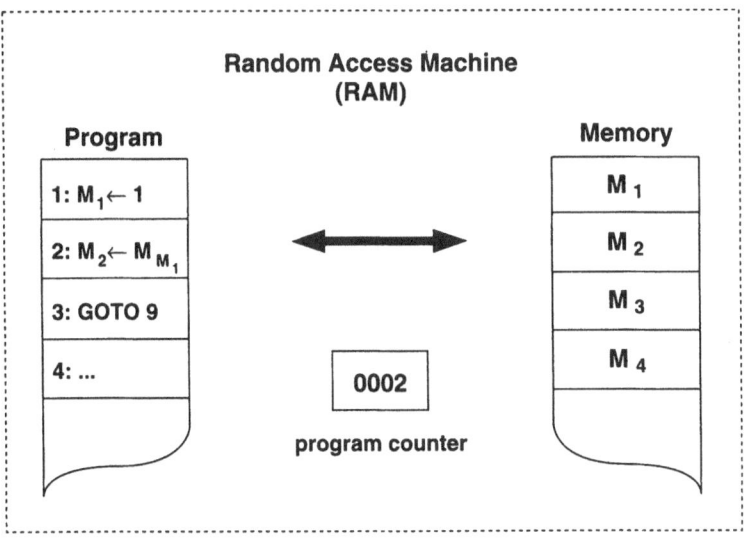

Figure 2.1 The Random Access Machine (RAM)

an "obvious", even though perhaps tedious, way to implement the algorithm in some high level programming language and hence also on a RAM.

Example 2.1 GRAPH CONNECTEDNESS. Assume we want to design an algorithm solving the following problem: *Given a graph $G = (V,E)$, is it connected?* There exists a beautiful and simple algorithm:

Algorithm 2.2 (GRAPH CONNECTIVITY)
Input: A graph $G = (V,E)$.
Output: TRUE if G is connected; otherwise FALSE.
(1) **for all** $v \in V$ **do** marked[v] := FALSE;
(2) Choose $v_0 \in V$ arbitrarily;
(3) $Q := \{v_0\}$; marked[v_0] := TRUE;
(4) **while** $Q \neq \emptyset$ **do**
(5)   Choose $w \in Q$ arbitrarily; $Q := Q \setminus \{w\}$;
(6)   **for all** $v \in \Gamma(w)$ **do**
(7)   **if** marked[v] = FALSE **then**
                marked[v] := TRUE; $Q := Q \cup \{v\}$;
(8) **if** marked[v] = TRUE for all $v \in V$ **then return** TRUE
                                    else return FALSE.

We postpone the analysis of the correctness and efficiency of this algorithm to Examples 2.3 and 2.4. Here we only add some general comments about the notation of this and future algorithms. First note that while we borrowed the

while and **for** instructions from Pascal we disregarded the block statements
begin ... **end**. Instead we use indentation. Next observe that we stated the input
of our algorithm rather abstractly: "a graph". Surely, there are many ways to
model this combinatorial structure with data structures like arrays or lists. We
will study these in more detail in Examples 2.5 and 2.6. Nevertheless, within the
context of algorithms we will usually restrict ourselves to the abstract level of the
above example and leave it to the reader to choose and implement an appropriate
data structure.

Having agreed on the definition of an algorithm, let us proceed to the next
question. What properties should an algorithm have? Clearly, it should be correct
and efficient. Let us first consider its correctness. Testing it on various examples
gives a good first indication — but doesn't rule out the possibility of wrong results for particular inputs. Strictly speaking, a formal proof of correctness would,
however, require a formal algorithm. And we just agreed that we do allow algorithms stated in colloquial English. But, putting this formal point of view aside,
it is nevertheless obvious what we mean if we claim that an algorithm is "correct". Usually, we will stick to this more intuitive notion of correctness. That is,
we restrict ourselves to showing that the general approach of the algorithm is
correct, but omit the precise verification of the implementation issues.

Example 2.3 CORRECTNESS ANALYSIS. Consider the algorithm in Example 2.1. To show
that it is correct it suffices to verify that at termination of the algorithm marked[w] =
TRUE for a vertex $w \in V$ if and only if there exists a v_0-w path in G. Assume first that
$w \in V$ is a vertex with marked[w] = TRUE. If $w = v_0$ there is nothing to show. Otherwise
marked[w] must have been set to true in statement (7) of the algorithm. Let w_1 be the
vertex w of that iteration of the **while** loop. Repeating the above argument iteratively
for w_1 instead of w (yielding a vertex w_2), and then for w_2 etc. we will finally end up
with a vertex $w_k = v_0$. Then $v_0 = w_k, w_{k-1}, \ldots, w_1, w$ defines a v_0-w path. The similar
proof of the other direction is left to the reader.

The second important property of an algorithm is its efficiency. Here we
distinguish between *space* and *time efficiency*. Let us, for the moment, restrict
our attention to algorithms on a RAM. Measuring space and time requirements
of such an algorithm for a *fixed input* is straightforward. We just check how many
different memory cells the algorithm accesses and how many instructions are
performed until the **halt** instruction is reached. But, naturally, these quantities
will vary with the input. In particular, if the "size" of the input tends to infinity
then we expect that the time and space requirements will also tend to infinity.

So, how should we define the requirements of an algorithm per se? Let us
first make precise what the size of an input is. Within the context of a RAM
we will always assume that the input consists of a finite sequence of 0's and 1's

stored in the first memory cell M_1. For the analysis of an algorithm we partition the set of all inputs into different classes according to their length and consider every such class separately. The (time or space) complexity T_A of an algorithm is then defined as a function in the input length n. Commonly, one sets $T_A(n)$ to the *maximum* of the corresponding values, where the maximum is taken over all inputs of length n. This is the so-called *worst-case analysis*. It is the measure which we will usually employ in this book. Other possibilities are *average-case analysis* (which we will study in more detail in Chapter 8) or the (rather seldomly used) *best-case analysis*.

Estimating $T_A(n)$ precisely is usually a very tedious, if not practically impossible, task. But in some sense the precise value of this function is not important anyway. What we are really interested in is how this function grows with n. Accordingly, we speak of

linear algorithms,	if $T_A(n) = \mathcal{O}(n)$,
quadratic algorithms,	if $T_A(n) = \mathcal{O}(n^2)$,
polynomial algorithms,	if $T_A(n) = \mathcal{O}(n^k)$ for some $k \in \mathbb{N}$,
exponential algorithms,	if $T_A(n) = 2^{\mathcal{O}(n)}$.

At that point we are now ready to drop our assumption that A should be an algorithm for a RAM. By our tacit assumption, all algorithms allow an "obvious" implementation on a RAM. While any two of these implementations might vary in the precise details of the implementation, the order of magnitude of the number of necessary instructions can usually easily be determined. And that is all what interests us. Note also, that this allows to add some flexibility to the underlying RAM. E.g., adding an instruction $M_i \leftarrow -M_i$ saves a constant number of instructions each time we alter the sign of a number, but won't change the order of magnitude of the running time.

Example 2.4 RUNTIME ANALYSIS. We consider again the algorithm in Example 2.1. The initialization in step (1) clearly needs $\mathcal{O}(n)$ time, while the steps in lines (2) and (3) can be done in constant time. Similarly, the final **if** statement in line (8) can easily be implemented in $\mathcal{O}(n)$ time. For the analysis of the **while** loop there are two possibilities. In the straightforward argument we note that the **while** loop is executed at most n times (once for every vertex) and that in each iteration we have to perform constant work for each of the at most $n - 1$ neighbors of the current vertex w. This gives an overall work of at most $\mathcal{O}(n) + n \cdot \mathcal{O}(n) + \mathcal{O}(n) = \mathcal{O}(n^2)$.

But we can do better by considering the statements within the **while** loop separately. First we note that the while condition in line (4) and the statement in line (5) both have to be executed exactly n times. Since each single of these executions can be performed in constant time, the total time spent for these operations can be bounded by $\mathcal{O}(n)$. (Admittedly, we are cheating a bit here: the time required for the test "Is $Q \neq \emptyset$?" depends on the data structure used for implementing Q. Implementing it as a list yields the claimed time bound. We will come back to this point at the end of this section.) Finally, consider the statement in line (6). By the definition of the **for** loop it is executed only for edges $\{w,v\} \in E$. On the other hand every edge $\{a,b\} \in E$ is considered at most twice (once for "w"$= a$ and once for "w"$= b$). Hence, the total running time can be bounded by $\mathcal{O}(n + m)$, which is strictly smaller than our first bound whenever $m \ll n^2$.

There is one problem which we did neglect so far. We did assume that every instruction of a RAM can be performed in one time step. But every memory cell of a RAM may contain an arbitrarily large number. So how realistic is this assumption?

Let's consider an example. Say we want to calculate $n!$. Clearly, this can be done with $n - 1$ multiplications. That is, we could say that the complexity of the algorithm is $\mathcal{O}(n)$. On the other hand, the number of bits in the result is already $\Omega(n \log n)$ — and it is somewhat unrealistic that these many bits can be written in just $\mathcal{O}(n)$ time steps. So, what is wrong in this example? We did violate the "unwritten rules of good behavior". More precisely, we allowed the numbers stored in the memory cells to be significantly larger than our input n (which can be stored using just $\lceil \log_2 n \rceil$ bits).

Thus, whenever we obey the "unwritten rules" (more precisely, if we ensure that at all times that the numbers written in the memory cells contain at most a constant times more bits than the largest number in the input) we may stick to the *unit cost model*, which counts every operation as one step, regardless of how many bits are involved. If, however, these rules are violated we should switch to the so-called *logarithmic cost model* which counts the number of bit operations. Besides a single exception in Chapter 8, we will always use the unit cost model.

Elementary Data Structures

We conclude this section with a quick overview of some well-known data structures. More specifically, we will look on arrays and lists. An *array* is just a contiguous block of memory cells. In the definition of an array we state its *dimension* and the size of each coordinate. For example in $C++$ the definition int $A[3][4]$; yields a two dimensional array with entries 0 to 2 in the first coordinate and entries 0 to 3 in the second coordinate. Upon seeing it the compiler reserves space for $3 \cdot 4 = 12$ integer values. In the program this block of memory cells can subsequently be accessed by the notation $A[i][j]$. The compiler maps that to the corresponding entry within the reserved memory block.

Example 2.5 STORING A GRAPH. Arrays are a very simple but nevertheless powerful way to store a graph. Namely, we use a two dimensional matrix A of size $n \times n$ (recall that n denotes the number of vertices of the graph), called the *adjacency matrix* of the graph, initialize it with zeros everywhere and, while parsing through the input, put a one at all positions $A[i,j]$ which correspond to an edge. For graphs this yields a symmetric matrix, and we could therefore save a factor of two in the memory requirements, by storing only the upper half. But as this has no effect within the big Oh notation we usually won't bother with these details.

Assume we have a graph G given by its adjacency matrix A and we want to access all neighbors of a vertex v_0. Clearly, this is easily done by parsing through the row corresponding to v_0: the neighbors of v_0 are given by the set of w's such that $A[v_0,w] = 1$. While this is easy it is not particularly fast. Assume for example that v_0 has just, say, two neighbors, it nevertheless takes time $\mathcal{O}(n)$ to actually find them. Here lists allow a much faster implementation.

Lists are a data structure which allow to store an a priori unknown number of items. This is achieved by storing one object that points to the beginning of the list and letting each item point to its successor, if there is one. Here is a schematic drawing of such a simple list:

We can make lists slightly more powerful, by adding additional pointers. E.g., we can let an item point not only to its successor but also to its predecessor, if there is one. This are the so-called doubly linked lists. Also, instead of having the last element point to nothing, or as it is usually called to *nil*, we can let it point back to the first element in the list, thus obtaining a so-called cyclic list. The precise way of implementing lists varies among different programming languages. We ignore this issue here.

Example 2.6 STORING A GRAPH (CONT.) Storing the graph via an adjacency matrix as indicated in Example 2.5 is a simple and concise way. For many applications it is, however, more useful to store the graph using lists, e.g., keeping for every vertex a list of all its neighbors. In particular, this allows scanning all neighbors of a vertex more quickly, namely, in $\mathcal{O}(\deg(v))$ time instead of $\mathcal{O}(n)$ time. On the other hand testing whether an edge $\{u,v\}$ belongs to $E(G)$ or not can be done in constant time with the help of an adjacency matrix, and is usually much more time consuming if we have to scan through the list of neighbors. That is, both ways of storing a graph have their advantages and disadvantages, and it depends on the context which is better.

The concept of a pointer is not restricted to lists. One can easily build up more complex data structures like trees and other graphs. This is very important for the design of efficient algorithms. In the next section we will see an example for a powerful data structure which is based on trees. Here we show that combining the two approaches of storing a graph from Examples 2.5 and 2.6 leads to a data structure which has surprising properties. To motivate this the reader should, before reading on, try to find a method for storing a graph, given as a list of its m edges, in $\mathcal{O}(m)$ time in such a way that subsequent queries of the form *"Is $\{x,y\}$ an edge of the graph?"* can be answered in constant time.

Example 2.7 STORING A GRAPH (CONT.) Assume we are dealing with a sparse graph, i.e., a graph satisfying $m \ll n^2$, and are trying to find an algorithm for some problem that has a running time which is linear in m. May we use the adjacency matrix to store the graph G? It seems not, as constructing this matrix needs $\Omega(n^2)$ time and this would crossly spoil our intended linear running time. Can we avoid this? What seems to be impossible at first sight is actually not difficult at all. Roughly speaking, we just leave out the initialization step during the generation of the matrix! In return we are a bit more careful while storing the edges: instead of simply putting a 1 at places corresponding to an edge we store a pointer to the edge within a separately kept edge list, as indicated in Figure 2.2. Clearly, generating the matrix in this way can be done in $\mathcal{O}(m)$. And checking whether an entry $A[u,v]$ represents an edge of the graph is easily done in $\mathcal{O}(1)$ time: we just test whether the entry $A[u,v]$ points to an entry in the edge list *and* whether this edge is indeed the edge $\{u,v\}$.

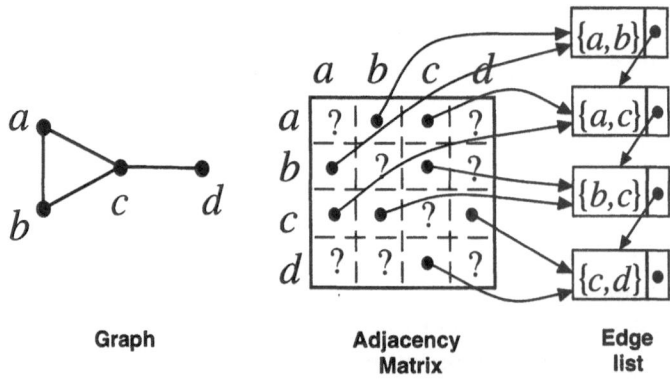

Graph   Adjacency   Edge
          Matrix   list

Figure 2.2 Adjacency matrix of a sparse graph.

We close this section with an illustration of how the use of appropriate data structures combined with some facts about properties of graphs from Chapter 1 can be used to design efficient algorithms.

Example 2.8 GRAPH COLORING. A coloring of a graph $G = (V,E)$ with k colors is a mapping $c : V \to \{1,\dots,k\}$ such that $c(u) \neq c(v)$ for all edges $\{u,v\} \in E$. Clearly, every graph permits a coloring with $k = |V|$ colors. For the complete graph this bound is obviously best possible, i.e., there exists no coloring using *less than* $|V|$ colors. For other graphs, however, colorings using fewer colors often do exists. Consider, for example, a bipartite graph $G = (A \uplus B; E)$. Clearly, G can be colored with just two colors by coloring every vertex in A with color 1 and every vertex in B with color 2. For planar graphs the famous *Four Color Theorem* (see page 5) states that every planar graph can be colored with four colors. While the proof of this theorem is very complicated, it is not difficult to see that planar graphs can be colored with *six* colors. To see how this is done, recall that we know from Problem 1.7 that every planar graph $G = (A \uplus B, E)$ contains a vertex v_0 of degree at most five. Now observe what happens if we delete such a vertex. The remaining graph G' is still planar and contains fewer vertices. So we can color G' recursively. Having found a coloring for G' we consider the colors of the neighbors of v_0. As v_0 has at most five neighbors, there has to exist at least one color (out of the total six colors) that is not used by any of the neighbors of v_0. We can therefore extend the coloring of G' to a coloring of G by assigning such a color to v_0.

We claim that the above coloring algorithm for planar graphs can be implemented in such a way that the algorithms has a running time of $\mathcal{O}(n)$. Observe that in every recursion the most time consuming step is finding the vertex v_0. Clearly, this can be done by parsing the adjacency lists of all vertices in V. Note however that this would require $\mathcal{O}(n)$ time in each recursion and would thus result in an $\mathcal{O}(n^2)$ algorithm. Alternatively, one can, however, distribute all vertices of G in "buckets" $B[k]$ so that, for each $0 \le k < n$, $B[k]$ points to a list that contains all vertices of degree k. With the help of these buckets the vertex v_0 can then be found in constant time. Note also that, because v_0 has degree at most five, updating the buckets $B[k]$ also takes only constant time. In total we thus have an algorithm with running time $\mathcal{O}(n)$.

2.2 Excursion: Fibonacci heaps and amortized time

The efficiency of algorithms manipulating large amounts of data depends heavily on the use of appropriate and clever data structures. Of particular importance is the manipulation of sets which dynamically grow and shrink during the execution of the algorithm. An element of such a set is usually characterized by a special value, called its *key*. Search trees are a fundamental data structure for supporting operations like insertion and deletion of elements and checking whether an element with a given key is currently contained in the set. It is not too difficult to design search trees that guarantee that any of these operations can be performed within $\mathcal{O}(\log n)$ time, if n denotes the number of elements in the set.

In algorithmic graph theory it often happens that one does not need the full strength of search trees. More precisely, it is often sufficient that the data structure supports the following operations:

Insert:	add a new element to the set,
Extract-Min:	return the element with minimum key among all elements and remove it from the set,
Decrease-Key:	change the key of an element to a new (smaller) value.

A data structure which supports these three operations is called a *priority queue*.

There are numerous ways to implement a priority queue. Straightforward approaches are to use (sorted or unsorted) arrays or lists. Also, search trees can be used. The following table states the time bounds which can be achieved with these data structures (assuming that the set will never contain more than n elements simultaneously):

	sorted array or list	unsorted array or list	search tree
Insert	$\mathcal{O}(n)$	$\mathcal{O}(1)$	$\mathcal{O}(\log n)$
Extract-Min	$\mathcal{O}(1)$	$\mathcal{O}(n)$	$\mathcal{O}(\log n)$
Decrease-Key	$\mathcal{O}(n)$	$\mathcal{O}(1)$	$\mathcal{O}(\log n)$

These bounds imply in particular that, using for example an array, any sequence of k of these operations can be performed in $\mathcal{O}(k \cdot n)$ steps. Moreover, it is not difficult to show that there also exists sequences of k operations which do need $\Omega(k \cdot n)$ steps, cf. Problem 2.1.

On the other hand, the so-called Fibonacci heaps we are going to describe in this section have the property that any sequence of k operations can be performed in $\mathcal{O}(k + l \cdot \log n)$ time, where $l \leq k$ is the number of Extract-Min operations. While this might give rise to the impression that each Insert and Decrease-Key operation can be performed in $\mathcal{O}(1)$ time and each Extract-Min in $\mathcal{O}(\log n)$ time, this is far from being true! There exists no data structure that is guaranteed to perform that well. All we can say is that over a sequence of operations the (extra) time spent for some operations can be saved during other operations. This calls for a new technique in analyzing algorithms. It is known as *amortized analysis*. Amortized analysis effectively bounds the "average time" required to performance a certain operation. Its main property, however, is that it can also be used to bound the time required to perform a sequence of operations in the *worst case*. This is best explained via an example.

Example 2.9 CLIMBING UP A FLIGHT OF STAIRS. Assume you are at the foot of a flight of stairs consisting of, say, n steps. At each turn you may perform any of the following two actions:

Up: Climb up one step (if not at the top already);

Down: Go all the way down to the bottom.

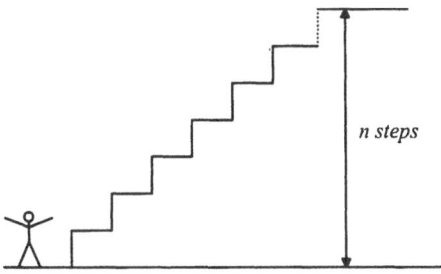

n steps

Under the assumption that mastering a single step (independent of the direction) requires one unit of time, try to answer the following questions:

1.) What is the maximum time spent for an action?

2.) What is the maximum time spent for an arbitrary sequence of k actions starting at the bottom of the stairs?

The answer to the first question is clearly n (achieved when you are at the top and choose the action Down). A correct answer for the second question is thus certainly $\mathcal{O}(k \cdot n)$. But this is far from best possible. Why? Intuitively, because we can only go down steps which we went up previously. For a formal proof that in fact $2k$ time units always suffice we introduce two functions:

$$\phi(i) \quad := \quad \text{height after the first } i \text{ actions, and}$$
$$t(i) \quad := \quad \text{time spent during the first } i \text{ actions.}$$

We claim that $t(i) + \phi(i) \leq 2i$ for all $i \geq 0$. Since ϕ is nonnegative, this implies that $t(k) \leq 2k$. We proceed by induction on i. For $i = 0$ there is nothing to show. So assume the claim holds for the first $i - 1$ actions and consider the ith action. If it is Up then $t(i) = t(i-1) + 1$ and $\phi(i) = \phi(i-1) + 1$, and hence $t(i) + \phi(i) = t(i-1) + \phi(i-1) + 2 \leq 2i$ by the inductive hypothesis. On the other hand, if the ith action is Down then $t(i) = t(i-1) + \phi(i-1)$ and $\phi(i) = 0$, and hence, again using the inductive hypothesis, $t(i) + \phi(i) = t(i-1) + \phi(i-1) \leq 2i - 2$.

Let us extract the main ideas in the proof of Example 2.9. Our aim was to show that "on average" every action needs two time units. But charging two time units to Up overcharges this particular operation by one. We introduced the function $\phi(i)$ to keep track of these "spare" time units. The function ϕ is also called a *potential* function. It accumulates time units and releases them whenever they are required. Defining and using an appropriate potential function is what is known as *amortized analysis*.

We now come back to our original aim: the design of Fibonacci heaps (or shortly F-heaps). F-heaps are a collection of ordered trees. An ordered tree is a rooted tree such that each vertex has a collection of successors, called its *children*, so that the key of each vertex is less than or equal to the keys of all its children. The roots of these trees are contained in a (doubly linked) cyclic list called the *root list*. At any time the pointer min[H] of an F-heap H points to the vertex with the smallest key in the root list (and thus to the vertex with the smallest key in the entire heap) and $n[H]$ denotes the number of vertices currently contained in the heap. Figure 2.3 shows an example of an F-heap containing 16 elements (ignore the shading of the vertices for the moment). The children of a vertex are contained in a (doubly linked) cyclic list, the vertex itself points to any one of its children. To simplify the drawings the cyclic lists for the children of a vertex are not shown.

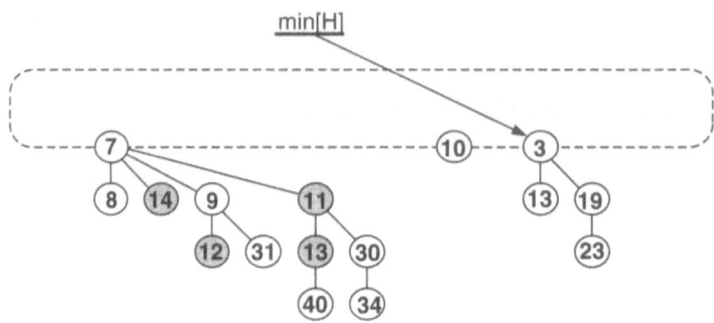

Figure 2.3 A Fibonacci heap containing 16 elements; the vertices with marked$[x]$ = TRUE are shaded.

Formally, to every vertex x in an F-heap the following fields and pointers are attached:

key$[x]$:   key of x,
parent$[x]$:   pointer to the parent of x,
child$[x]$:   pointer to an (arbitrary) child of x,
left$[x]$:   pointer to the left sibling of x,
right$[x]$:   pointer to the right sibling of x,
rank$[x]$:   number of children of x,
marked$[x]$:   boolean field indicating whether x has lost a child already.

What do we need the field marked$[x]$ for? Assume an F-heap is given. By applying the function **Decrease-Key** to the child of a vertex it might become necessary to cut the child from its parent as otherwise the order of the tree would be violated. Moreover, if this happens repeatedly the tree will get sparser and sparser – in the extreme case the heap might even transmute into a collection of simple lists. And as lists do not support the desired functions sufficiently fast it is necessary to reorganize the heap once in a while. The threshold which we will use is the following: each vertex might loose one child without any reorganization of the heap taking place but no more. This is the rôle of the field marked$[x]$: it memorizes whether x has lost a child since the last time it became a child of another vertex.

We are now all set to precisely state the implementations of the desired three functions. For sake of completeness we start with a fourth one.

Initialize. Generate an empty heap H by letting min$[H] :=$ **nil** and $n[H] := 0$.

Insert. Insert the new vertex x into the root list of H, changing $\min[H]$ to x if $key[x]$ is smaller than $key[\min[H]]$:

> `Insert(H,x)`
>
> $parent[x] := \mathbf{nil}$; $child[x] := \mathbf{nil}$;
> $marked[x] := \text{FALSE}$; $rank[x] := 0$;
> add x to the root list of H, updating $\min[H]$ if necessary;
> $n[H] := n[H] + 1$;

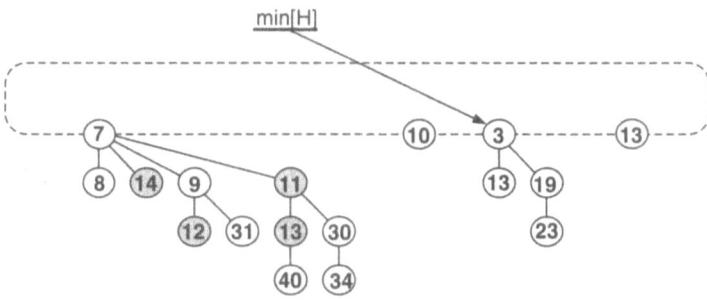

Figure 2.4 Insertion of a vertex with key 13.

Extract-Min. Insert all children of $\min[H]$ in the root list and remove vertex $\min[H]$ from the root list. Then reorganize the heap so that the root list does not contain two vertices of the same rank. This is done by repeatedly making one of two vertices of the same rank a child of the other. Formally, this step is implemented in the procedure `Consolidate`. (In the algorithm $A[\cdot]$ denotes an array of size $\lfloor \log_{3/2} n[H] \rfloor$.)

> `Extract-Min(H)`
>
> $z := \min[H]$;
> **if** $z \neq \mathbf{nil}$ **then**
>   **for** each child x of z **do**
> $parent[x] := \mathbf{nil}$; $marked[x] := \text{FALSE}$;
>   add x to the root list of H;
>   remove z from the root list of H;
> $n[H] := n[H] - 1$;
>   **if** root list of H empty
>   **then** $\min[H] := \mathbf{nil}$
>   **else** `Consolidate(H)`;
> **return** z;

> `Consolidate(H)`
>
> **for** $i := 1$ **to** $\lfloor \log_{3/2} n[H] \rfloor$ **do** $A[i] := \mathbf{nil}$;
> **repeat**
>   remove an arbitrary vertex x from the root list of H;
>   **while** $A[rank[x]] \neq \mathbf{nil}$ **do**
> $y := A[rank[x]]$; $A[rank[x]] := \mathbf{nil}$;
>   **if** $key[x] > key[y]$ **then** exchange $x \leftrightarrow y$;
>   make y a child of x, incrementing $rank[x]$;

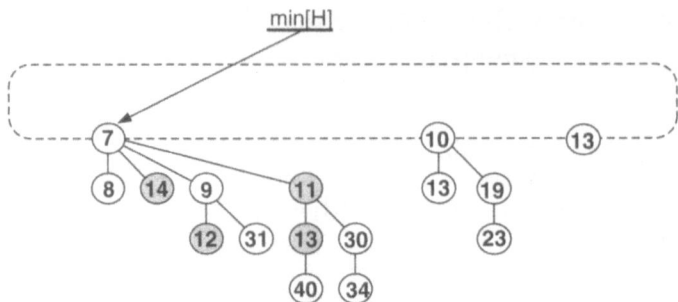

Figure 2.5 *F*-heap after performing an Extract-Min operation.

> $A[\mathrm{rank}[x]] := x;$
> **until** root list of H is empty;
> $\min[H] := \mathbf{nil};$
> **for** $i := 1$ **to** $\lfloor \log_{3/2} n[H] \rfloor$ **do**
>   **if** $A[i] \neq \mathbf{nil}$ **then** add $A[i]$ to the root list of H, updating $\min[H]$ if necessary;

Decrease-Key. First we remove the edge between the vertex and its parent and insert the vertex (with its new key) in the root list. If the parent hasn't previously lost a child already (i.e., if its marked-value is FALSE) set its marked-value to TRUE. Otherwise remove the edge between the parent and its parent as well, and repeat the above procedure for its parent, continuing in this way until a vertex is reached which either belongs to the root list already or hasn't yet lost a child. Note that a priori there no upper bound for number of cut operation which will be performed. These operations are therefore also called *cascading cuts*.

> Decrease-Key(H,x,k)
> **if** $k \geq \mathrm{key}[x]$ **then error** ("new key is not smaller than current key");
> $\mathrm{key}[x] := k;$
> **if** $\mathrm{parent}[x] \neq \mathbf{nil}$ **and** $\mathrm{key}[x] < \mathrm{key}[\mathrm{parent}[x]]$ **then** Cut(H,x);

> Cut(H,x)
> $y := \mathrm{parent}[x];$
> $\mathrm{parent}[x] := \mathbf{nil};$ $\mathrm{marked}[x] := \mathrm{FALSE};$
> remove x from the child list of y, decrementing $\mathrm{rank}[y];$
> add x to the root list of H, updating $\min[H]$ if necessary;
> **if** $\mathrm{parent}[y] \neq \mathbf{nil}$ **then**
>   **if** $\mathrm{marked}[y] = \mathrm{TRUE}$ **then**
>   Cut(H,y);
>   **else**
> $\mathrm{marked}[y] := \mathrm{TRUE};$

Before we proceed to the analysis of the (amortized) running times of these procedures, we collect some properties of *F*-heaps.

Lemma 2.10 *Let x be a vertex in an F-heap and suppose that $\mathrm{rank}[x] = k$. Let $y_1, \ldots, y_k$ be the children of x in the order in which they were linked to x, starting with the earliest. Then $\mathrm{rank}[y_i] \geq i - 2$ for all $i = 2, \ldots, k$.*

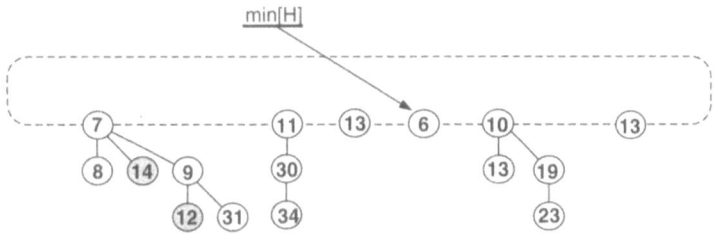

Figure 2.6 *F*-heap after performing a `Decrease-Key` operation for the vertex with old key 40 and new key 6.

Proof. Consider the moment when y_i became a child of x. As the procedure `Consolidate` is the only place where one vertex becomes a child of another, this implies that at that moment y_i and x must have had the same rank. And, since $y_1,\ldots,y_{i-1}$ were linked to x before y_i, at that time $\text{rank}[x] \geq i-1$. That is, when y_i became a child of x we must have had $\text{rank}[y_i] \geq i-1$. Since then, vertex y_i can have lost at most one child (as it would have been cut from x as soon as it lost a second child). □

Lemma 2.11 *Let x be any vertex in an F-heap and let $k = \text{rank}[x]$. Then the subtree rooted at x has size at least $(\frac{3}{2})^k$.*

Proof. Let S_k denote the minimum number of vertices that a subtree of an *F*-heap rooted at a vertex of rank at least k can have. It suffices to show that $S_k \geq (\frac{3}{2})^k$. We proceed by induction on k. Trivially, $S_0 = 1$ and $S_1 \geq 2$, so the base of the induction holds. For the inductive step consider an arbitrary integer $k \geq 2$. From Lemma 2.10 we know that $S_k \geq 2 + \sum_{i=2}^{k} S_{i-2}$. Thus, using the inductive hypothesis we obtain

$$S_k \geq 2 + \sum_{i=2}^{k}(\tfrac{3}{2})^{i-2} = 2 + (1 - (\tfrac{3}{2})^{k-1})/(1 - \tfrac{3}{2}) = \tfrac{4}{3} \cdot (\tfrac{3}{2})^k \geq (\tfrac{3}{2})^k,$$

as claimed. □

Corollary 2.12 *Let H be an F-heap containing at most n elements. Then the rank of each vertex x in H satisfies $\text{rank}[x] \leq \log_{3/2} n$.*

Proof. Trivially, the subtree rooted at x can contain at most n vertices. Using Lemma 2.11 we thus obtain that $(\frac{3}{2})^{\text{rank}[x]}$ is at most n. □

With these lemmas at hand we are now ready to prove our main theorem of this section.

Theorem 2.13 *Starting with an empty heap any k of the operations* Insert, Decrease-Key, *and* Extract-Min *can be performed in* $\mathcal{O}(k + l \cdot \log n)$ *time, where l is the number of* Extract-Min *operations and n is the maximum number of elements contained in the heap at any time.*

Proof.   We use the idea of a potential function which was introduced in Example 2.9. Let C be a fixed, sufficiently large integer (see below) and let

$$
\begin{aligned}
r(i) &:= \text{number of vertices in the root list after } i \text{ operations,} \\
m(i) &:= \text{number of marked vertices after } i \text{ operations,} \\
\phi(i) &:= C \cdot (r(i) + 2\,m(i)), \\
t(i) &:= \text{time spent during the first } i \text{ operations, and} \\
a(i) &:= [\phi(i) + t(i)] - [\phi(i-1) + t(i-1)].
\end{aligned}
$$

Then $a(i)$ is the amortized time for the ith operation. It suffices to show that $a(i)$ is constant whenever the ith operation is an Insert or a Decrease-Key operation and bounded by $\mathcal{O}(\log n)$ for an Extract-Min operation. To see this we consider the three cases separately.

Insert: Then
$$
\begin{aligned}
r(i) &= r(i-1) + 1 \\
m(i) &= m(i-1), \\
t(i) &\leq t(i-1) + C,
\end{aligned}
$$
and hence $a(i) \leq 2C$.

Extract-Min: Then
$$
\begin{aligned}
r(i) &\leq \log_{3/2} n, \\
m(i) &\leq m(i-1), \\
t(i) &\leq t(i-1) + C(r(i-1) + \log_{3/2} n).
\end{aligned}
$$
(To see the last inequality observe first that by Corollary 2.12 the vertex with minimum key has at most $\log_{3/2} n$ children. Secondly, observe that each linking of two vertices in the root list reduces the number of vertices in the root list by one.) Hence, $a(i) \leq 2C \log_{3/2} n$.

Decrease-Key: Let x denote the number of calls of the procedure Cut. Then
$$
\begin{aligned}
r(i) &= r(i-1) + x, \\
m(i) &\leq m(i-1) - (x-1) + 1, \\
t(i) &\leq t(i-1) + Cx,
\end{aligned}
$$
and hence $a(i) \leq 4C$.

(Note that the above inequalities for $t(i)$ hold whenever C is chosen large enough.)

$\square$

Remark.   There is a question a curious reader might be wondering about: *Why are Fibonacci heaps called Fibonacci heaps?* – Originally, the exponential growth of the subtrees rooted at vertices of rank k wasn't shown directly, as we have done in Lemma 2.11. Rather they were shown to grow at least as fast as the Fibonacci numbers – which were known to grow exponentially, see Problem 2.11.

Problems

2.1 Consider an implementation of a set data structure by an array or a list. Show that for all $k \geq n$ there exist sequences of k operations which need a total of $\Omega(k \cdot n)$ steps.

2.2 Assume $\mathcal{A}$ is a program for a random access machine. Show that there exists a program $\mathcal{A}'$ that computes for every input the same result as $\mathcal{A}$ and satisfies the additional property that it accesses a memory cell M_i if and only if it also accesses the cell M_{i-1}. The running time of algorithm $\mathcal{A}'$ should satisfy $T_{\mathcal{A}'}(n) = \mathcal{O}(T_{\mathcal{A}}(n)^2)$.

2.3 Design an algorithm for the following problem. Input: positive integers $m \leq n$, real numbers $b_1 \leq b_2 \leq \cdots \leq b_m$, and n not necessarily different integers $a_1, \ldots, a_n \in \{1, \ldots, n\}$ in arbitrary order. Output: $c_1, \ldots, c_{m+1}$ such that $c_i := |\{a_j \mid b_{i-1} < a_j \leq b_i\}|$ (where $b_0 = -\infty$ and $b_{m+1} = +\infty$). The running time of your algorithm should not exceed $\mathcal{O}(n)$.

2.4 Design an algorithm which checks in time $\mathcal{O}(n+m)$ whether a given graph $G = (V,E)$ is bipartite.

2.5 Design an algorithm for the following problem. Input: a graph $G = (V,E)$ and a vertex $s \in V$. Output: a spanning tree T of G such that for all vertices $x \in V$ the length of the (unique) s-x-path in T is equal to the length of a shortest path from s to x in G. The running time of your algorithm should not exceed $\mathcal{O}(n+m)$.

2.6 Design an efficient algorithm that checks whether a graph $G = (V,E)$ contains an Euler circuit and, if so, finds one. The running time of your algorithm should not exceed $\mathcal{O}(n+m)$.

2.7 Design an algorithm which finds in connected bipartite graphs $G = (A \uplus B, E)$ a maximum matching. The running time of your algorithm should not exceed $\mathcal{O}(nm)$. (Hint: consider an arbitrary matching M. Show that, if M is not yet a maximum matching, then there exists a path P which contains edges in and not in M alternatingly and which starts and ends with edges not in M.)

2.8 Design an algorithm which finds in trees $T = (V,E)$ in time $\mathcal{O}(n)$ two vertices $x,y \in V$ that have maximum distance. (The distance of two vertices is the length of a shortest path connecting them.)

2.9 Design an algorithm which decides in time $\mathcal{O}(m)$ whether a planar graph $G = (V,E)$ is triangle-free.

2.10 Let $x_1, \ldots, x_n$ be n points in the Euclidean plane. The convex hull for these points can be constructed as follows. First sort the points according to their x coordinate and then determine the convex hull successively for the first i points (in the sorted order). Describe an algorithm for the second step, which runs in $\mathcal{O}(n)$ time. (Hint: use amortized cost analysis.)

2.11 The Fibonacci numbers are defined by $F_0 := 0$, $F_1 := 1$ and the recursion
$F_k = F_{k-1} + F_{k-2}$ for $k \geq 2$. Show that

a) $F_k = 1 + \sum_{i=0}^{k-2} F_i$ for all $k \geq 2$.

b) [Alternative version of Lemma 2.11] If x is a node of rank k in an
F-heap then the subtree rooted at x has size at least F_{k+2}.

c) $F_k \geq \left(\frac{1+\sqrt{5}}{2} \right)^{k-2}$ for all $k \geq 2$.

2.12 Design an implementation of the operation $\texttt{Delete}(H,x)$ which removes a
node x from an F-heap H. What can you say about the amortized running
time of your procedure?

2.13 Design an implementation of the operation $\texttt{Merge}(H_1,H_2)$ which should
merge two F-heaps H_1 and H_2 (containing different nodes) into a single
F-heap. What can you say about the amortized running time of your procedure?

2.14 Design a modification of the Fibonacci heaps for situations in which the
keys come from a known (small) universe U. Your data structure should
still support each $\texttt{Insert}$, and $\texttt{Decrease-Key}$ operation in constant amortized time. $\texttt{Extract-Min}$ operations should be performed in amortized time
$\mathcal{O}(\log(\min\{n,|U|\}))$, where n is the maximum number of elements stored in
the heap simultaneously.

2.15 Prove or disprove. There exists a sequence of n operations such that, starting
from an empty heap, after performing these operations the Fibonacci heap
will consist of exactly one list of size $\theta(n)$.

Notes

There are almost uncountably many books on algorithms. Two classical ones are "The Art
of Computer Programming, Vol. 1, Fundamental Algorithms" by Knuth (1973) and "The
Design and Analysis of Computer Algorithms" by Aho, Hopcroft, and Ullman (1974).
In the latter book, as well as e.g. in the book by Papadimitriou (1994), the reader also
finds more detailed information on the Turing machine and its relation to random access machines. Two recent, highly recommendable textbooks on algorithms are Cormen,
Leiserson, Rivest (2001) and Sedgewick (1998).

Fibonacci heaps were invented by Fredman and Tarjan (1987). Tarjan (1985) presents
a nice survey on the concept of amortized analysis and its applications.

3

Basics III: Complexity

Complexity theory has many facets. Their motivations and goals, however, are similar: to determine the computational "difficulty" or "complexity" of a problem. Clearly, the computational complexity of problems might differ widely. For some the reader will find a (fast) algorithm immediately. For some it might require some efforts to design just any (finite) algorithm. And for some even that might not be possible at all. These fundamental differences (and associated interesting problems) are best illustrated via some examples:

1. GRAPH CONNECTIVITY: *Given a graph $G = (V,E)$, is G connected?* In Section 2.1 we designed an algorithm whose running time is linear in the size of the input (i.e., in the number of vertices and edges). Surely, in the worst case we have to consider (read) every edge before we can decide whether the graph is connected. That is, our algorithm is best possible up to a multiplicative constant.

2. SORTING: *Given n integers $a_1,\ldots,a_n$, return them in ascending order.* Here it is not difficult to design an $\mathcal{O}(n^2)$ algorithm. With some more effort (and appropriate data structures) the running time can be decreased to $\mathcal{O}(n \log n)$ and it can be shown that every comparison based algorithm requires $\Omega(n \log n)$ steps.

3. STEINER PROBLEM: *Given a network $N = (V,E,\ell)$ and a terminal set $K \subseteq V$, return a Steiner minimum tree.* Clearly, for a finite network there exist only finitely many different Steiner trees for K. Thus, enumerating all of them yields a finite – even though inefficient – algorithm. A main issue of this book is whether and under which conditions there exist more efficient algorithms.

4. LINEAR PROGRAMMING: *Given a real $m \times n$ matrix A, a vector $b \in \mathbb{R}^m$, and a vector $c \in \mathbb{R}^n$, return a vector $x \in \mathbb{R}^n$ such that $Ax \le b$ and $c^T x$ is minimum among all such vectors.* This is one of the central problems in operations research. At first sight it is not even obvious that there exists a finite algorithm for solving linear programming problems: it takes some thinking (or a theorem) to show that an optimal solution can always be found among the *corners* of the polyhedron $P = \{x \mid Ax \le b\}$. In addition, it can be shown that there exists an algorithm whose running time is polynomially

bounded in n and m (and of course the size of the numbers in A, b and c). This is one of the most famous and significant results in combinatorial optimization.

5. GLOBAL MINIMIZATION: *Given a function $f : \mathbb{R}^n \to \mathbb{R}$, find a vector $x \in \mathbb{R}^n$ such that $f(x)$ is minimum.*This is now a problem which, at the abstract level at which we posed it, does not necessarily have a finite algorithm. Note, however, that for this problem it is not even clear what the "input" of the problem is: in general it is not possible to encode such a function f by a finite sequence of zeros and ones. In this book we will not consider such general problems.

In general there are two fundamental problems in complexity theory which we will encounter in this book. Namely, does a problem have a polynomial time algorithm solving it and, if so, what is the fastest possible algorithm (up to a multiplicative constant).

3.1   Introduction to complexity theory

Before we continue, we would like to remind the reader that we already convinced ourselves in Section 2.1 that for the answer to these problems it does not matter what kind of machine our algorithms are designed for. Concerning the "input" and the "output" of an algorithm we have been rather sloppy so far. In the definition of a RAM we assumed that the input and the output is a number contained in the first memory cell at the beginning respectively at the end of the algorithm. In the above examples, however, the inputs and outputs of the problems were allowed to be much more general: *e.g.*, a graph or a sequence of numbers. But, clearly, there are straightforward ways to transform these "general" inputs into a string of zeros and ones which we may in turn interpret as a number (and vice versa for the output). In this book we will always tacitly assume these kind of transformations. More precisely, in the abstract notions of complexity theory, which we are about to define now, the inputs, also called the *instances* of a problem, are always finite strings of zeros and ones. On the other hand, in the examples of problems which we consider the instances will always be graphs, networks or a finite sequence of integers, etc. Bridging this gap is routine but tedious and usually left to the reader.

Decision Problems

Let $\Sigma = \{0,1\}$ and let Σ^* denote all finite words over the alphabet Σ (that is, all finite strings of zeros and ones). An *instance* of a problem is a word $x \in \Sigma^*$. The set $\mathcal{I}$ of all instances of a problem will usually form a *proper* subset of Σ^*. Thus every algorithm has to verify first whether a given input is a *valid* input. This is a common problem for every computer programmer: its solution is usually straightforward even though perhaps tedious. Formally, we call a set $\mathcal{I} \subseteq \Sigma^*$ *recognizable in polynomial time*, if there exists an algorithm $\mathcal{A}$ that stops for every string $x \in \Sigma^*$ after at most $\mathcal{O}(|x|^k)$ steps, for an appropriate $k \in \mathbb{N}$, and returns "1" or ACCEPT if $x \in \mathcal{I}$ and "0" or REJECT if $x \in \Sigma^* \setminus \mathcal{I}$. In this book we will only consider sets of instances $\mathcal{I} \subseteq \Sigma^*$ that are recognizable in polynomial time.

Note that in some sense the problem whether a word x belongs to some set $\mathcal{I} \subseteq \Sigma^*$ is quite a simple problem: there are just two possible answers. That is, the algorithm has only to decide between YES instances and NO instances. Such kind of problems are called decision problems and play a fundamental rôle in the theory of computational complexity. Decision problems come in various shapes. For example, deciding whether a given string of 0's and 1's is a valid encoding of a graph is a decision problem. But it is, of course, a rather boring one. Let us look at another, more interesting, example: given a graph $G = (V,E)$ and a terminal set K. Does there exist a Steiner tree for K containing, say, at most $3|K|$ edges? Clearly, this is also a decision problem. But somehow it has a bit more structure. Namely, for every instance consisting of the graph G and the terminal set K we can attribute a set of possible "solutions", e.g. the set of all Steiner trees for K which contain at most $3|K|$ edges. The decision problem then is to decide whether the solution set is nonempty. We will formalize this now. A *decision problem* Π is a tuple $\langle \mathcal{I}, \mathrm{Sol} \rangle$ such that

- $\mathcal{I} \subseteq \Sigma^*$ is a set of *instances* that is recognizable in polynomial time.
- For every instance $I \in \mathcal{I}$, $\mathrm{Sol}(I) \subseteq \Sigma^*$ denotes the set of *solutions* of I.

An algorithm $\mathcal{A}$ is said to *solve* a decision problem $\Pi = \langle \mathcal{I}, \mathrm{Sol} \rangle$ if the algorithm stops for all instances $I \in \mathcal{I}$ and returns ACCEPT if $\mathrm{Sol}(I) \neq \emptyset$ and REJECT otherwise. Note that we *do not* require that the algorithm *finds* a member of $\mathrm{Sol}(I)$. It simply has to decide whether $\mathrm{Sol}(I)$ is empty or not. We will come back to the problem of finding solutions later in this chapter.

Example 3.1 THE SATISFIABILITY PROBLEM. Let $X = \{x_1, x_2, \ldots, x_n\}$ be a set of n Boolean variables. A truth assignment to X is a mapping $\tau : X \longrightarrow \{\text{TRUE}, \text{FALSE}\}$ which assigns to every variable the Boolean value TRUE or FALSE. We freely identify TRUE with the (numerical) value 1 and FALSE with 0. The negation $\overline{x}_i$ of x_i evaluates to 1 if and only if $x_i = 0$. The logical conjunction $x_i \wedge x_j$ evaluates to 1 if and only if $x_i = x_j = 1$ and the disjunction $x_i \vee x_j$ evaluates to 1 if and only if $x_i = 1$ or $x_j = 1$. The set $\{\wedge, \vee, ^-\}$ is already a complete basis, i.e., every Boolean function $f : \{0,1\}^n \longrightarrow \{0,1\}$ can be expressed using only conjunctions, disjunctions and negations. The most straightforward way of achieving this is as a disjunction of at most 2^n conjunctions. For example, the Boolean function $f : \{0,1\}^3 \longrightarrow \{0,1\}$, where $f(x_1, x_2, x_3) = 1$ if and only if $x_1 + x_2 + x_3 \equiv 1 \bmod 2$ can be expressed as $F_0 = (x_1 \wedge x_2 \wedge x_3) \vee (x_1 \wedge \overline{x_2} \wedge \overline{x_3}) \vee (\overline{x_1} \wedge x_2 \wedge \overline{x_3}) \vee (\overline{x_1} \wedge \overline{x_2} \wedge x_3)$. This leads naturally to the following formalization.

A *Boolean formula* can be any Boolean variable, such as x_i, or the negation $\overline{F}$ of any Boolean formula F, or an expression of the form $F_1 \vee F_2$, where F_1 and F_2 are Boolean formulae, or an expression of the form $F_1 \wedge F_2$, where F_1 and F_2 are Boolean formulae.

A *literal* is a formula consisting of just one variable x_i or its negation $\overline{x_i}$.

A Boolean formula F is said to be in *disjunctive normal form* if it is a disjunction of clauses D_i such that $F = D_1 \vee D_2 \vee \cdots \vee D_m$. Each of these clauses D_i is a conjunction of literals so that $D_i = y_{i_1} \wedge y_{i_2} \wedge \cdots \wedge y_{i_k}$. Obviously, the Boolean formula F_0, described above, is in disjunctive normal form. A Boolean formula F is said to be in *conjunctive normal form* if it is a conjunction of clauses C_i such that $F = C_1 \wedge C_2 \wedge \cdots \wedge C_m$. Each of these clauses C_i, in turn, is a disjunction of literals so that $C_i = y_{i_1} \vee y_{i_2} \vee \cdots \vee y_{i_\ell}$. It can easily be seen that every Boolean formula in disjunctive normal form can be transformed in polynomial time into a Boolean formula in conjunctive normal form. In fact, *every* Boolean formula can be transformed in polynomial time into a Boolean formula in conjunctive normal form, cf. Problem 3.1. For this reason and because of its structure, Boolean formulae in conjunctive normal form are of particular interest.

A clause in a Boolean formula in conjunctive normal form is satisfied by a truth assignment if at least one literal included in it evaluates to TRUE. The Boolean formula F is satisfied if all its clauses are satisfied. The SATISFIABILITY problem is the following decision problem for Boolean formulae in conjunctive normal form.

SATISFIABILITY:

Instance: A Boolean formula F in conjunctive normal form.
Question: Is $\mathrm{Sol}(F) \neq \emptyset$, i.e., does there exist a truth assignment which satisfies F?

Sometimes it is convenient to work with Boolean formulae which carry more structure. This leads to the definition of the k-SATISFIABILITY problem.

k-SATISFIABILITY (kSAT):

Instance: A Boolean formula F in conjunctive normal form such that each clause of F contains *at most* k literals.
Question: Is $\mathrm{Sol}(F) \neq \emptyset$, i.e., does there exist a truth assignment which satisfies F?

It is not difficult to show that every Boolean formula F in conjunctive normal form can be expressed as a 3SAT-formula of length at most 3 times the length of F, cf. Problem 3.1. Thereby, the *length* of a formula is the number of literals which occur in the formula. Note, however, that there are Boolean formulae which cannot be expressed as a 2SAT-formula.

Optimization Problems

In the above definition of a decision problem, $\mathrm{Sol}(I)$ is the set of certificates which "prove" that the instance $I \in \mathcal{I}$ has to be accepted. Recall the problem to decide whether a given graph $G = (V,E)$ with terminal set $K \subseteq V$ contains a Steiner tree for K with at most $3|K|$ edges. Here $\mathrm{Sol}(G,K)$ was exactly the set of all those Steiner trees for K which contain at most $3|K|$ edges.

In the context of optimization problems, Sol(I) has to contain all possible candidates for an optimal solution for the given instance $I \in \mathcal{I}$. Thus, asking for a Steiner minimum tree, given some graph $G = (V,E)$ with terminal set $K \subseteq V$, Sol(G,K) naturally contains *all* Steiner trees for K in G. Formally, an optimization problem Π is a four-tuple $\langle \mathcal{I}, \text{Sol}, \text{val}, \text{goal} \rangle$ such that

- $\mathcal{I} \subseteq \Sigma^*$ is the set of *instances* that is recognizable in polynomial time;
- For every instance $I \in \mathcal{I}$, Sol(I) $\subseteq \Sigma^*$ denotes the set of *solutions* of I and is nonempty.
- For every instance I and solution $x \in$ Sol(I), the value val(I,x) is a positive integer. The function val($\cdot,\cdot$) is called the *objective function*.
- goal $\in \{\min, \max\}$.

The aim of an optimization problem is to find for a given instance I a solution $x_{\text{opt}} \in$ Sol(I) such that

$$\text{val}(I,x_{\text{opt}}) = \begin{cases} \min\{\text{val}(I,x) \mid x \in \text{Sol}(I)\} & \text{if goal} = \min, \\ \max\{\text{val}(I,x) \mid x \in \text{Sol}(I)\} & \text{if goal} = \max. \end{cases}$$

We abbreviate the value of the optimal solution by opt(I) := val(I,x_{opt}). Note that opt(I) is well defined, as Sol(I) is, by definition, nonempty for all $I \in \mathcal{I}$.

Most problems reflecting "real world" problems are optimization problems. Decision problems occur rather rarely. There is, however, a very neat and easy way of transforming an optimization problem in a decision problem. As an example consider the MINIMUM STEINER PROBLEM IN GRAPHS introduced in Chapter 1. This is an optimization problem and asks for a Steiner minimum tree in a graph (and not just for a Steiner tree with at most $3|K|$ edges, as we did above). We can, however, easily turn it into a decision problem by just querying whether there exists a Steiner tree of a given length:

STEINER PROBLEM IN GRAPHS:

Instance: A graph $G = (V,E)$, a set $K \subseteq V$ of terminals, and an integer B.
Question: Does there exist a Steiner tree T for K in G such that $|E(T)| \leq B$?

Clearly, if we can solve the decision problem we can also solve the optimization problem by repeatedly solving the decision problem for various bounds B and graphs G'. Namely, in a first phase we let $G' = G$ and decrease B successively, starting with $B = |E|$, until the answer to the decision problem changes from YES (= ACCEPT) to NO (= REJECT). This determines the length B_0 of the Steiner minimum tree. In the second phase we compute the Steiner minimum tree. Starting with $G' = G$ we repeatedly temporarily remove an edge from G' and check whether there still exists a Steiner tree of length B_0. If yes then we make the deletion of the edge permanent, if not we reinsert the edge. We continue in this way until the graph G' is a Steiner tree for K. It is worthwhile to note that while this process is finite it is far from being efficient. Implementing the first phase by a binary search will usually drastically decrease the running time.

The necessity of a clever implementation becomes even clearer if we consider the STEINER PROBLEM IN NETWORKS instead, where ℓ might be a function which takes exponentially large values compared to the number of vertices in the underlying graph.

> STEINER PROBLEM IN NETWORKS:
>
> *Instance:* A network $N = (V,E,\ell)$, a set $K \subseteq V$ of terminals, and an integer B.
> *Question:* Does there exist a Steiner tree T for K in N such that $\ell(T) \leq B$?

Efficient algorithms

When is an algorithm efficient? There is no natural answer. Usually, a linear time algorithm is, but an exponential time algorithm is not. On the other hand, consider two algorithms $\mathcal{A}$ and $\mathcal{B}$ whose running times are

$$T_{\mathcal{A}}(n) = 2^{2^{2^{2^2}}} \cdot n \qquad \text{and} \qquad T_{\mathcal{B}}(n) = 2^{\frac{1}{1000} \cdot n},$$

respectively. Clearly, according to our conventions to "hide" constants in the $\mathcal{O}$-notation, algorithm $\mathcal{A}$ is a linear time algorithm while $\mathcal{B}$ is an exponential time algorithm. But for small n, in fact for $n < 2^{25}$ — and so in particular for all practical purposes — algorithm $\mathcal{B}$ is much faster than algorithm $\mathcal{A}$. Nevertheless one has agreed to call an algorithm *efficient* if its running time is polynomial in the size of the input. With very few exceptions this convention has turned out to be very fruitful.

The classes $\mathcal{P}$ and $\mathcal{NP}$

A decision problem Π belongs to the class $\mathcal{P}$ if and only if there exists a polynomial time algorithm which solves Π. That is, the class $\mathcal{P}$ contains all problems which can be solved by an efficient algorithm. Examples of problems in $\mathcal{P}$ are GRAPH CONNECTIVITY and 2SAT, cf. Problem 3.2. Observe that even though SORTING can be solved by a polynomial time algorithm it is not contained in $\mathcal{P}$ since it is not a decision problem.

For many decision problems (like, e.g., STEINER PROBLEM IN GRAPHS and SATISFIABILITY) no polynomial time algorithm is known. Nevertheless some of these problems have a property which is not inherent to every decision problem: there exist algorithms which, if presented with an instance of the problem (i.e., a graph G, a terminal set K, and a bound B, respectively a Boolean formula F) and, in addition with a potential solution x (i.e., a subgraph T of G, respectively a truth assignment τ for the variables in F) these algorithms *verify* in polynomial time whether x is a valid solution (i.e., whether T is a Steiner tree for K containing at most B edges, respectively whether τ satisfies F). The decision problems with this property form the class $\mathcal{NP}$. This abbreviation comes from *Nondeterministic Polynomial time* – essentially meaning that it is "easy" to verify that a given string x is a solution, but possibly "hard" to find such a string. Formally, a decision problem $\Pi = \langle \mathcal{I}, \text{Sol} \rangle$ belongs to the class $\mathcal{NP}$, if and only if :

- The size of the solutions is polynomially bounded in the length of I, i.e., there exists a polynomial p such that

$$|x| \le p(|I|) \qquad \text{for all } I \in \mathcal{I} \text{ and } x \in \text{Sol}(I).$$

- There exists an algorithm $\mathcal{A}$ and a polynomial q such that for every $I \in \mathcal{I}$ the following is true:

  - If $\text{Sol}(I) \ne \emptyset$ then there exists an $x_0 \in \text{Sol}(I)$ such that algorithm $\mathcal{A}$ accepts the input (I, x_0) in time $q(|I|)$.
  - If $\text{Sol}(I) = \emptyset$ then for every $x \in \Sigma^*$ the algorithm $\mathcal{A}$ rejects the input (I, x) in time $q(|I|)$.

From this definition it is obvious that $\mathcal{P} \subseteq \mathcal{NP}$. The question whether the converse inclusion is also true, i.e., the question

$$\mathcal{P} \overset{?}{=} \mathcal{NP},$$

is one of the most central topics in complexity theory. Practical evidence seems to point towards $\mathcal{P} \ne \mathcal{NP}$. There are thousands of important and well studied problems in $\mathcal{NP}$ and so far no polynomial time algorithm is known which solves a single one of them. But a proof that indeed $\mathcal{P} \ne \mathcal{NP}$ seems to be out of reach in the near future.

A reasonable strategy to attack the $\mathcal{P}$ versus $\mathcal{NP}$ question is to study the most difficult problems in $\mathcal{NP}$, viz. the $\mathcal{NP}$-complete problems.

A first glance at $\mathcal{NP}$-completeness

A very important concept in complexity theory is the concept of reducibility. It allows to show that one problem is at least as "difficult" as another one. The first appearance of this concept which we meet in this book is polynomial reducibility.

Definition 3.2 *A decision problem* $\Pi = \langle \mathcal{I}, \text{Sol} \rangle$ *is said to be polynomially reducible to another decision problem* $\Pi^* = \langle \mathcal{I}^*, \text{Sol}^* \rangle$, *written as* $\Pi \le_p \Pi^*$, *if and only if there exists a function* $f : \mathcal{I} \to \mathcal{I}^*$ *such that* f *is computable in polynomial time and*

$$\text{Sol}(I) \ne \emptyset \quad \Longleftrightarrow \quad \text{Sol}^*(f(I)) \ne \emptyset \qquad \text{for all } I \in \mathcal{I}.$$

Observe that the relation $\le_p$ is transitive. The notion of polynomial reducibility allows us to classify the "most difficult" problems in $\mathcal{NP}$.

Definition 3.3 *A decision problem* Π^* *is called* $\mathcal{NP}$-*complete if and only if* $\Pi^* \in$ $\mathcal{NP}$ *and* $\Pi \leq_p \Pi^*$ *for all problems* $\Pi \in \mathcal{NP}$. *If* Π^* *satisfies the latter condition, but is not contained in* $\mathcal{NP}$, *it is said to be* $\mathcal{NP}$-*hard.*

In order to prove that $\mathcal{P} = \mathcal{NP}$, it is enough to identify just *one* $\mathcal{NP}$-complete problem which admits a polynomial time algorithm for its solution. On the other hand, if we can prove that there exists a problem in $\mathcal{NP}$ which does not allow any polynomial time algorithm for its solution then all $\mathcal{NP}$-complete problems do not have polynomial time algorithms. Therefore it is sensible to collect many problems which are $\mathcal{NP}$-complete.

In order to show that a certain problem $\Pi^* \in \mathcal{NP}$ is $\mathcal{NP}$-complete, we just have to pick one problem Π which is already known to be $\mathcal{NP}$-complete and to show that $\Pi \leq_p \Pi^*$. But so far we have not yet seen a single problem which is $\mathcal{NP}$-complete. And to prove that there exist a problem Π^* which is $\mathcal{NP}$-complete one has to show for *every* problem $\Pi \in \mathcal{NP}$ that $\Pi \leq_p \Pi^*$. This was done in a seminal paper by Stephen Cook in 1971, who showed that the SATISFIABILITY problem is $\mathcal{NP}$-complete.

Theorem 3.4 (COOK)   SATISFIABILITY *is* $\mathcal{NP}$-*complete.*

Sketch of proof.   Let $\Pi = \langle \mathcal{I}, \text{Sol} \rangle$ be an arbitrary problem in $\mathcal{NP}$ and let $\mathcal{A}$ be a polynomial time algorithm which decides for every pair $I \in \mathcal{I}$ and $x \in \Sigma^*$ whether $x \in \text{Sol}(I)$. The idea of the proof of Cook's theorem is to define for every I an instance of the SATISFIABILITY-problem which "simulates" the algorithm $\mathcal{A}$. How can this be done? The principal ideas are quite simple, but the precise formulation of all details is rather tedious. So here we will restrict ourselves to presenting the main ideas.

We shall assume that $\mathcal{A}$ is a program for a RAM. Let n denote the length of the input, i.e., the length of the sequence which is contained in the memory cell M_1 (which contains I and x) when the program starts. Since $\mathcal{A}$ is polynomially bounded in n, there are polynomials $p(\cdot), q(\cdot), r(\cdot)$, and $s(\cdot)$ such that $\mathcal{A}$ has only access to the memory cells $M_1, M_2, \ldots, M_{p(n)}$, the longest sequence which occurs in one memory cell has at most $q(n)$ digits, the length of the program is $r(n)$, and the running time is at most $s(n)$. The existence of the polynomials $q(n), r(n)$ and $s(n)$ should be obvious. The argument for $p(n)$ is slightly more complicated. Clearly, $\mathcal{A}$ can access only polynomially many memory cells. But, due to the possibility of indirect addressing, these need not necessarily be the first $p(n)$ memory cells. One can, however, show (cf. Problem 2.2) that for every polynomial time algorithm $\mathcal{A}$ there exists a polynomial time algorithm $\mathcal{A}'$ which computes the same results and accesses the memory cells consecutively.

For every memory cell M_i, $i \leq p(n)$, for every digit $j \leq q(n)$, and for every time step $t \leq s(n)$ we introduce a variable $m(i,j,t)$. The intended interpretation is that $m(i,j,t)$ is TRUE if and only if the cell M_i contains at position j a "1" at time t. Moreover, we introduce variables $\ell(i,t)$ for $i \leq r(n)$ and $t < s(n)$ with the intention that $\ell(i,t)$ should be TRUE if and only if the value of the program counter at time t is equal to i. Any truth assignment according to the intended interpretation gives a complete description of the algorithm $\mathcal{A}$, requiring only polynomially many variables.

What about the clauses? The clauses now will be forced to stick to the intended interpretation, i.e., the resulting formula will only be satisfied by a truth assignment if this assignment describes indeed an execution of the algorithm $\mathcal{A}$ and, additionally, if $\mathcal{A}$ accepts the given input. To simplify notation, we introduce the exclusive or (XOR)-function denoted by $x_i \oplus x_j$ and its negation $x_i \Leftrightarrow x_j$, called *equivalence*. The expression $x_i \oplus x_j$ evaluates to 1 if and only if exactly one of the variables equals 1. Thus, $x_i \oplus x_j$ can be expressed as $x_i \oplus x_j = (x_i \vee x_j) \wedge (\overline{x}_i \vee \overline{x}_j)$ and, hence, $(x_i \Leftrightarrow x_j) = (x_i \wedge x_j) \vee (\overline{x}_i \wedge \overline{x}_j)$.

We have to ensure that for each t exactly one $\ell(i,t)$ becomes true and that start and stop configurations are described properly. This can easily be done with polynomially many clauses, the details are left to the reader. Then we have to ensure that for every $t < s(n)$ the $(t+1)$st configuration is a successor of the tth configuration. Assume for example that the tth step of the program is an addition $M_{i_0} \leftarrow M_{i_1} + M_{i_2}$. First we ensure that changes only occur in memory cell M_{i_0} by adding clauses $m(i,j,t) \Leftrightarrow m(i,j,t+1)$ for every $i \neq i_0$ and j. Next, we have to ensure that the content of cell M_{i_0} changes correctly.

We start considering the last digit, i.e., $j = 0$. Of course, $x_0 := m(i_0,0,t+1)$ should be true if and only if exactly one of the variables $x_1 := m(i_1,0,t)$ and $x_2 := m(i_2,0,t)$ i.e., $x_1 \oplus x_2$ is true. Hence, we just have to add the clause $x_0 \Leftrightarrow (x_1 \oplus x_2)$. For arbitrary j_0, the situation becomes slightly more complicated because we have to deal with carry-overs. Surely, the carry-over at j_0 depends only on $y_1 := m(i_1,j_0 - 1,t), y_2 := m(i_2,j_0 - 1,t), y_3 := m(i_0,j_0 - 1,t+1)$ and is given by $C := (y_1 \wedge y_2) \vee (\overline{y}_3 \wedge (y_1 \oplus y_2))$, i.e., if C becomes true, a carry-over occurs, otherwise not. Hence, with the same reasoning as before, we have to add the clause $y_0 \Leftrightarrow y_1 \oplus y_2 \oplus C$, where $y_0 := m(i_0,j_0,t+1)$.

Similarly, one can simulate every step of the algorithm $\mathcal{A}$ by a clause of polynomial length. Finally, one obtains a formula, which can easily be converted into an equivalent formula in conjunctive normal form, such that this formula is satisfiable if and only if $\mathcal{A}$ accepts (I,x), i.e., $x \in \mathrm{Sol}(I)$.   □

Observe that the reduction shown in the proof of Cook's theorem gives actually more than asked for. Namely, the definition of $\Pi \leq_p \Pi^*$ just requires that the function f maps instances I of Π with nonempty solution sets onto instances $f(I)$ of Π^* with nonempty solution set. The transformation f in the proof of Cook's theorem has the additional property that a satisfying assignment for the Boolean formula $f(I)$ contains also an encoding of an actual solution $x \in \mathrm{Sol}(I)$. We summarize this observation in the following corollary.

Corollary 3.5 *Let $\mathcal{F}$ denote the set of instances of* SATISFIABILITY, *i.e., the set of Boolean formulae in conjunctive normal form. Then for every* $\Pi = \langle \mathcal{I}, \text{Sol} \rangle \in \mathcal{NP}$ *there exist mappings* f *and* g *which are computable in polynomial time such that* $f(I) \in \mathcal{F}$ *for every* $I \in \mathcal{I}$ *and* $g(I, \tau) \in \text{Sol}(I)$ *for every* $\tau \in \text{Sol}(f(I))$. $\square$

Corollary 3.6 3SAT *is* $\mathcal{NP}$-*complete.*

Proof.   In the light of Theorem 3.4 we have only to show that SATISFIABILITY $\leq_p$ 3SAT. This follows immediately from Problem 3.1. $\square$

Surely, as a consequence of Corollary 3.6, kSAT is $\mathcal{NP}$-complete for every $k \geq 3$. On the other hand, Problem 3.2 provides a polynomial time algorithm for 2SAT. Thus one can say that the step from 2 to 3 variables per clause crosses the border to $\mathcal{NP}$-completeness.

The classes $\mathcal{PO}$ *and* $\mathcal{NPO}$

As a final step, we now introduce complexity classes for optimization problems. These classes will be studied in more detail in Chapter 7.

An optimization problem belongs to the class $\mathcal{NPO}$, if and only if :

• The size of the solutions is polynomially bounded in the length of I, i.e., there exists a polynomial p such that

$$|x| \leq p(|I|) \qquad \text{for all } I \in \mathcal{I} \text{ and } x \in \text{Sol}(I).$$

• The question "*Is* $x \in \text{Sol}(I)$?" is decidable in polynomial time.
• The function val$(\cdot, \cdot)$ is computable in polynomial time.

An optimization problem in $\mathcal{NPO}$ belongs to the class $\mathcal{PO}$ if it is solvable in polynomial time.

Example 3.7 MAXkSAT. The problem kSAT is a decision problem asking for the existence of an assignment satisfying all clauses. It is reasonable to ask more generally for truth assignments that satisfy (not necessarily all clauses but) as many clauses as possible. This leads us to the following maximization version of the kSAT problem:

> MAXkSAT:
>
> *Given:* A Boolean formula F in conjunctive normal form such that each clause of F contains at most k literals.
> *Find:*   A truth assignment which maximizes the number of satisfied clauses.

Note that knowing a polynomial time algorithm to solve the kSAT problem does not lead to a polynomial time algorithm for the MAXkSAT problem. In fact, for 2SAT such an algorithm would even prove $\mathcal{P} = \mathcal{NP}$. More precisely, in Problem 3.2 the reader is asked to show that there exists a polynomial time algorithm to solve 2SAT. But it can be shown, cf. Problem 3.5, that the following decision variant of MAX2SAT is $\mathcal{NP}$-complete: given a Boolean formula F in conjunctive normal form such that each clause of F contains at most 2 literals and a positive integer B, does there exist a truth assignment such that at least B clauses of F become true?

Every optimization problem $\Pi = \langle \mathcal{I},\text{Sol},\text{val},\text{goal}\rangle$ has a naturally associated decision problem, called the *underlying decision problem*, $\Pi_D = \langle \mathcal{I}_D,\text{Sol}_D\rangle$ defined by $\mathcal{I}_D := \mathcal{I} \times \mathbb{Z}$ and

$$\text{Sol}(I,B) := \begin{cases} \{x \in \text{Sol}(I) \mid \text{val}(I,x) \geq B\}, & \text{if goal} = \max \\ \{x \in \text{Sol}(I) \mid \text{val}(I,x) \leq B\}, & \text{if goal} = \min. \end{cases}$$

It should be immediately clear, that the optimization version is no easier than the decision version.

Theorem 3.8 *If Π is an optimization problem such that the underlying decision problem Π_D is $\mathcal{NP}$-complete, then $\Pi \notin \mathcal{PO}$, unless $\mathcal{P} = \mathcal{NP}$.*   □

The converse is not necessarily true. If, however, we do know an a priori bound for the optimal value, then a binary search procedure similar to the one outlined on page 45 can usually be used to solve the optimization problem based on an algorithm for the decision problem.

3.2   Excursion: More $\mathcal{NP}$-complete problems

Proving $\mathcal{NP}$-completeness results somehow serves as an excuse for not giving a polynomial time algorithm. On the other hand, it is an important ingredient of our methodology for studying computational problems. By now there are several thousands of problems known to be $\mathcal{NP}$-complete. In this section we will mention just a few of them which are important for the course of this book. But they will nevertheless give a good impression of the art of proving $\mathcal{NP}$-completeness results.

Steiner problem in graphs

The first $\mathcal{NP}$-completeness result we are going to prove is a key result for this book. Assuming the widely believed conjecture that $\mathcal{P} \neq \mathcal{NP}$, it answers the question whether there exists an algorithm which solves the Steiner problem for every terminal set K in polynomial time. And the answer is: no!

Theorem 3.9 STEINER PROBLEM IN GRAPHS *is* $\mathcal{NP}$-*complete*.

Proof. As obviously STEINER PROBLEM IN GRAPHS $\in \mathcal{NP}$, it suffices to show that STEINER PROBLEM IN GRAPHS is in fact $\mathcal{NP}$-complete. To see this, we reduce 3SAT to STEINER PROBLEM IN GRAPHS. Let $x_1, \ldots, x_n$ be the variables and $C_1, \ldots, C_m$ the clauses in an arbitrary instance of 3SAT. Our aim is to construct a graph $G = (V, E)$, a terminal set K, and a bound B such that G contains a Steiner tree T for K of size at most B if and only if the given 3SAT instance is satisfiable.

The graph G is constructed as follows. First we connect two vertices u and v by a *variable path* as shown in Figure 3.1. Second we create for every clause C_i a

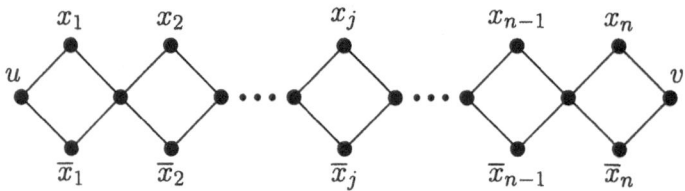

Figure 3.1 Transforming 3SAT to STEINER PROBLEM IN GRAPHS: the variable path.

clause gadget consisting of a vertex C_i that is connected to the literals contained in the clause C_i by paths of length $t = 2n + 1$. As terminal set we choose $K = \{u, v\} \cup \{C_1, \ldots, C_m\}$ and set B to $B = 2n + t \cdot m$.

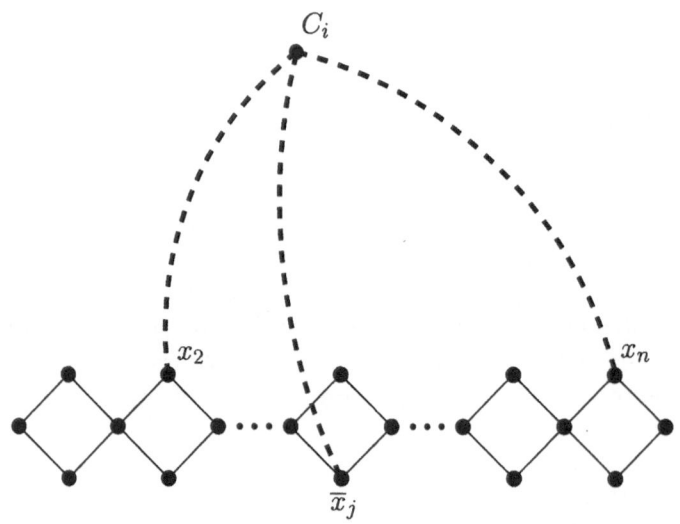

Figure 3.2 The clause gadget for the clause $C_i = x_2 \vee \overline{x}_j \vee x_n$. The dashed lines indicate paths of length $t = 2n + 1$ from C_i to the appropriate vertices on the variable path.

Assume first that the 3SAT instance is satisfiable. To construct a Steiner tree for K we start with a u-v path P reflecting a satisfying assignment. That is, we let $x_i \in P$ if x_i is set to true in this assignment, and $\overline{x}_i \in P$ otherwise. Next observe that for every clause the vertex C_i can be connected to P by a path of length t. In this way we obtain a Steiner tree for K of length $2n + t \cdot m = B$.

To see the other direction assume now that T is a Steiner tree for K of length at most B. Trivially, for each clause the vertex C_i has to be connected to the variable path. Assume for the moment that there exists a clause C_{i_0} for which C_{i_0} is connected to the variable path by at least *two* of its paths. Then $|E(T)| \geq (m+1) \cdot t > B$, so this can't be. This shows that u and v can only be connected along the variable path, which requires at least $2n$ edges. As each clause needs at least t edges to connect C_i to the variable path, we conclude that the u-v path contains *exactly* $2n$ edges and that each clause gadget is connected to *this* path using exactly t edges. Thus the u-v path reflects a satisfying assignment.

The observation that this construction can easily be obtained in polynomial time concludes the proof of Theorem 3.9. □

The restricted Steiner problem

In this and the next paragraph we will show that two seemingly easier special cases of the Steiner problem in graphs are nevertheless still $\mathcal{NP}$-complete. First we will prove that the Steiner problem stays $\mathcal{NP}$-complete even if we know in advance that the structure of the Steiner minimum tree is "simple", i.e., if the Steiner minimum tree is decomposable in a rather easy way. Then, in the next paragraph, we show that the Steiner problem remains $\mathcal{NP}$-complete if the graph under consideration is planar. In each case we first prove that a certain variant of the 3SAT problem is $\mathcal{NP}$-complete.

RESTRICTED3SAT:

Instance: A 3SAT-formula F such that every variable occurs in at most 3 clauses. Moreover, it is required that each variable appears both negated and unnegated.

Question: Does there exist a satisfying assignment for F?

Theorem 3.10 RESTRICTED3SAT *is* $\mathcal{NP}$-*complete.*

Proof. Obviously, RESTRICTED3SAT is in $\mathcal{NP}$. To show that the restricted problem is $\mathcal{NP}$-complete we reduce 3SAT to RESTRICTED3SAT. Given a 3SAT formula F, let z be a variable with exactly k occurrences in F. Let $z_1, \dots, z_k$ be k new variables. Then we replace the ith occurrence of z with z_i and append the 2SAT formula

$$(\overline{z}_1 \vee z_2) \wedge (\overline{z}_2 \vee z_3) \wedge \cdots \wedge (\overline{z}_k \vee z_1).$$

Observe that this formula forces all the z_i to have the same truth value in any satisfying assignment, and in the resulting expression there are exactly three occurrences of each z_i, at least one of them is unnegated and at least one of them is negated. We proceed the same way for all variables. Since this construction can obviously be obtained in polynomial time, we have shown that $3\textsc{Sat} \leq_p$ $\textsc{Restricted3Sat}$. □

Assume that a Steiner minimum tree for K is given such that some of the terminals are interior points. Then we can decompose this tree (by splitting terminals) into components so that terminals only occur as leaves of these components. Such a component is called a *full component*. Figure 3.3 illustrates the decomposition of a Steiner tree into full components.

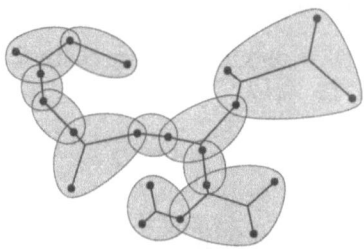

Figure 3.3 The full components of a Steiner tree

The reader is invited to prove that if it is known in advance that for a given Steiner problem there exists a Steiner minimum tree such that every full component contains at most two terminals then there exists a polynomial time algorithm to determine a Steiner minimum tree. We will exploit this idea in detail in Chapters 6 and 7. Though there is some evidence for it, cf. Chapter 8, no polynomial time algorithm is known for the Steiner problem if the existence of a Steiner minimum tree is guaranteed having only full components containing at most three terminals. But in general the situation is even worse.

R-RESTRICTED STEINER PROBLEM:

Instance: A graph $G = (V,E)$, a set $K \subseteq V$, and an integer B.
Question: Does there exist a Steiner tree T for K such that every full component contains at most r terminals and such that $|E(T)| \leq B$?

Theorem 3.11 *For every* $r \geq 4$, R-RESTRICTED STEINER PROBLEM *is $\mathcal{NP}$-complete.*

Proof. We adopt the proof of Theorem 3.9 reducing RESTRICTED3SAT to the 4-RESTRICTED STEINER PROBLEM. In order to do so we construct a graph as in the proof of Theorem 3.9 but, additionally, we include all vertices of the variable path different from the x_i's and $\overline{x_i}$'s to the terminal set. Then the fact that every variable occurs at most twice in unnegated form and at most twice in negated form implies that there exists a Steiner minimum tree for the new terminal set, in which every full component contains at most 4 terminals. The remaining arguments remain the same.   □

The Steiner problem in planar graphs

Planar graphs are in many respects an important class of graphs. Restricting the Steiner problem to this class, one might hope that its solution becomes considerably easier. Unfortunately, this is not the case. To prove that the Steiner problem remains $\mathcal{NP}$-complete also in planar graphs, we first prove that a variant of the 3SAT problem, called PLANAR3SAT, is also $\mathcal{NP}$-complete.

PLANAR3SAT:

Instance: A 3SAT-formula $F = C_1 \wedge \cdots \wedge C_m$ using variables $x_1, \ldots, x_n$ such that the following graph $G = (V,E)$ is planar:

$$V = \{x_i, \overline{x_i}, y_i, z_i \mid 1 \leq i \leq n\} \cup \{C_j \mid 1 \leq j \leq m\}, \text{ and}$$
$$E = \{\{y_i, x_i\}, \{y_i, \overline{x_i}\}, \{x_i, z_i\}, \{\overline{x_i}, z_i\} \mid 1 \leq i \leq n\}$$
$$\cup \{\{z_i, y_{i+1}\} \mid 1 \leq i < n\} \cup \{\{z_n, y_1\}\}$$
$$\cup \{\{\lambda, C_j\} \mid 1 \leq j \leq m, \lambda \text{ a literal of } C_j\}.$$

Question: Does there exist a satisfying assignment for F?

Note that the graph associated to a 3SAT formula in problem PLANAR3SAT is very similar to the one constructed in the proof of Theorem 3.9.

Theorem 3.12 PLANAR3SAT *is $\mathcal{NP}$-complete.*

Proof. We shall reduce 3SAT to PLANAR3SAT in several steps. First we show how to transform a given 3SAT instance into an equivalent one such that its occurrence graph, which has as vertices all variables and clauses and edges from a variable to a clause if and only if the variable or its negation appears in the clause, is planar. Second we show how to add a "line" through all vertices without destroying planarity. Lastly, we show how to "pull apart" each variable vertex into a negated and an unnegated vertex.

Let us start with the first step. Assume a 3SAT instance $F = C_1 \wedge \cdots \wedge C_m$ using n variables $x_1, \ldots, x_n$ is given. Embed its occurrence graph in the plane as follows: put all variable vertices on the x-axis and all clause vertices on the y-axis and connect two adjacent vertices by a "hook" as shown in Figure 3.4. (Ignore the bold-faced line for the moment.)

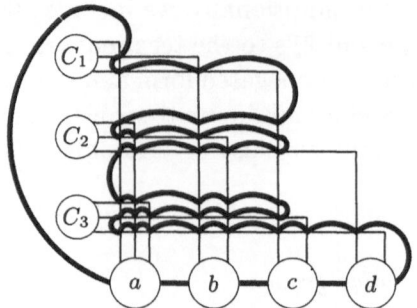

Figure 3.4 Occurrence graph of the instance $(a \vee \bar{b} \vee \bar{c}) \wedge (\bar{a} \vee b \vee d) \wedge (a \vee c \vee \bar{d})$.

In general this graph will of course not be planar. On the other hand, it will certainly not contain more than $9m^2$ crossings as there are at most $3m$ such hooks. We will now show how to remove these crossings successively. Consider a crossing

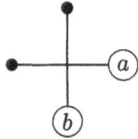

We replace it by the subgraph shown in Figure 3.5.

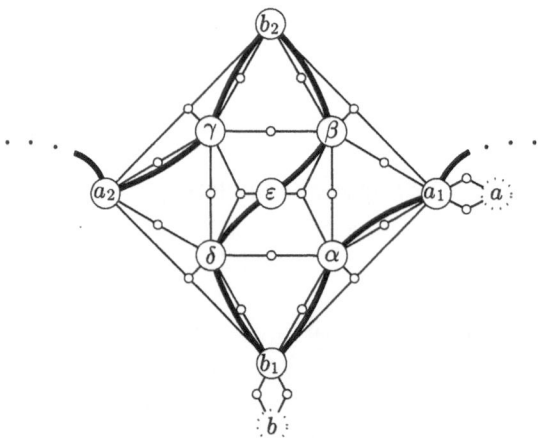

Figure 3.5 Gadget replacing a crossing.

Here, a_1, a_2, b_1, b_2, and α, β, γ, δ, and ϵ are new variables and the little circles represent the following clauses:

$\bar{a}_1 \vee \bar{b}_1 \vee \alpha,\ a_1 \vee \bar{\alpha},\ b_1 \vee \bar{\alpha},$	*(i.e.,*	$a_1 \wedge b_1 \Leftrightarrow \alpha$ *)*
$\bar{a}_1 \vee b_2 \vee \beta,\ a_1 \vee \bar{\beta},\ \bar{b}_2 \vee \bar{\beta},$	*(i.e.,*	$a_1 \wedge \bar{b}_2 \Leftrightarrow \beta$ *)*
$a_2 \vee b_2 \vee \gamma,\ \bar{a}_2 \vee \bar{\gamma},\ \bar{b}_2 \vee \bar{\gamma},$	*(i.e.,*	$\bar{a}_2 \wedge \bar{b}_2 \Leftrightarrow \gamma$ *)*
$a_2 \vee \bar{b}_1 \vee \delta,\ \bar{a}_2 \vee \bar{\delta},\ b_1 \vee \bar{\delta},$	*(i.e.,*	$\bar{a}_2 \wedge b_1 \Leftrightarrow \delta$ *)*
$\alpha \vee \beta \vee \epsilon,\ \gamma \vee \delta \vee \bar{\epsilon},$	*(i.e.,*	$\alpha \vee \beta \vee \gamma \vee \delta$ *)*
$\bar{\alpha} \vee \bar{\beta},\ \bar{\beta} \vee \bar{\gamma},\ \bar{\gamma} \vee \bar{\delta},\ \bar{\delta} \vee \bar{\alpha},$	*(i.e.,*	$\alpha \Rightarrow \bar{\beta} \wedge \bar{\delta},\ \beta \Rightarrow \bar{\alpha} \wedge \bar{\gamma},$
	and,	$\gamma \Rightarrow \bar{\beta} \wedge \bar{\delta},\ \delta \Rightarrow \bar{\alpha} \wedge \bar{\gamma}$ *)*
$a \vee \bar{a}_1,\ \bar{a} \vee a_1,$	*(i.e.,*	$a \Leftrightarrow a_1$ *)*
$b \vee \bar{b}_1,\ \bar{b} \vee b_1$	*(i.e.,*	$b \Leftrightarrow b_1$ *)*

Straightforward case checking verifies that these clauses ensure that in every satisfying assignment $a \Leftrightarrow a_1 \Leftrightarrow a_2$ and $b \Leftrightarrow b_1 \Leftrightarrow b_2$. Using these gadget successively for each crossing (with different variables) we will finally end up at an equivalent 3SAT instance whose occurrence graph is planar.

The remaining two steps are easy. The bold-faced line in Figures 3.4 and 3.5 shows how a line through all variable vertices may be found (use some extra crossing gadgets for intersections with vertical lines). Finally, we need to replace each variable vertex, a say, by two vertices, one corresponding to the literal a, the other to $\bar{a}$. At that point it is convenient to label all incoming edges with "+" or "−" according to whether the edge corresponds to the use of a or $\bar{a}$. In case the bold-faced line separates the plus from the minus edges, the vertex can easily be pulled apart:

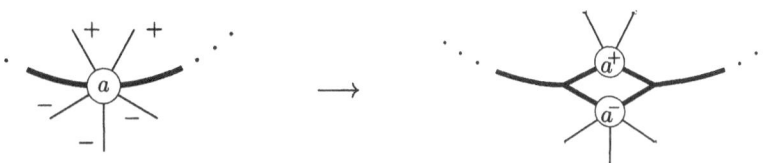

If not, we replace the vertex a by an appropriate number of vertices $a_1, \ldots, a_k$ (degree of a many will always suffice), and add the clauses $(\bar{a}_i \vee a_{i+1}) \wedge (a_i \vee \bar{a}_{i+1})$ for all $i = 1, \ldots, k-1$. (Note that these clauses guarantee that $a_1 \Leftrightarrow a_2 \Leftrightarrow \cdots \Leftrightarrow a_k$ in every satisfying truth assignment.) As indicated in the following figure

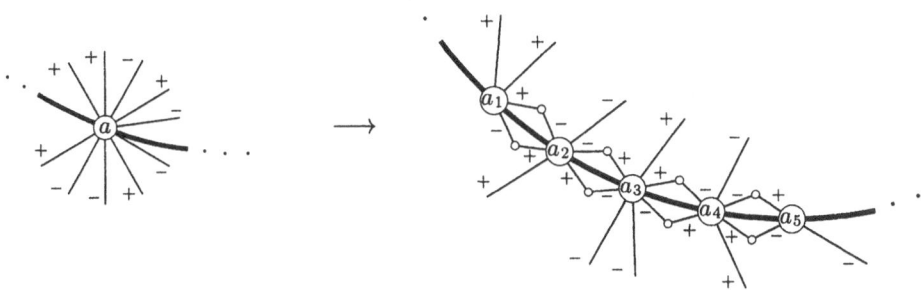

the clause vertices which were previously connected to a can now be connected to the new vertices a_i in such a way that the bold-faced line separates the plus from the minus edges in each of the vertices a_i.   □

The same reduction as in the proof of Theorem 3.9 but using only instances of PLANAR3SAT yields the $\mathcal{NP}$-completeness of STEINER PROBLEM IN PLANAR GRAPHS.

STEINER PROBLEM IN PLANAR GRAPHS:

Instance: A planar graph $G = (V,E)$, a set $K \subseteq V$ of terminals, an integer B.
Question: Does there exist a Steiner tree T for K in G such that $|E(T)| \leq B$?

Corollary 3.13 STEINER PROBLEM IN PLANAR GRAPHS *is $\mathcal{NP}$-complete.*   □

Tripartite matching

The next two problems are not directly related to the Steiner problem but will turn out to be useful in further investigations on the difficulty of several decision and optimization problems studied in this text book.

First assume that two sets are given: a set of girls and a set of boys each of the same cardinality. Moreover, there is a relation between some girls and some boys which might be interpreted as "girl A wants to live with boy B". For obvious reasons, the question is to find a pairing between girls and boys such that each girl is related to exactly one boy with whom she wants to live and vice versa. Translated into a graph theoretic setting this problem is exactly that of finding a perfect matching in a bipartite graph. From Problem 2.7 we know that there exists a polynomial time algorithm which solves this problem.

Now, consider what happens if we add one more restriction to our problem. Assume that there are three sets of the same size given: a set of girls, a set of boys, and a set of cities. Moreover, there is a triple relation which might be interpreted as "girl A wants to live with boy B in city C". The task is to find a matching according to the given triple relation such that each girl, each boy, and each city is covered by exactly one such triple. Unfortunately, preferences for certain cities complicates the situation quite a bit: we will shown that this new problem is $\mathcal{NP}$-complete.

TRIPARTITE MATCHING:

Instance: Three pairwise disjoint sets X, Y and Z, each containing n elements, and a set $S \subseteq X \times Y \times Z$.
Question: Does there exist a *perfect matching*, i.e., a subset $M \subseteq S$ such that every element $v \in X \cup Y \cup Z$ is contained in exactly one element of M?

Theorem 3.14 TRIPARTITE MATCHING *is $\mathcal{NP}$-complete.*

Proof. We shall reduce 3SAT to TRIPARTITE MATCHING. Let $x_1,\ldots,x_n$ be the variables and $C_1,\ldots,C_m$ the clauses in an arbitrary instance of 3SAT. We construct an instance of TRIPARTITE MATCHING as follows. We replace every variable x_i by $4m$ variables $x_i^j,\bar{x}_i^j,y_i^j,z_i^j$, $1 \le j \le m$. Then we add some triples in such a way (see set S_1 below) such that each element y_i^j and z_i^j is contained in exactly two triples and such that the choice of any one triple covering, say, y_i^1 automatically also fixes the choice of the triples covering the remaining elements y_i^j and z_i^j and such that by this choice either all elements x_i^j or all elements $\bar{x}_i^j$ are covered. Figure 3.6 illustrates how this is achieved.

Next we add for each clause C_j two elements a^j and b^j and three triples (see set S_2 below) each containing a^j and b^j and exactly one of the three literals of C_j. Finally, we add some extra triples (see set S_3 below) that ensure that every partial matching which covers all elements y_i^j, z_j^j, a^j and b^j can be extended to a perfect matching.

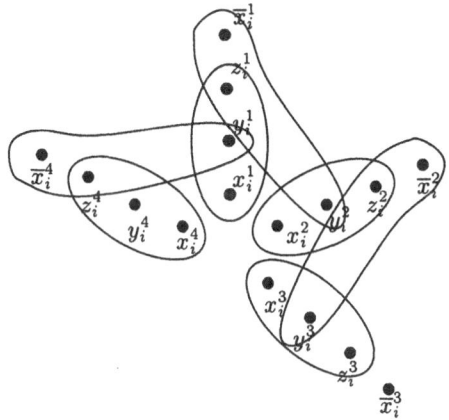

Figure 3.6 Gadget for variable x_i

Formally, the desired instance of TRIPARTITE MATCHING is defined as follows:

$$
\begin{aligned}
X &:= \{x_i^j,\bar{x}_i^j \mid 1 \le i \le n, 1 \le j \le m\} \\
Y &:= \{y_i^j \mid 1 \le i \le n, 1 \le j \le m\} \cup \{a^j \mid 1 \le j \le m\} \\
&\quad \cup \{c_k \mid 1 \le k \le (n-1)m\} \\
Z &:= \{z_i^j \mid 1 \le i \le n, 1 \le j \le m\} \cup \{b^j \mid 1 \le j \le m\} \\
&\quad \cup \{d_k \mid 1 \le k \le (n-1)m\} \\
S_1 &:= \{(x_i^j,y_i^j,z_i^j),(\bar{x}_i^j,y_i^{j+1},z_i^j) \mid 1 \le i \le n, 1 \le j \le m-1\} \\
&\quad \cup \{(x_i^m,y_i^m,z_i^m),(\bar{x}_i^m,y_i^1,z_i^m) \mid 1 \le i \le n\} \\
S_2 &:= \{(\lambda^j,a^j,b^j) \mid 1 \le j \le m,\ \lambda \text{ a literal of } C_j\} \\
S_3 &:= \{(x,c_k,d_k) \mid x \in X,\ 1 \le k \le (n-1)m\}
\end{aligned}
$$

and $S := S_1 \cup S_2 \cup S_3$.

We claim that the given 3SAT instance is satisfiable if and only if the constructed instance of TRIPARTITE MATCHING contains a perfect matching.

Assume a perfect matching M is given. Every element y_i^j is contained in only two elements of S. Hence, for each $1 \le i \le n$, $M \cap S_1$ either contains all x_i^j (meaning that x_i is set to FALSE) or none (meaning that x_i is set to TRUE). As each element of $M \cap S_2$ picks up one of the remaining elements x_i^j (one for each clause), this truth assignment is a satisfying assignment. As outlined above the elements in S_3 allow the extension of this matching to a perfect matching.

Using the same ideas, the construction of a perfect matching from a satisfying assignment is straightforward.   □

The subset sum problem

Finally, we will cross the border between $\mathcal{P}$ and $\mathcal{NP}$ once more. Let B be a fixed integer. Given any sequence $a_1, \dots, a_n$ of integers from $\{1, \dots, B\}$ and an arbitrary integer K the reader will easily find a polynomial time algorithm to decide whether there is a subsequence $I \subseteq \{1, \dots, n\}$ such that $\sum_{i \in I} a_i = K$, cf. Problem 3.8. But as soon as we drop the assumption that each of the integers $a_1, \dots, a_n$ is bounded by B, i.e., if we allow $a_1, \dots, a_n$ to be arbitrary integers, the problem becomes difficult. More precisely:

SUBSET SUM:

Instance: Integers $a_1, \dots, a_n$, and K.
Question: Does there exist $I \subseteq \{1, \dots, n\}$ such that $\sum_{i \in I} a_i = K$?

Theorem 3.15 SUBSET SUM *is* $\mathcal{NP}$*-complete.*

Proof.   Surely SUBSET SUM $\in \mathcal{NP}$: given $I \subseteq \{1, \dots, n\}$ it is easy to check in polynomial time that $\sum_{i \in I} a_i = K$.

We show that the problem is $\mathcal{NP}$-complete by polynomially reducing TRIPARTITE MATCHING to SUBSET SUM. Given X, Y, Z, each containing n elements, and $S \subseteq X \times Y \times Z$ where $S = \{S_1, \dots, S_m\}$. We can think of each triple $S_i \in S$ as a 0-1 vector of length $3n$, containing exactly three 1's, corresponding to the three elements in S_i. Now interpret these 0-1 vectors as integers written to the base $m + 1$. Say, a_i is the integer associated to S_i in this way. Moreover, let K be the integer corresponding to the constant 1 vector of length $3n$. We claim that there exists a subset $M \subseteq S$ which is a perfect matching if and only if there exists a subset $I \subseteq \{1, \dots, m\}$ such that $\sum_{i \in I} a_i = K$. First assume that $M = \{S_i \mid i \in I\}$ for some $I \subseteq \{1, \dots, m\}$ is a perfect matching. Then it is obvious that $\sum_{i \in I} a_i = K$.

So assume now that there exists $I \subseteq \{1,\ldots,m\}$ such that $\sum_{i \in I} a_i = K$. Note that there are at most m summands and, expanding each summand again into its base $m+1$ representation, we observe that only the digits 0 and 1 appear in the summands. Thus, there is no carry-over in this addition. Hence, $\sum_{i \in I} a_i = K$ means that to each of the $3n$ positions there corresponds a unique S_i having a "1" in this position. Therefore, the family $M = \{S_i \mid i \in I\}$ is a perfect matching.

$\square$

Problems

3.1 Show that every Boolean formula F can be expressed as a 3SAT-formula of length at most 3 times the length of F. Show that there exists a Boolean formula which can not be expressed as a 2SAT-formula.

3.2 Design an algorithm which decides in polynomial time whether a given 2SAT formula is satisfiable, i.e., show that 2SAT $\in \mathcal{P}$.

3.3 An algorithm is called *nondeterministic* if in each step it first "guesses" a value $v \in \{0,1\}$ and then performs an instruction depending on the value v. A set $L \subseteq \Sigma^*$ is called recognizable by a nondeterministic algorithm $\mathcal{A}$ if for all $x \in L$ there exist appropriate guesses so that $\mathcal{A}$ returns TRUE, while for all $x \notin L$ the algorithm returns FALSE regardless of the guesses for the values v. Show that a decision problem $\Pi = \langle \mathcal{I}, \text{Sol} \rangle$ belongs to $\mathcal{NP}$ if and only if the set $L_\Pi = \{I \in \mathcal{I} \mid \text{Sol}(I) \neq \emptyset\}$ can be recognized by a nondeterministic polynomial time algorithm.

3.4* Let $\Pi = \langle \mathcal{I}, \text{Sol} \rangle$ be a decision problem. Then Π is said to belong to the class $co\mathcal{NP}$ whenever $\Sigma^* \setminus L_\Pi$ can be decided by a nondeterministic polynomial time algorithm. Show that $\langle \mathbb{N}, Sol \rangle$ where $\text{Sol}(n) = \{i \in \mathbb{N} \mid 2 < i < n \text{ and } i \mid n\}$ belongs to the class $\mathcal{NP} \cap co\mathcal{NP}$.

3.5 Show that the following variant of the MAX2SAT problem is $\mathcal{NP}$-complete. Given a Boolean formula F in conjunctive normal form such that each clause of F contains at most 2 literals and a positive integer B. Does there exist a truth assignment such that at least B clauses of F are true?

3.6 EXACT3SAT is the subproblem of 3SAT where only instances are considered which have exactly three pairwise different literals per clause. Show that EXACT3SAT is $\mathcal{NP}$-complete.

3.7 Show that the restriction of EXACT3SAT in which each variable may appear at most three times is in $\mathcal{P}$. (Hint: Use Problem 2.7.)

3.8 Let B be a fixed integer. Show that there exists a polynomial time algorithm which decides for every input $a_1,\ldots,a_n$ of integers from $\{1,\ldots,B\}$ and every integer K whether there is some $I \subseteq \{1,\ldots,n\}$ such that $\sum_{i \in I} a_i = K$.

3.9* Show that deciding whether a graph $G = (V,E)$ has a Hamilton cycle is $\mathcal{NP}$-complete.

3.10 PARTITION is the following problem. Given integers $a_1, \ldots, a_n$. Does there exist $I \subseteq \{1, \ldots, n\}$ such that $\sum_{i \in I} a_i = \sum_{j \notin I} a_j$? Show that PARTITION is $\mathcal{NP}$-complete.

3.11 3-PARTITION is the following problem. Given $3n$ integers $a_1, \ldots, a_{3n}$ and a bound $B \in \mathbb{Z}^+$ such that $\frac{B}{4} < a_i < \frac{B}{2}$ for all $1 \leq i \leq 3n$ and $\sum_{i=1}^{3n} a_i = n \cdot B$. Does there exist a partition of the elements a_i into n disjoint sets $A_1, \ldots, A_n$ such that for each $1 \leq i \leq n$, $\sum_{a \in A_i} a = B$? Show that 3-PARTITION is $\mathcal{NP}$-complete.

3.12 An *independent set* in a graph $G = (V, E)$ is a set $X \subseteq V$ such that $|X \cap e| \leq 1$ for all $e \in E$. Show that deciding whether a graph $G = (V, E)$ has an independent set of size B is in $\mathcal{P}$ if B is a fixed constant, and is $\mathcal{NP}$-complete if B is part of the input.

Notes

There is a vast literature concerned with complexity theory. The reader might *e.g.* consult the excellent book by Papadimitriou (1994) on *Computational Complexity* to find a thorough treatment of the field and many references. An easily readable introduction to the theory of complexity is the textbook by Bovet and Crescenzi (1994).

Nearly every textbook on algorithms contains procedures for SORTING n integers in a comparison based model in $\mathcal{O}(n \log n)$ time. Within this model $\mathcal{O}(n \log n)$ is also best possible, see *e.g.* Cormen, Leiserson, and Rivest (2001). Changing the model slightly, i.e., considering so-called conservative or word-based algorithms, Fredman and Willard (1993, 1994) broke through the $\mathcal{O}(n \log n)$ barrier and proved that n integers can be sorted in $\mathcal{O}(n\sqrt{\log n})$ time. Hagerup (1998) provides a survey for algorithms for sorting and search problems on the word RAM.

The classical, and empirically very successful method to solve LINEAR PROGRAMMING is the simplex method due to Dantzig (1951). Khachiyan (1979) was the first to state the polynomial time solvability of LINEAR PROGRAMMING using the so-called ellipsoid method. For an excellent account of this method and its consequences see Grötschel, Lovász, and Schrijver (1988).

Cook's theorem (Theorem 3.4) is, as the name indicates, due to Cook (1971). In a subsequent paper, Karp (1972) revealed the true wealth of $\mathcal{NP}$-completeness by showing that many natural problems are indeed $\mathcal{NP}$-complete. Karp's paper includes $\mathcal{NP}$-completeness proofs for the STEINER PROBLEM IN GRAPHS (Theorem 3.9), TRIPARTITE MATCHING (Theorem 3.14), SUBSET SUM (Theorem 3.15), and PARTITION (Problem 3.10). The proof that PLANAR3SAT is $\mathcal{NP}$-complete (Theorem 3.12) is due to Lichtenstein (1982).

The bible of $\mathcal{NP}$-complete problems and reduction techniques is Garey and Johnson (1979), which listed over 300 $\mathcal{NP}$-complete problems already in 1979. In addition to this book many new interesting results concerning $\mathcal{NP}$-completeness can be found in "The on-going $\mathcal{NP}$-completeness column" by Johnson which appeared over many years in the *Journal of Algorithms*.

4

Special Terminal Sets

In this chapter we begin our study of Steiner trees. We start by considering two well known graph problems which occur as special cases of the Steiner problem. Let $N = (V,E,\ell)$ be a network and $K \subseteq V$ be a terminal set. If K is restricted to be of cardinality two, say $K = \{s,t\}$, then the Steiner problem reduces to the problem of finding a shortest path between the vertices s and t. If on the other hand K is the whole vertex set of N, i.e., $K = V$, then the Steiner problem reduces to the problem of finding a minimum spanning tree in a network. Both of these problems are very important graph problems and, as we will see later, both of them will also play an important rôle in studying the general Steiner tree problem.

4.1 The shortest path problem

We will present a labeling and scanning procedure to solve the problem of finding a shortest path between two vertices, say s and t, which was proposed by Dijkstra in 1959. This algorithm can be viewed as a wave which propagates through the graph, starting at s and phasing out at t. There are three kinds of vertices: those which are already scanned, they are behind the wave in silent water, those which are labeled but not yet scanned, they can be considered as on the crest of the wave, and finally those which are not yet labeled. In the beginning, only the vertex s is scanned, all neighbors of s are labeled and all other vertices are unlabeled. Then at each stage of the algorithm one vertex drops from the crest of the wave into silent water and some new vertices move onto the crest, i.e., they receive a label. As soon as t is scanned the algorithm stops.

The reader should have in mind that weight functions in networks are by definition nonnegative functions. The proof that Dijkstra's algorithm works correctly depends heavily on this fact.

Theorem 4.1 *Dijkstra's algorithm is correct and can be implemented with running time $\mathcal{O}(m + n \log n)$.*

Algorithm 4.2 (DIJKSTRA'S ALGORITHM)
Input :   A connected network $N = (V,E,\ell)$ and vertices $s,t \in V$.
Output : A shortest s-t path in N.
{ *Initialization* }
$W := \emptyset$; $\rho[s] := 0$; pred$[s] := $ **nil**;
for all $v \in V \setminus \{s\}$ **do** $\rho[v] := \infty$;
{ *Scanning and Labeling* }
while $t \notin W$ **do**
   { *Scanning* }
   Find $x_0 \in V \setminus W$ such that $\rho[x_0] = \min\{\rho[v] \mid v \in V \setminus W\}$;
   $W := W \cup \{x_0\}$;
   { *Labeling* }
   for all $v \in \Gamma(x_0) \cap (V \setminus W)$ such that $\rho[v] > \rho[x_0] + \ell(x_0,v)$ **do**
     $\rho[v] := \rho[x_0] + \ell(x_0,v)$; pred$[v] := x_0$;

Proof.   To see the correctness, we show that the following statements hold after each iteration of the while-loop:

(i) For all $v \in W$ the length of a shortest s-v path is $\rho[v]$ and (if $s \neq v$) there exists a shortest s-v path which ends with the edge $\{\text{pred}[v],v\}$.

(ii) For all $v \in V \setminus W$ with $\rho[v] < \infty$ the length of a shortest s-v path in the network $N[W \cup \{v\}]$ induced by the vertex set $W \cup \{v\}$ is $\rho[v]$ and there exists a shortest such s-v path which ends with the edge $\{\text{pred}[v],v\}$.

Trivially, (i) and (ii) hold after the initialization step. Thus, it suffices to show that if (i) and (ii) hold at the beginning of an iteration of the while loop, they also hold at the end (with respect to the new set W). So consider the vertex x_0 which will be added to the set W. By assumption (ii) there exists an s-x_0 path P in $N[W \cup \{x_0\}]$ ending with $\{\text{pred}[x_0],x_0\}$ such that its length satisfies $\ell(P) = \rho[x_0]$. Let P' be an arbitrary s-x_0 path which is not totally contained in $N[W \cup \{x_0\}]$. Then P' contains a first vertex $y \notin W \cup \{x_0\}$. By choice of x_0 and the fact that all edge lengths are nonnegative we conclude $\ell(P) = \rho[x_0] \leq \rho[y] \leq \ell(P')$. That is, (i) holds for the new set $W \cup \{x_0\}$. Also, from the way the ρ-values are updated in the for-loop it follows immediately that (ii) holds as well.

Now consider the running time. One easily observes that the most time consuming step is hidden in the line "*Find $x_0 \in V \setminus W$ such that ...*". A straightforward implementation would be to store the ρ-values in an array. Finding an appropriate $x_0 \in V \setminus W$ takes then $\mathcal{O}(n)$ time, as we have to run over the whole

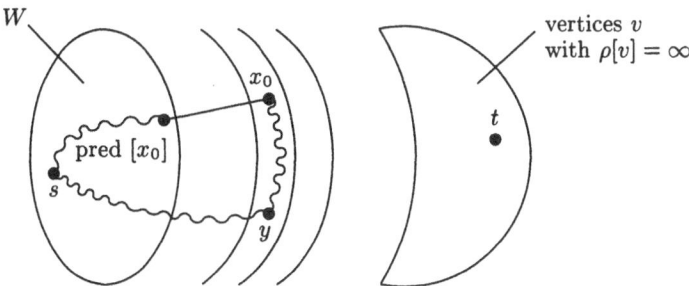

Figure 4.1 Dijkstra's scanning procedure.

array. As the while loop might be executed up to $n - 1$ times, this yields a total running time of $\mathcal{O}(n^2)$. We obtain a better running time, by observing that we can use a priority queue in order to store the ρ-values. That is, we can use the Fibonacci heaps which were introduced in Section 2.2. to store the ρ-values.

The initialization clearly takes at most $\mathcal{O}(n)$ time. Then, throughout the labeling and scanning phase, there are at most $n-1$ executions of the **Extract-Min** operation (one at each iteration of the while-loop) and at most m executions of the **Decrease-Key** operation. According to Theorem 2.13, these can be performed in $\mathcal{O}(m + n \log n)$ time. Furthermore, the labeling of all neighbors can certainly be done in $\mathcal{O}(m)$ time, as each edge is considered at most twice (once from each endpoint). Here we assume that the data structure for the graph is such that every vertex contains a list with all its incident edges. □

What has to be changed, if we require not just a shortest path from s to *some* vertex t, but shortest paths from s to *all* other vertices? After a moment's thought the reader will probably guess the correct answer: essentially nothing. By changing the condition of the while-loop from $t \notin W$ to $W \neq V$, Dijkstra's algorithm computes the desired shortest paths within the same time bound.

Theorem 4.3 *Given a network $N = (V, E, \ell)$ and a vertex $s \in V$, one can compute in $\mathcal{O}(m + n \log n)$ time shortest paths from s to all vertices $v \in V$.* □

We close this section on the shortest path problem by considering the case of unit costs, i.e., the problem of finding shortest paths in connected graphs. How does Dijkstra's algorithm behave in this special situation? The following two observations are obvious and left to the reader as an easy exercise, cf. Problem 4.2. First, the label $\rho[v]$ of a vertex v will change at most once: from infinity to some finite value. Second, if v is assigned its finite label $\rho[v]$ before w has got its finite label $\rho[w]$, then $\rho[v] \leq \rho[w]$. Hence, when choosing $x_0 \in V \backslash W$ such that $\rho[x_0]$ is minimum in the scanning phase of Dijkstra's algorithm, the algorithm can just pick the vertex which was labeled first. Keeping the vertices which are labeled in a list, adding vertices at the end and removing vertices from the front, x_0 can be found in constant time. Dijkstra's algorithm has thus a running time of only $\mathcal{O}(m)$ when applied to (unweighted) graphs.

Dijkstra's algorithm for the special case of graphs is often called *breadth-first search algorithm*. It can, of course, also be applied to decide in $\mathcal{O}(m + n)$ steps whether a given graph is connected or to find in time $\mathcal{O}(m)$ a spanning tree in a connected graph.

4.2   The minimum spanning tree problem

The minimum spanning tree problem is among the first network optimization problems ever studied. It dates back to the middle of the 1920s. The ideas and developments to solve this simple problem have played a central rôle in the design of computer algorithms.

Here we will present two algorithms for solving this important problem. The first one is the Jarník-Prim algorithm. The basic idea of this algorithm is very close to the one we have seen in the previous section. Again, we distinguish three kinds of vertices: those which are already scanned, those which are labeled but not yet scanned and, finally, those vertices which are not yet labeled. In the beginning, only one (arbitrarily chosen) vertex, say s, is scanned, all neighbors of s are labeled. Then, starting from this root vertex s, at every step the tree is enlarged by one edge (and so also by one vertex) until it contains all vertices. More precisely, at every step the shortest edge connecting a vertex from the tree with a vertex outside is added.

> **Algorithm 4.4** (JARNÍK-PRIM ALGORITHM)
> **Input:**   A connected network $N = (V,E,\ell)$.
> **Output:** A minimum spanning tree $T = (V,F)$ in N.
>
> *{ Initialization }*
> Choose $s \in V$ arbitrarily;
> $W := \{s\}$; $F := \emptyset$; $x_0 := s$;
> $\sigma[s] := 0$; pred$[s] := $ **nil**;
> **for all** $v \in V \setminus \{s\}$ **do** $\sigma[v] := \infty$;
> *{ Scanning and Labeling }*
> **while** $W \neq V$ **do**
>   Find $x_0 \in V \setminus W$ such that $\sigma[x_0] = \min\{\sigma[v] \mid v \in V \setminus W\}$;
> $W := W \cup \{x_0\}$; $F := F \cup \{\{x_0,\text{pred}[x_0]\}\}$;
>   **for all** $v \in \Gamma(x_0) \cap (V \setminus W)$ such that $\sigma[v] > \ell(x_0,v)$ **do**
> $\sigma[v] := \ell(x_0,v)$; pred$[v] := x_0$;

Theorem 4.5 *The Jarník-Prim algorithm is correct and can be implemented with running time $\mathcal{O}(m + n \log n)$.*

Proof. To see the correctness we show, similarly as in the proof of Theorem 4.3, that a set of statements hold after each iteration of the while-loop. Namely,

(i) The edges in F form a spanning tree for the vertices in W.
(ii) For all $v \in V \setminus W$ the value $\sigma[v]$ satisfies

$$\sigma[v] = \min\{\ell(x,v) \mid x \in W \text{ and } \{x,v\} \in E\},$$

and

$$\sigma[v] = \ell(\text{pred}[v],v).$$

In other words, $\sigma[v]$ is the length of the shortest edge that connects v to the vertices in W, and $\{\text{pred}[v],v\}$ is such an edge.
(iii) There exists a minimum spanning tree $T^* = (V,F^*)$ such that $F \subseteq F^*$.

Trivially, (i) - (iii) hold after the initialization. So assume now, (i) - (iii) hold at the beginning of some iteration. Then one easily observes, by the way the algorithm is designed, that (i) and (ii) also hold at the end of the iteration with respect to the new set W. To see (iii) let $T^* = (V,F^*)$ be a minimum spanning tree such that F^* contains (the old) tree F. If $\{x_0,\text{pred}[x_0]\}$ is contained in F^* there is nothing to show. So assume not. Then there exists a path P in T^* that connects $\text{pred}[x_0]$ and x_0. As $\text{pred}[x_0]$ is in W, but x_0 is not, the path P has to contain at least one edge, say $\{y,z\}$, such that $y \in W$ and $z \notin W$. By choice of x_0 and assumption (ii), we know that

$$\ell(\text{pred}[x_0],x_0) = \sigma[x_0] \leq \sigma[z] \leq \ell(y,z).$$

Now consider $\widetilde{F} := (F^* \setminus \{\{y,z\}\}) \cup \{\{\text{pred}[x_0],x_0\}\}$. Then $\widetilde{T} = (V,\widetilde{F})$ is a spanning tree of length

$$\ell(\widetilde{T}) = \ell(T^*) - \ell(y,z) + \ell(\text{pred}[x_0],x_0) \leq \ell(T^*).$$

That is, $\widetilde{T}$ is a minimum spanning tree such that its edge set $\widetilde{F}$ contains the new set $F := F \cup \{\{x_0,\text{pred}[x_0]\}\}$. This concludes the proof of the correctness of the algorithm.

With the same arguments as in the proof of Theorem 4.3 we obtain an implementation having complexity $\mathcal{O}(m + n \log n)$ if we store the σ-values as Fibonacci heaps: the initialization clearly takes at most $\mathcal{O}(n)$ time. Then, throughout the labeling and scanning phase, there are at most $n - 1$ executions of the **Extract-Min** operation (one at each iteration of the while-loop) and at most m executions of the **Decrease-Key** operation. According to Theorem 2.13, this can be performed in $\mathcal{O}(m + n \log n)$ time. □

Whereas the Jarník-Prim algorithm starts growing a minimum spanning tree at one (arbitrarily chosen) root and picks at every step the shortest edge incident to the already existing tree, Kruskal's algorithm, which we will present next, behaves differently. Here, one grows a forest which finally connects to a minimum spanning tree. The initial forest has as components all vertices. Then, at every step one greedily picks the shortest remaining edge and adds this edge to the already existing forest if this edge does not create a cycle. As soon as the forest becomes a tree no more edge can be added and the algorithm stops.

> **Algorithm 4.6** (KRUSKAL'S ALGORITHM)
> Input: A connected network $N = (V,E,\ell)$.
> Output: A minimum spanning tree $T = (V,F)$ in N.
> { *Initialization* }
> Sort the edges such that $\ell(e_1) \leq \ldots \leq \ell(e_m)$.
> Let $F := \emptyset$.
> { *Building the tree* }
> for $i := 1$ to m do
>   if $(V,F \cup \{e_i\})$ is acyclic then $F := F \cup \{e_i\}$.

As we will see in the next section, the idea of choosing at every step "greedily" the globally cheapest element has led to a very fruitful concept in combinatorics. For the moment, we leave it as an exercise to the reader to verify that Kruskal's algorithm works correctly. The correctness then will also follow from a more general result in Section 4.3. Moreover, one can show (cf. Problem 4.6) that Kruskal's algorithm can be implemented in such a way that its running time is dominated by the initialization step, i.e., by sorting the m edges. Without any additional assumptions, this sorting requires by standard techniques $O(m \log n)$ time, cf. notes to Chapter 3.

4.3   Excursion: Matroids and the greedy algorithm

In his seminal paper "On the abstract properties of linear dependence" Whitney (1935) introduced the concept of a *matroid*. His intention was to elaborate fundamental properties of dependence which are common to graphs and matrices.

A *matroid* $\mathcal{M}$ is an ordered pair $(S,\mathcal{I})$ consisting of a finite ground set S and a collection $\mathcal{I}$ of subsets of S satisfying the following three axioms:

(M1) $\emptyset \in \mathcal{I}$.
(M2) If $A \in \mathcal{I}$ and $B \subseteq A$, then $B \in \mathcal{I}$.
(M3) If $A,B \in \mathcal{I}$ and $|A| > |B|$, then there exists an element $x \in A \setminus B$ such that $B \cup \{x\} \in \mathcal{I}$.

The members of $\mathcal{I}$ are called the *independent sets* of the matroid $\mathcal{M}$. A subset $A \subseteq S$ that is not in $\mathcal{I}$ is called *dependent*. An (inclusion) minimal dependent set is called a *circuit* of $\mathcal{M}$.

The following examples indicate that the concept of a matroid indeed generalizes important properties of graph theory and linear algebra.

Example 4.7 FOREST MATROID. Let $G = (V,E)$ be a graph and let $\mathcal{I}$ be the collection of subsets of E which are acyclic, i.e., $\mathcal{I}$ is the collection of forests in G. Then $(E,\mathcal{I})$ is a matroid, called the *forest matroid* of the graph G.

LINEAR MATROID. Let S be a finite set of vectors of an arbitrary vector space and let $\mathcal{I}$ be the collection of subsets of S which are linearly independent. Then $(S,\mathcal{I})$ is a matroid.

Many notions in matroid theory are coined after similar properties in graphs. We have already defined *circuits* of a matroid. The following two propositions reflect well-known properties of cycles in a graph. The similar proofs for matroids are left to the reader as an exercise.

Proposition 4.8 *If C_1 and C_2 are two distinct circuits of a matroid and $e \in C_1 \cap C_2$, then there exists a circuit $C_3 \subseteq (C_1 \cup C_2) \setminus \{e\}$.*   □

Proposition 4.9 *Suppose A is an independent set in a matroid and e is an element such that $A \cup \{e\}$ is dependent. Then there exists a unique circuit C in $A \cup \{e\}$ and this circuit contains e.*   □

Let $\mathcal{M} = (S,\mathcal{I})$ be a matroid and X be an arbitrary subset of S. The *rank* $r(X)$ of X is defined as the cardinality of the largest independent set contained in X:

$$r(X) = \max\{|A| \mid A \subseteq X, A \in \mathcal{I}\}.$$

The function r is called the *rank function* of the matroid $\mathcal{M}$. Note that the definition of the rank function captures the concepts of "dimension" and "components" from linear algebra and graph theory respectively.

Proposition 4.10 *If r is the rank function of a matroid $\mathcal{M} = (S,\mathcal{I})$, then a subset $X \subseteq S$ is independent if and only if $r(X) = |X|$.*   □

Matroids can be defined in many different but equivalent ways, cf. *e.g.*, Problems 4.8 and 4.10. The perhaps most surprising one is that matroids are exactly those combinatorial structures for which a certain optimization problem can be solved by the greedy algorithm.

Algorithm 4.11 (GREEDY ALGORITHM)
Input: A finite set S, a collection $\mathcal{I}$ of subsets of S, and a nonnegative weight
function $w : S \to \mathbb{N}_0$.
Output: A set $A \in \mathcal{I}$.
{ *Initialization* }
Sort the elements of S such that $w(e_1) \leq \ldots \leq w(e_{|S|})$.
Let $A := \emptyset$.
{ *Build set* }
for $i := 1$ **to** $|S|$ **do**
  if $A \cup \{e_i\} \in \mathcal{I}$ **then** $A := A \cup \{e_i\}$.

Ideally, the greedy algorithm is expected to output a set A in $\mathcal{I}$ that is inclusion maximal and has minimum weight among all inclusion maximal elements in $\mathcal{I}$. We say that the greedy algorithm is *optimal* if it stops with such a set $A \in \mathcal{I}$, i.e., if the greedy algorithm terminates with some $A \in \mathcal{I}$ satisfying

$$w(A) = \min\{w(B) \mid B \in \mathcal{I} \text{ and } B \cup \{x\} \notin \mathcal{I} \text{ for all } x \in S \setminus B\}.$$

With regard to the implementation of the greedy algorithm, one word of caution might be appropriate. In general, it would be rather unwise to actually require that the input contains an explicit list of all independent sets $I \in \mathcal{I}$, as there might well be exponentially many of such sets. Instead one usually assumes that the elements of $\mathcal{I}$ are implicitly given by an *oracle*, which when presented with a subset $X \subseteq S$ determines (for example in time propotional to the size of $|X|$) whether X is a member of $\mathcal{I}$ or not.

Theorem 4.12 *Let $(S,\mathcal{I})$ be a system which satisfies axioms* (M1) *and* (M2). *Then $(S,\mathcal{I})$ is a matroid (i.e., satisfies axiom* (M3) *as well), if and only if the greedy algorithm is optimal for all nonnegative weight functions $w : S \to \mathbb{N}_0$.*

Proof. First assume that $(S,\mathcal{I})$ is a matroid, but that there exists a weight function $w : S \to \mathbb{N}_0$ for which the greedy algorithm is not optimal. That is, the greedy algorithm stops with a set $A = \{a_1,\ldots,a_k\}$, but there exists a maximal independent set $B = \{b_1,\ldots,b_{k'}\}$ such that $w(A) > w(B)$. Observe first that in fact $k = k'$. Indeed, by construction, the greedy algorithm will always return a maximal independent set. Furthermore, as $(S,\mathcal{I})$ is a matroid, axiom (M3) implies that all maximal independent sets have the same cardinality.

We assume that the elements of A are ordered in the way they were added, that is that $w(a_1) \leq \cdots \leq w(a_k)$. By suitably renumbering the elements of B we may similarly suppose that $w(b_1) \leq \cdots \leq w(b_k)$. Then the assumption $w(A) > w(B)$ implies that there exists an $1 \leq i \leq k$ such that $w(a_i) > w(b_i)$. For such an i let

$$A' = \{a_1,\ldots,a_{i-1}\}, \quad \text{and}$$
$$B' = \{b_1,\ldots,b_{i-1},b_i\}.$$

As A' and B' are both independent and $|B'| > |A'|$ the fact that $(S,\mathcal{I})$ is a matroid implies that there exists an $x \in B' \setminus A'$ such that $A' \cup \{x\} \in \mathcal{I}$. But, by the choice of i and the definition of B', we have $w(x) \leq w(b_i) < w(a_i)$. This, however, cannot be, as in this case the greedy algorithm would have chosen x.

To show the other direction assume now that the greedy algorithm is optimal for all weight functions, but that (M3) is not fulfilled in $(S,\mathcal{I})$. Then there exist two sets $A, B \in \mathcal{I}$ such that $|A| > |B|$ and $B \cup \{x\} \notin \mathcal{I}$ for all $x \in A \setminus B$. We claim that we may assume that A is a maximal independent set. Indeed, if it is not, we add elements to A and if possible also to B until A is a maximal independent set. Let $a = |A \setminus B|$ and $b = |B \setminus A|$. Then $a > b$. We distinguish two cases.

CASE 1. B is a maximal independent set. Consider the weight function

$$w(e) = \begin{cases} b+1 & \text{if } e \in A \setminus B, \\ 0 & \text{if } e \in A \cap B, \\ a+1 & \text{otherwise.} \end{cases}$$

Then $w(A) = a(b+1) > b(a+1) = w(B)$. But the greedy algorithm is forced to choose A.

CASE 2. B is not a maximal independent set. Consider the weight function

$$w'(e) = \begin{cases} 0 & \text{if } e \in B, \\ a & \text{if } e \in A \setminus B, \\ a^2 + 1 & \text{otherwise.} \end{cases}$$

Then $w'(A) = a^2$. But the greedy algorithm is forced to choose first all elements in B and then an element in $S \setminus (A \cup B)$ of weight $a^2 + 1$.

In both cases the greedy algorithm therefore does not stop with a maximal independent set of minimum weight, contradicting our assumption. □

At first sight it might look rather obscure to consider optimization problems where a set of *maximum* size and *minimum* weight is sought. On the other hand, the alert reader might have already observed the striking similarity between the greedy algorithm and the algorithm of Kruskal from the previous section. This is of course no coincidence: Kruskal's algorithm is just the greedy algorithm for the forest matroid of the graph G.

Problems

4.1   Let $N = (V,E,\ell)$ be a network and $s \in V$. Apply Dijkstra's algorithm in
      the "one to all" version as indicated in the paragraph before Theorem 4.3.
      Now consider the union of all edges $\{v,\text{pred}[v]\}$ for all $v \in V \setminus \{s\}$ (or,
      alternatively, the union of all shortest s-v paths for $v \in V$). What can you
      say about this graph?

4.2   Apply Dijkstra's algorithm to an unweighted connected graph $G = (V,E)$
      and vertices $s,t \in V$. Show that the label $\rho[v]$ of a vertex $v \in V$ will change
      at most once: namely, from infinity to a finite value. Prove further that if
      $v \in V$ is assigned its finite label $\rho[v]$ before $w \in V$ is assigned its finite label
      $\rho[w]$, then $\rho[v] \leq \rho[w]$.

4.3   Let $N = (V,E,\ell)$ be a connected network in which no two edges have the
      same length. Prove that N contains a unique minimum spanning tree.

4.4   Let $N = (V,E,\ell)$ be a connected network. A set of edges $F \subseteq E$ is called a
      cut if there exists a nonempty set of vertices $W \subset V$ such that the set F is
      given by $F = \{e \in E \mid |e \cap W| = 1\}$. We color the edges of N successively
      using the following two rules:
      BLUE RULE. Select a cut F that does not contain a blue edge. Among the
      uncolored edges in F choose $e_0 \in F$ such that $\ell(e_0) = \min\{\ell(e) \mid e \in F,$
      e uncolored$\}$ and color e_0 blue.
      RED RULE. Select a cycle C in N that does not contain a red edge. Among
      the uncolored edges in C choose $e_1 \in C$ such that $\ell(e_1) = \max\{\ell(e) | e \in$
      C, e uncolored$\}$ and color e_1 red.
      Start with an uncolored connected network and apply at each step one of
      the two rules, in arbitrary order, until all edges are colored. Show that this is
      always possible and that at the end, the edges colored blue form a minimum
      spanning tree.

4.5   Show that Kruskal's algorithm as well as the Jarník-Prim algorithm are
      special cases of the red-blue algorithm introduced in the previous problem.

4.6   Show that there exists an implementation of Kruskal's algorithm which
      solves the minimum spanning tree problem in $\mathcal{O}(m + n \log n)$ time plus the
      time for sorting the edges.

4.7   Prove Propositions 4.8 and 4.9.

4.8   Assume that $\mathcal{C}$ is a collection of subsets of a finite set S. Show that $\mathcal{C}$ is the
      collection of circuits of a matroid with ground set S if and only if $\mathcal{C}$ satisfies
      the following three conditions:
      (C1) $\emptyset \notin \mathcal{C}$.
      (C2) If $C_1,C_2 \in \mathcal{C}$ and $C_1 \subseteq C_2$, then $C_1 = C_2$.
      (C3) If $C_1,C_2 \in \mathcal{C}$, $C_1 \neq C_2$, and $e \in C_1 \cap C_2$, then there exists a set
      $\quad C_3 \in \mathcal{C}$ such that $C_3 \subseteq (C_1 \cup C_2) \setminus \{e\}$.

4.9 Show that the rank function r of a matroid $\mathcal{M} = (S,\mathcal{I})$ satisfies the following conditions:

(R1) If $X \subseteq S$, then $0 \leq r(X) \leq |X|$.
(R2) If $X \subseteq Y \subseteq S$, then $r(X) \leq r(Y)$.
(R3) If X and Y are subsets of S, then $r(X \cup Y) + r(X \cap Y) \leq r(X) + r(Y)$.

4.10 Let S be a finite set and r be a function that maps all subsets of S into the set of nonnegative integers and satisfies (R1) - (R3) of the previous problem. Let $\mathcal{I}$ denote the collection of subsets of S for which $r(X) = |X|$. Show that $(S,\mathcal{I})$ is a matroid.

4.11 Show that a system $(S,\mathcal{I})$ that satisfies axioms (M1) and (M2) is a matroid if and only if the following algorithm MAX-GREEDY is optimal for every weight function $w : S \to \mathbb{N}_0$, i.e., outputs a set $A \in \mathcal{I}$ such that $w(A) = \max\{w(B) \mid B \in \mathcal{I}\}$:

> Sort the elements of S such that $w(e_1) \geq \ldots \geq w(e_{|S|})$.
> Let $A := \emptyset$.
> **for** $i := 1$ **to** $|S|$ **do**
>   **if** $A \cup \{e_i\} \in \mathcal{I}$ **then** $A := A \cup \{e_i\}$.

(Note that this statement is *not* identical to Theorem 4.12!)

Notes

Several papers on the shortest path problem were published around the time Dijkstra (1959) proposed his algorithm. Dreyfus (1969) presents a nice survey of the early work on this problem. Dijkstra's algorithm for finding a shortest s-t path in a network relies on the fact that all weights of the edges are nonnegative. Bellman (1958) and Ford (1956) present early algorithms which allow edges with negative weights. In fact, the goal of these algorithms is either to find a shortest path between s and t or to detect a negative cycle. This is essentially best possible, as finding a shortest s-t path in the presence of negative cycles is known to be $\mathcal{NP}$-complete, see Garey and Johnson (1979).

The straightforward implementation of Dijkstra's algorithm using arrays runs in $\mathcal{O}(n^2)$ time which is already optimal for dense networks, i.e., networks with $m = \Omega(n^2)$. The best time bound for solving the shortest path problem with nonnegative edge lengths using Fibonacci-heaps is $\mathcal{O}(m + n \log n)$ and was obtained by Fredman and Tarjan (1987). By making stronger assumptions on the underlying RAM-machine Fredman and Willard (1994) were able to design a variation of Fibonacci-heaps which can be used to implement Dijkstra's algorithm with complexity $\mathcal{O}(m + n \log n / \log \log n)$. Using a different approach, Thorup (1997) provided even a linear time algorithm for the shortest path problem.

A history of the minimum spanning tree problem was written by Graham and Hell (1985). Tarjan (1983) provides a good survey which compares the different methods to solve this problem. The first algorithm known to solve the minimum spanning tree problem already dates back to 1926 and is due to the Czech mathematician Borůvka (1926).

His motivation to consider this problem was that it arose in connection with the rural electrification of South Moravia. The Jarník-Prim algorithm was invented by Jarník (1930) in a paper which uses the same title as Borůvka's paper does, and was rediscovered by Prim (1957). Dijkstra's shortest path algorithm is almost identical to the Jarník-Prim minimum spanning tree algorithm. In fact, Dijkstra discussed both algorithms in the same paper, apparently unaware as well as of Jarník's as of Prim's paper. Kruskal's algorithm appeared in Kruskal (1956).

Kruskal's algorithm can easily be implemented in time $\mathcal{O}(m + n \log n)$ plus the time required for sorting the edges. An implementation which requires $\mathcal{O}(m \cdot \alpha(n,m))$ time, provided that the edges are already sorted was given by Tarjan (1975). Here, $\alpha(n,m)$ denotes the functional inverse of the Ackermann function. For all practical purposes, this number is not larger than 5. In general, sorting the m edges requires $\mathcal{O}(m \log n)$ time. But, for instance, when the weights are small integers radix sort can be used to sort in linear time. For further comments on sorting, see the notes of Chapter 3. Within their RAM-model, Fredman and Willard were able to devise a linear-time algorithm for finding a minimum spanning tree. A randomized linear-time algorithm within the usual RAM-model was given by Karger, Klein, and Tarjan (1995).

For further reading on the rich and well-studied body of shortest path and minimum spanning tree algorithms we recommend Tarjan (1983), van Leeuwen (1990), and Ahuja, Magnanti, and Orlin (1993).

The concept of matroids dates back to 1931, when van der Waerden (1931) first studied linear and algebraic dependence axiomatically, and Whitney (1935) coined the term matroid. Meanwhile, matroid theory has developed into an important part of discrete mathematics. The extension of Kruskal's greedy algorithm to matroids was first carried out by Rado (1957). The books by Welsh (1976) and Oxley (1992), provide excellent mathematical accounts of the field.

5

Exact Algorithms

The STEINER PROBLEM IN NETWORKS deals with *finite* objects only. So, in principle, the problem can easily be solved: just enumerate all subsets of the edges, check whether they form a Steiner tree which span the given terminal set, and keep the smallest one. Although such an approach leads to a finite algorithm, it is certainly not a very efficient one. Its complexity is exponential in the number of edges of the network. On the other hand, as we have seen in Chapter 3, so far there is no polynomial time algorithm known for the STEINER PROBLEM IN GRAPHS and, hence, in particular none for the STEINER PROBLEM IN NETWORKS.

In this chapter we discuss two algorithms for attacking the STEINER PROBLEM IN NETWORKS. In principle, they behave like the algorithm sketched above, but are slightly more efficient. Nevertheless, they will usually require exponentially many steps.

These algorithms also have the nice additional feature that they considerably extend the class of polynomially solvable special cases studied in the last chapter. As we have observed there, if the terminal set K coincides with the whole vertex set V, then the Steiner problem coincides with the minimum spanning tree problem and, hence, can be solved efficiently. In this chapter we will see that this is also true if K almost coincides with the whole vertex set. More precisely, we will show that if $|V| - |K|$ is constant, one can still solve the Steiner problem in polynomial time. We also saw in the previous chapter that if K has cardinality two, then the Steiner problem reduces to the shortest path problem and, hence, allows a fast solution. Here, we will show that if $|K|$ is constant then there exists a polynomial time algorithm which generates a solution to the Steiner problem.

5.1   The enumeration algorithm

Let $N = (V,E,\ell)$ be a network and $K \subseteq V$ be a terminal set. Let T be any Steiner tree for K.

Surely, T may contain up to $|V \setminus K|$ Steiner points. (Recall that, by definition, all vertices of a Steiner tree T that do not belong to K are called Steiner points.) Using that, also by definition, every leaf in a Steiner tree must be a terminal we can considerably restrict the number of Steiner points in T which are branching points, i.e., which have degree at least 3.

Lemma 5.1 *A Steiner tree T for a terminal set K with $|K| = k$ contains at most $k - 2$ Steiner points of degree at least 3.*

Proof.   Let s_2 denote the number of Steiner points in T whose degree is two and let s_3 denote the number of Steiner points which have degree at least three. As, by definition, all leaves of a Steiner tree belong to the terminal set K, we obtain $|V(T)| = k + s_2 + s_3$. As T is a tree, this implies

$$\sum_{v \in V(T)} d_T(v) = 2|E(T)| = 2(|V(T)| - 1) = 2(k + s_2 + s_3 - 1).$$

On the other hand, by definition of s_2 and s_3,

$$\sum_{v \in V(T)} d_T(v) = \sum_{v \in K} d_T(v) + \sum_{v \in V(T) \setminus K} d_T(v) \geq k + 2s_2 + 3s_3.$$

Thus $k - 2 \geq s_3$.   □

Consider two branching points u and w of T, i.e., two vertices u and w of degree at least 3, which are connected by a path P. If all interior vertices of P have degree 2 in T and none of them is a terminal then the fact that T is a Steiner minimum tree implies that P is a shortest $u - w$ path in N. To collect all shortest paths in N we denote by $D(N) = (V, E_D, \ell_D)$ the so-called *(full) distance network* of N given by $E_D = \begin{bmatrix} V \\ 2 \end{bmatrix}$ and $\ell_D(u,v) = p(u,v)$, the length of a shortest $u - v$ path in N. Note that $D(N)$ can be computed, for example, by calling Dijkstra's Algorithm 4.2 once for each vertex in V (cf. Theorem 4.3).

Let S be the set of all branching points in a Steiner minimum tree T. By the remarks above, T corresponds to a minimum spanning tree in the subnetwork of $D(N)$ induced by the set $K \cup S$. Moreover, by Lemma 5.1, $|K \cup S| \leq 2k - 2$. These are the ideas behind the following enumeration algorithm.

Algorithm 5.2 (ENUMERATION ALGORITHM)

Input: A network $N = (V,E,\ell)$ and a terminal set $K \subseteq V$.

Output: A Steiner minimum tree T for K.

(1) Compute the distance network $D(N) = (V,E_D,\ell_D)$ and store for each edge $\{u,w\}$ of E_D a shortest u-w path in N.

(2) Compute for all subsets S of $V \setminus K$ of size $|S| \le k - 2$ a minimum spanning tree for the subnetwork of $D(N)$ induced by $K \cup S$.

(3) Transform the shortest of the spanning trees encountered in step (2) into a Steiner tree T of the network N.

Theorem 5.3 *The enumeration algorithm computes a Steiner minimum tree. It can be implemented such that its running time is bounded by* $\mathcal{O}(n^2 \log n + nm + \min\{n^{k-2}, 2^{n-k}\} \cdot k^2)$.

Proof. The correctness of the algorithm is immediate by the remarks above. To see the claimed complexity observe first that the distance network $D(N)$ can be computed in time $\mathcal{O}(n^2 \log n + nm)$ by applying Dijkstra's algorithm n times, cf. Theorem 4.3. In step (2) we have to compute a minimum spanning tree (on at most $2k-2$ vertices) for at most $\Sigma_{i=0}^{k-2} \binom{n-k}{i} \le \min\{n^{k-2}, 2^{n-k}\}$ different sets $K \cup S$. As each such computation can easily be done in $\mathcal{O}(k^2)$ time using the algorithm of Jarník and Prim, cf. Theorem 4.5, step (2) needs at most $\mathcal{O}(\min\{n^{k-2}, 2^{n-k}\} \cdot k^2)$ time. In step (3) only the shortest of the spanning trees encountered in step (2), say T_D, is transformed into a Steiner tree and, hence, every edge of the original network N can be contained in at most one of the paths in N which serve as edges in T_D. Hence, step (3) can easily be done in $\mathcal{O}(m)$ time. □

As remarked in the introduction to this chapter, the running time of the algorithm is in fact bounded polynomially in the size of the graph if we only consider instances where the number of non-terminals is bounded by a fixed constant.

Corollary 5.4 *Let $C \in \mathbb{N}$ be arbitrary but fixed. Then the enumeration algorithm computes for all networks $N = (V,E,\ell)$ and terminal sets $K \subseteq V$ with $|V \setminus K| \le C$ a Steiner minimum tree in $\mathcal{O}(n^2 \log n + nm)$ time.* □

In fact, one can do even better. Using the Jarník-Prim algorithm for finding a minimum spanning tree, one can easily construct a procedure which computes a Steiner minimum tree in $\mathcal{O}(n \log n + m)$ time if $V \setminus K$ is of constant size, cf. Problem 5.1.

Observe that the enumeration algorithm reduces also to a polynomial time algorithm if we bound the number of terminals by a constant, but in this case the algorithm which we will present in the next section is more efficient.

5.2   The Dreyfus-Wagner algorithm

The algorithm of Dreyfus and Wagner is an example of a method known as *dynamic programming*. Its essential ingredient is a recursion formula which computes the length of a Steiner minimum tree for a given terminal set from the lengths of the Steiner minimum trees for *all* proper subsets of this terminal set.

More precisely, we will proceed as follows. In a first step we compute the Steiner minimum trees for all 2-element subsets of K, in the second step we use those to compute the Steiner minimum trees for all 3-element subsets of K, and so one. We continue this procedure until we find a Steiner minimum tree for the entire terminal set K.

We start with some notation. For $X \subseteq K$ and $v \in V \setminus X$ let $s(X \cup \{v\})$ denote the length of a Steiner minimum tree for $X \cup \{v\}$ and let $s_v(X \cup \{v\})$ denote the length of a Steiner minimum tree for $X \cup \{v\}$ in which v has degree at least 2. Note that these definitions imply in particular that $s_v(X \cup \{v\}) \geq s(X \cup \{v\})$. While the definition of $s(X \cup \{v\})$ and, in particular, $s_v(X \cup \{v\})$ might look a bit awkward at first sight, their usefulness should become obvious from the following lemma: it implies that these values can easily be computed recursively.

Lemma 5.5 *Let $X \subseteq K$, $X \neq \emptyset$, and $v \in V \setminus X$. Then*

$$s_v(X \cup \{v\}) = \min_{\emptyset \neq X' \subsetneq X} \{s(X' \cup \{v\}) + s((X \setminus X') \cup \{v\})\} \qquad (5.1)$$

and

$$s(X \cup \{v\}) = \min \Big\{ \min_{w \in X}\{p(v,w) + s(X)\}, \min_{w \in V \setminus X}\{p(v,w) + s_w(X \cup \{w\})\} \Big\}.$$
$$(5.2)$$

(Recall that $p(v,w)$ denotes the length of a shortest v-w path.)

Proof.   To see (5.1), assume T is a Steiner tree for $X \cup \{v\}$ of length $s_v(X \cup \{v\})$ in which v has degree at least 2. Clearly, we can then decompose T into two subtrees T_1 and T_2 such that T_1 is a Steiner tree for $X' \cup \{v\}$ and T_2 is a Steiner tree for $(X \setminus X') \cup \{v\}$, where X' is a suitable nonempty subset of X. Figure 5.1 illustrates theses facts. (Note that in this figure as well as those in the remainder of the book terminals are denoted by squares, while ordinary vertices are denoted by circles.) That is, if we minimize over all proper subsets X' we obtain (5.1). (Strictly speaking we obtain only "$\geq$", but the other direction is obvious.)

To see (5.2), assume now that T is a Steiner minimum tree for $X \cup \{v\}$ with no restriction on the degree of v. For a leaf u in T, we let P_u denote the longest path in T starting in u in which all interior points have degree 2 in T and do not belong to K. We distinguish three cases. Minimizing over all three cases will then imply the validity of recursion (5.2).

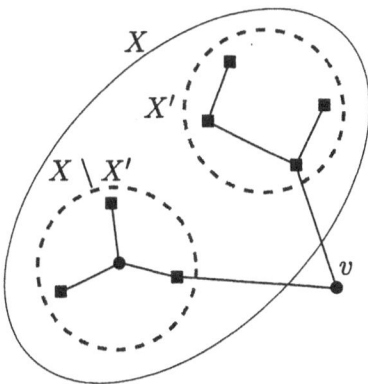

Figure 5.1 The Dreyfus-Wagner recursion (1).

CASE 1. v has degree at least 2 in T. Then $s(X \cup \{v\}) = s_v(X \cup \{v\})$. (Observe that this case is included by choosing $w = v$.)

CASE 2. v is a leaf in T and P_v ends at a vertex $w \in X$. In this case T is the union of a Steiner minimum tree for X and a shortest path from v to w. Hence, $s(X \cup \{v\}) = s(X) + p(v,w)$.

CASE 3. v is a leaf in T and P_v ends at a vertex $w \notin X$. Then w has degree at least 3 in T. That is, in this case T is the union of a Steiner minimum tree for $X \cup \{w\}$ in which w has degree at least 2 and a shortest path from v to w. Hence, $s(X \cup \{v\}) = s_w(X \cup \{w\}) + p(v,w)$. □

With Lemma 5.5 at hand, the design of an algorithm for computing the length of a Steiner minimum tree (which is equal to $s(K) = s((K \setminus \{v\}) \cup v)$ for an arbitrary $v \in K$) is now straightforward.

Theorem 5.7 *The Dreyfus-Wagner algorithm computes the length of a Steiner minimum tree in $\mathcal{O}(3^k n + 2^k n^2 + n^2 \log n + nm)$ steps.*

Proof. The correctness of the algorithm follows immediately from Lemma 5.5. To see that the complexity of the algorithm is as claimed, observe first that the initialization (the computation of the shortest paths) can be done in $\mathcal{O}(n^2 \log n +$

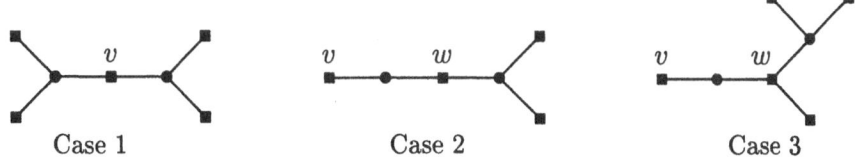

Figure 5.2 The Dreyfus-Wagner recursion (2).

Algorithm 5.6 (DREYFUS-WAGNER ALGORITHM)
Input: A network $N = (V,E,\ell)$ and a terminal set $K \subseteq V$.
Output: The length of a Steiner minimum tree T for K.
{ *Initialization* }
for all $v,w \in V$ **do**
    compute $p(v,w)$;
for all $\{x,y\} \subseteq K$ **do**
    $s(\{x,y\}) := p(x,y)$;
{ *Perform recursion according to Lemma 5.5* }
for $i = 2$ **to** $k - 1$ **do**
    for all $X \subseteq K$ with $|X| = i$ and all $v \in V \setminus X$ **do**
        $s_v(X \cup \{v\}) := \min_{\emptyset \neq X' \subsetneq X} \{s(X' \cup \{v\}) + s((X \setminus X') \cup \{v\})\}$;
    for all $X \subseteq K$ with $|X| = i$ and all $v \in V \setminus X$ **do**
        $s(X \cup \{v\}) := \min \big\{ \min_{w \in X} \{p(v,w) + s(X)\}, \min_{w \in V \setminus X} \{p(v,w) + s_w(X \cup \{w\})\} \big\}$;

nm) by using Dijkstra's algorithm n times, cf. Theorem 4.3. The number of additions and comparisons in recursion (1) of Lemma 5.5 is of the same order as the number of possibilities to choose v, X' and X. As every terminal of K belongs to exactly one of the sets X', $X \setminus X'$ and $K \setminus X$ the number of steps in recursion (1) is bounded by $\mathcal{O}(3^k n)$.

For each tuple (X,v) the computation of recursion (2) needs $\mathcal{O}(n)$ additions and comparisons. The number of tuples for which recursion (2) is carried out is bounded by $\mathcal{O}(2^k n)$, the number of choices of X and v.   □

Observe that the algorithm, as presented here, only computes the *length* of a Steiner minimum tree. It is, however, not difficult to modify the algorithm in such a way that it also constructs such a tree explicitly. One simply has to keep track of the sets and vertices for which the minima in the computation of $s(X \cup \{v\})$ and $s_v(X \cup \{v\})$ were attained and add a backtracking procedure at the end, cf. Problem 5.2.

Combining this observation with Theorem 5.7 yields a polynomial time algorithm for instances in which the number of terminals is bounded by a fixed constant.

Corollary 5.8 *Let $C \in \mathbb{N}$ be arbitrary but fixed. Then the Dreyfus-Wagner algorithm computes for all networks $N = (V,E,\ell)$ and terminal sets $K \subseteq V$ with $|K| \leq C$ a Steiner minimum tree in $\mathcal{O}(n^2 \log n + nm)$ time.*

5.3   Excursion: Dynamic programming

Dynamic programming is a rather general technique for solving combinatorial optimization problems which have the property that their optimal solution can be computed recursively from solutions to subproblems.

For the Steiner tree problem such a recursion is given in Lemma 5.5: in order to compute a Steiner minimum tree for a given terminal set $K \subseteq V$, it is enough to know the answer for all subsets $K' \subset K$. Another way of looking at this technique is to consider it as a recursive table-filling strategy, where we fill a table $T[i,j]$ row by row, starting with $T[0,j]$ for all j, continuing with $T[1,j]$, and so on. Dynamic programming is effective when this table is small, i.e., when the solutions to subproblems can be used many times throughout the recursive process.

This table-filling strategy can be best visualized by a standard example of dynamic programming, viz. by computing binomial coefficients. Pascal's formula states that

$$\binom{n}{k} = \binom{n-1}{k-1} + \binom{n-1}{k} \qquad \forall n,k \geq 1.$$

Hence, in order to compute $\binom{n}{k}$ it suffices to know this recursion and, additionally, the initial conditions that $\binom{m}{m} = \binom{m}{0} = 1$ for all $m > 0$. The table which has to be filled in order to compute $\binom{n}{k}$ is known as Pascal's triangle. Note that Pascal's triangle is of polynomial size, in fact even of size $\mathcal{O}(nk)$, whereas the number of Steiner trees which have to be computed during the Dreyfus-Wagner algorithm is exponential in $k = |K|$.

We will illustrate the dynamic programming approach with one more problem, known as the *knapsack problem*. This is a classical problem in the operations research literature.

> Maximum Knapsack:
>
> *Given:* A set of n objects with weights $w_1,\ldots,w_n$ and nonnegative integral profits $p_1,\ldots,p_n$, and a capacity $B \geq 0$ of the knapsack.
> *Find:*   A subset of weight at most B that maximizes the profit, i.e., a subset $I \subseteq \{1,\ldots,n\}$ such that $\sum_{i \in I} w_i \leq B$ and
> $$\sum_{i \in I} p_i = \max \Big\{ \sum_{i \in I'} p_i \mid I' \subseteq \{1,\ldots,n\} \text{ and } \sum_{i \in I'} w_i \leq B \Big\}.$$

In order to solve the knapsack problem using a dynamic programming approach, we need to introduce a suitable function that can be computed recursively. A good choice is:

$f(i,t) :=$ minimum possible weight, if only the first i objects are available and the total profit should be at least t. We set $f(i,t)$ to infinity if the total profit of the first i objects is less than t.

The value of an optimal packing therefore is the maximal t such that $f(n,t)$ is at most B.

Proposition 5.9 $f(i,t) = \min\{f(i-1,t), f(i-1,t-p_i) + w_i\}$.

Proof. Consider a subset of the first i objects that has total profit at least t and minimum weight among all such subsets. There are only two possibilities for the i-th object: either it does not belong to this subset (then all elements come from the first $i-1$ objects and achieve a total profit of t) or it belongs to this subset (then the remaining elements come from the first $i-1$ elements and achieve a total profit of at least $t - p_i$). □

With this proposition at hand, it is quite easy to compute the value of an optimal packing. One just has to recursively fill in an array $f[.,.]$ of size $n \times \sum_{i=1}^{n} p_i$ and to return $\max\{t \mid f[n,t] \leq B\}$.

Algorithm 5.10 (KNAPSACK PACKING)
Input: Weights $w_1, \ldots, w_n$, profits $p_1, \ldots, p_n \in \mathbb{N}$ and capacity B.
Output: Value $\max\{\sum_{i=1}^{n} p_i x_i \mid \sum_{i=1}^{n} w_i x_i \leq B, x \in \{0,1\}^n\}$ of the profit in
              an optimum packing of the knapsack.
{ *Initialization* }
$p := \sum_{i=1}^{n} p_i$;
for $t := 1$ **to** p_1 **do** $f[1,t] := w_1$;
for $t := p_1 + 1$ **to** p **do** $f[1,t] := \infty$;
{ *Recursion* }
for $i := 2$ **to** n **do**
 for $t := 1$ **to** p **do**
  if $t \leq p_i$ **then**
   $f[i,t] := \min\{f[i-1,t], w_i\}$
  else
   $f[i,t] := \min\{f[i-1,t], f[i-1,t-p_i] + w_i\}$;
{ *Determine optimal value* }
return $\max\{t \mid f[n,t] \leq B\}$.

Theorem 5.11 *The knapsack packing algorithm is correct and has complexity* $\mathcal{O}(n^2 p_{\max})$, *where* $p_{\max} = \max\{p_i \mid 1 \leq i \leq n\}$.

Proof. The correctness is an immediate consequence of Proposition 5.9. To verify the claimed bound on the complexity, observe that, trivially, $p \leq n p_{\max}$. Thus, the initialization and the computation of the output need $\mathcal{O}(n p_{\max})$ steps, while the recursion needs $\mathcal{O}(n^2 p_{\max})$ steps. □

Even though the statement about the complexity of the knapsack packing algorithm in Theorem 5.11 might rise the impression that Algorithm 5.10 is a polynomial time algorithm, it is important to realize that this is *not* the case. To see why, observe that the input for the knapsack packing problem consists just of

$$\lceil \log_2 B \rceil + \sum_{i=1}^{n} (\lceil \log_2 w_i \rceil + \lceil \log_2 p_i \rceil)$$

bits. The running time of a polynomial time algorithm must therefore be bounded by a polynomial in this value, but *not* by a polynomial in $p_{\max}$ itself.

On the other hand, it is actually not at all surprising that the knapsack packing algorithm is not a polynomial time algorithm. Consider the following decision version of the knapsack problem.

KNAPSACK:

Instance: A set of n objects with weights $w_1, \ldots, w_n$ and nonnegative integral profits $p_1, \ldots, p_n$, a capacity $B \geq 0$ of the knapsack and, additionally, an integer $K \geq 0$ for the required profit.

Question: Does there exist $I \subseteq \{1, \ldots, n\}$ such that $\sum_{i \in I} w_i \leq B$ and $\sum_{i \in I} p_i \geq K$?

Surely, this problem is $\mathcal{NP}$-complete because it is a generalization of SUBSET SUM (consider $w_i = p_i$ for every i and $B = K$). Hence, the $\mathcal{NP}$-completeness of KNAPSACK follows from Theorem 3.15.

Nevertheless, the knapsack packing algorithm seems to be "closer" to a polynomial time algorithm than, *e.g.*, the Dreyfus-Wagner algorithm. More precisely, the running time of the knapsack algorithm depends only polynomially on the number of objects to be packed. The reason for not being a polynomial time algorithm is just that some of the profits might be huge, or, in other words, that huge numbers may occur in the input. The running time of the Dreyfus-Wagner algorithm on the other hand is already exponential in the cardinality of the terminal set.

In general, an algorithm which solves a (decision or optimization) problem Π is called a *pseudopolynomial time* algorithm for Π if its running time is bounded from above by a polynomial function in the length of the input and in the largest number which occurs in the input. Thus, the knapsack packing algorithm is a pseudopolynomial time algorithm. Of course, every polynomial time algorithm for a problem Π is also a pseudopolynomial time algorithm.

We close this chapter by introducing a notion which captures the dependence of the difficulty of a problem on the use of large numbers.

Definition 5.12 *Let $\Pi = \langle \mathcal{I}, \mathrm{Sol} \rangle$ be an $\mathcal{NP}$-complete decision problem. As usual we let $|I|$ denote the length of the input I. Furthermore we use $Max(I)$ to denote the largest integer occurring in I. Then the problem Π is called strongly $\mathcal{NP}$-complete if there exists a polynomial p such that the restriction of Π to instances satisfying $Max(I) \leq p(|I|)$ is still $\mathcal{NP}$-complete.*

Obviously, STEINER PROBLEM IN GRAPHS, SATISFIABILITY and 3SAT are strongly $\mathcal{NP}$-complete, as the input of these problems does not contain any integer at all. Unless $\mathcal{P} = \mathcal{NP}$, the KNAPSACK problem, however, is not strongly $\mathcal{NP}$-complete.

Theorem 5.13 *If* $\Pi = \langle \mathcal{I}, \mathrm{Sol} \rangle$ *is strongly* $\mathcal{NP}$-*complete, then* Π *cannot be solved by a pseudopolynomial time algorithm, unless* $\mathcal{P} = \mathcal{NP}$.

Proof. As Π is strongly $\mathcal{NP}$-complete, there exists a polynomial p such that $\Pi_p = \langle \mathcal{I}_p, \mathrm{Sol} \rangle$, where $\mathcal{I}_p := \{I \in \mathcal{I} \mid Max(I) \leq p(|I|)\}$, is $\mathcal{NP}$-complete. Now consider what happens if we apply a pseudopolynomial time algorithm only to instances from $\mathcal{I}_p$. Clearly, in this case the running time can be bounded in a polynomial in $|I|$ only. That is, restricted to instances from $\mathcal{I}_p$ the algorithm is in fact a polynomial time algorithm. As Π_p is $\mathcal{NP}$-complete, this can only happen if $\mathcal{P} = \mathcal{NP}$.   □

Problems

5.1 Let $C \in \mathbb{N}$ be arbitrary but fixed. Design an algorithm with complexity $\mathcal{O}(n \log n + m)$ that solves the Steiner problem optimally for all networks $N = (V, E, \ell)$ and terminal sets $K \subseteq V$ such that $|V \setminus K| \leq C$. How does the constant hidden in the $\mathcal{O}(\cdot)$-term depend on C?

5.2 Modify the Dreyfus-Wagner algorithm in such a way that it computes not only the length of a Steiner minimum tree for T, but also the tree itself. (Hint: Add some kind of backtracking procedure.)

5.3 Suppose that we have a set with a product operation that is not associative. Therefore, brackets are needed to indicate the order in which the operations are to be carried out. Let a_n denote the number of ways to do this if there are n factors. For example one has $a_4 = 5$, corresponding to the products $(a(b(cd)))$, $(a((bc)d))$, $((ab)(cd))$, $((a(bc))d)$, and $(((ab)c)d)$. Design an algorithm which computes a_n efficiently. (Hint: First prove a recurrence relation. Then use dynamic programming.)

5.4 Let p_n denote the number of ways to cut an n-sided convex polygon into $n-2$ triangles using diagonal lines that do not cross. For example, a pentagon can be cut in 5 different ways, as shown in Figure 5.3. Therefore $p_5 = 5$. Design an algorithm which computes p_n efficiently.

5.5 Let $N = (V, E, \ell)$ be a network with vertices $V = \{v_1, \ldots, v_n\}$ and let (ℓ_{ij}) be a $n \times n$ matrix where $\ell_{ij} = \ell(v_i, v_j)$ if $\{v_i, v_j\} \in E$ and $\ell_{ij} = \infty$ otherwise. Prove that if the following operation

$$\ell_{ij} := \min\{\ell_{ij}, \ell_{ik} + \ell_{kj}\} \text{ for all } i,j = 1, \ldots, n \text{ but } i,j \neq k$$

    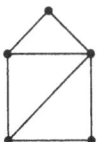

Figure 5.3 Cutting a pentagon into triangles.

is performed successively for $k = 1,2,\ldots,n$, then at the end each entry ℓ_{ij} will be equal to the length of the shortest path from v_i to v_j.

5.6 Dijkstra's algorithm for finding a shortest $s - t$ path in a network relies on the fact that all weights of the edges are nonnegative. Use the result of the previous problem to design an algorithm for finding shortest paths in networks with negative weights, assuming that there is no negative cycle (a cycle where the sum of the weights of its edges is negative). What is the running time of your algorithm?

5.7 Modify the knapsack packing algorithm of Section 5.3 in such a way that
   a) it computes not only the maximum profit, but also returns an index set $I \subseteq \{1,\ldots,n\}$ that attains this profit, and
   b) it has complexity $\mathcal{O}(nB)$ instead of $\mathcal{O}(n^2 p_{\max})$.

5.8 Let P_{opt} denote the maximum profit of a knapsack problem and let P_{greedy} denote the profit attained by using the following greedy algorithm:

> **Algorithm 5.14** (GREEDY KNAPSACK ALGORITHM)
> (1) Sort the objects such that $\frac{p_1}{w_1} \geq \ldots \geq \frac{p_n}{w_n}$; Let $I := \emptyset$.
>   Compute j_0 such that $p_{j_0} = \max\{p_j \mid 1 \leq j \leq n, w_j \leq B\}$.
> (2) **for** $i := 1$ **to** n **do**
>   **if** $w_i + \sum_{j \in I} w_j \leq B$ **then** $I := I \cup \{i\}$.
> (3) **if** $\sum_{j \in I} p_j > p_{j_0}$ **then return** $\sum_{j \in I} p_j$ **else return** p_{j_0}.

   a) Show that $\frac{p_1}{w_1} \geq \cdots \geq \frac{p_n}{w_n}$ implies

$$P_{\text{opt}} \leq \sum_{j=1}^{k} p_j + \frac{p_{k+1}}{w_{k+1}}\left(B - \sum_{j=1}^{k} w_j\right) \qquad \text{for all } k = 1,\ldots,n.$$

   b) Show that $P_{\text{greedy}} \geq \frac{1}{2} P_{\text{opt}}$.
   c) Show that the constant $1/2$ in b) is best possible.

5.9 Show that step (3) of the greedy algorithm of the previous problem is essential. More precisely, show that if (3) is replaced by

   (3') **return** $\sum_{j \in I} p_j$

   then the ratio $P'_{\text{greedy}}/P_{\text{opt}}$ can be arbitrarily large.

5.10 Find a pseudopolynomial time algorithm for the problem PARTITION, cf. Problem 3.10.

5.11* Show that the decision problem 3-PARTITION introduced in Problem 3.11 is strongly $\mathcal{NP}$-complete.

Notes

The enumeration algorithm was first developed by Hakimi (1971) and later streamlined by Lawler (1976). The dynamic programming approach is due to Dreyfus and Wagner (1972). Essentially the same algorithm was obtained by Levin (1971). Erickson, Monma, and Veinott, Jr. (1987) gave a similar dynamic programming algorithm for the more general problem of finding a minimum cost flow in a network with many demand nodes and concave cost functions on the edges. This algorithm can be specialized to the case of the Steiner problem in planar networks in which all terminals lie on the boundary of a single face yielding an algorithm of complexity $\mathcal{O}(nk^3 + (n \log n)k^2)$. Further improvements of exact algorithms for the Steiner problem in planar networks can be found in Bern (1990).

Optimization techniques using elements of dynamic programming have been in use a long time. The word "programming" refers here to the use of a tabular solution method. The book by Bellman (1957) was the first which studied dynamic programming systematically. Two other classical books on dynamic programming are: Nemhauser (1966) and Bellman and Dreyfus (1962). Since then, dynamic programming has proven to be a very successful algorithmic tool. For many applications of this general method we refer, *e.g.*, to Ahuja, Magnanti, and Orlin (1993) and Cormen, Leiserson, and Rivest (2001). Complexity aspects of the knapsack problem are covered in Garey and Johnson (1979). The concept of strong $\mathcal{NP}$-completeness was introduced by Garey and Johnson (1978). We will study its influence on the approximability of certain problems in Section 7.3.

The enumeration of binary trees, the number of ways to insert parentheses into a formula (Problem 5.3), and the number of ways to divide a polygon into triangles (Problem 5.4) belong to a huge class of counting problems, which all have essentially the same solution. The history of these problems goes back as far as 1760 when von Segner and Euler posed and solved an equivalent problem. This class of problems is known as the Catalan family (and the corresponding numbers as Catalan numbers) in recognition of the Belgian mathematician E.C. Catalan (1814-1894), who studied the bracketing problem. For more information on the Catalan numbers we refer the reader to Graham, Knuth, Patashnik (1989).

The use of the triangle operation for finding shortest paths in Problem 5.6 was observed by Floyd (1962), generalizing the transitive closure algorithm of Warshall (1962).

6

Approximation Algorithms

In Chapter 3 we have seen that the Steiner problem in networks is $\mathcal{NP}$-complete. It is thus unlikely that there exists an efficient (polynomial time) algorithm for its solution.

In this and the following chapters we will consider so-called *approximation algorithms*. As the name indicates, an approximation algorithm is not required to find an optimum solution exactly. Instead one only requires that it returns a solution whose value is not "too far off" from the value of an optimum solution. A widely used measure for the *quality* of an approximation algorithm is the maximum ratio between the value of the solution returned by the algorithm and the value of an optimum solution, where the maximum is taken over all admissible instances of the optimization problem under consideration. This ratio is also called the *performance ratio* of the approximation algorithm. In Chapter 7.3 we will define the notion of approximation algorithms for optimization problems in general. For now we restrict out considerations to the Steiner problem in networks. For that problem the performance ratio of an approximation algorithm is given by the maximum ratio of the length of the returned Steiner tree and the length of a Steiner minimum tree, where the maximum is taken over all admissible instances of Steiner problems in networks.

In this chapter we will develop a simple approximation algorithm with performance ratio two. This algorithm is based on the idea of computing a minimum spanning tree in an appropriately defined related network. In the next chapter we will then review this idea in a broader context. This will enable us to improve the quality of the approximation algorithm.

To shorten notation we will use the following notations in this and the following chapters. $N = (V, E, \ell; K)$ stands for a Steiner problem in a (connected) network with (nonnegative) length function $\ell \geq 0$ and terminal set K. To each such Steiner problem $N = (V,E,\ell; K)$ we associate a *complete distance network* $N_D = (K,E_D,\ell_D)$ as follows. The vertex set of N_D is equal to the terminal set K, the edge set is given by $E_D = [\binom{K}{2}]$, and the length function ℓ_D assigns to each edge $\{x,y\} \in E_D$ the length of a shortest $x - y$ path in N. With $smt(N)$ we denote the length of a Steiner minimum tree in N for the terminal set K.

6.1   A simple algorithm with performance ratio 2

The crucial idea of this chapter is the following lemma which relates the length of a Steiner minimum tree in a network to the length of a minimum spanning tree in the corresponding complete distance network.

Lemma 6.1 *Let $N = (V,E,\ell; K)$ be a Steiner problem. Then every minimum spanning tree T in the complete distance network N_D satisfies*

$$\ell_D(T) \leq (2 - \frac{2}{k}) \cdot smt(N),$$

where $k = |K|$ denotes the cardinality of the terminal set.

Proof.   Let S_{opt} be an arbitrary Steiner minimum tree in N. Embed the tree S_{opt} in the plane and consider a walk W along the outer region. This walk visits every terminal exactly once and traverses every edge twice. Its length is therefore exactly twice the length of S_{opt}.

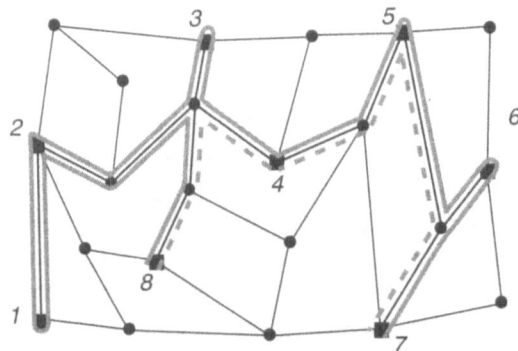

Figure 6.1   Illustration of the proof of Lemma 6.1. The walk W consists of the paths $1 - 2, 2 - 3, \ldots, 7 - 8$, and $8 - 1$. To obtain the walk W' we remove the path from 7 to 8.

Let t be the number of leaves in S_{opt}. Then W consists of $t \leq k$ paths between successive leaves in S_{opt}. Remove the longest of these paths from W. The length of the remaining walk W' is at most $(1 - \frac{1}{t})$ times that of W. Now observe that following the walk W' one can easily construct a spanning tree (in fact, even a path) in N_D whose length is at most that of W'. This observation concludes the proof of Lemma 6.1.   □

Example 6.2 The network $N = (V, E, \ell; K)$ with

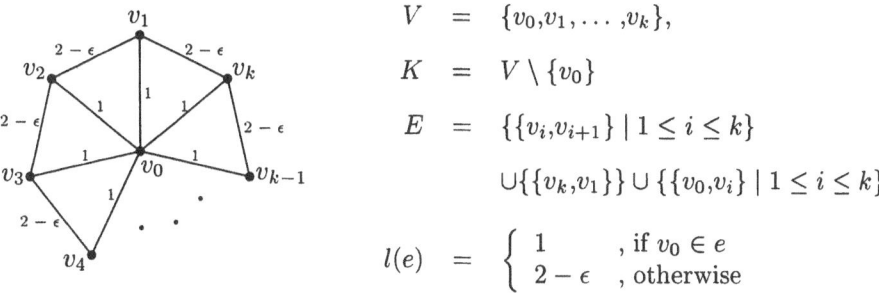

$$V = \{v_0, v_1, \ldots, v_k\},$$

$$K = V \setminus \{v_0\}$$

$$E = \{\{v_i, v_{i+1}\} \mid 1 \leq i \leq k\}$$
$$\cup \{\{v_k, v_1\}\} \cup \{\{v_0, v_i\} \mid 1 \leq i \leq k\},$$

$$\ell(e) = \begin{cases} 1 & \text{, if } v_0 \in e \\ 2 - \epsilon & \text{, otherwise} \end{cases}$$

shows that the bound of Lemma 6.1 is best possible.

We already know from Chapter 4.2 that minimum spanning trees are easy to compute. In Lemma 6.1 we saw that the minimum spanning tree in the complete distance network approximates the length of a Steiner minimum tree up to a factor of less than two. In fact, a minimum spanning tree in the complete distance network can also be used to actually *construct* a Steiner tree whose length exceeds that of a Steiner minimum tree by at most a factor of two.

Algorithm 6.3 (MST-ALGORITHM)
Input: Network $N = (V, E, \ell; K)$.
Output: Steiner tree S_K for N.
(1) Compute the distance network $N_D = (K, E_D, \ell_D)$.
(2) Compute a minimum spanning tree T_D in N_D.
(3) Transform T_D into a subnetwork $N[T_D]$ of N by replacing every edge of T_D by the corresponding shortest path.
(4) Compute a minimum spanning tree T for the subnetwork $N[T_D]$.
(5) Transform T into a Steiner tree S_K for N by successively deleting leaves which are no terminals.

The example in Figure 6.2 illustrates the various steps of the algorithm. In particular, it should make clear why steps (4) and (5) are necessary.

Theorem 6.4 *Let $N = (V, E, \ell; K)$ be a network. Then the MST-algorithm computes in polynomial time a Steiner tree S_K for N such that*

$$\ell(S_K) \leq \left(2 - \frac{2}{k}\right) \cdot smt(N).$$

Proof. According to Lemma 6.1, $\ell(T_D) \leq (2 - \frac{2}{k}) \cdot smt(N)$. Surely, the subsequent steps (3), (4) and (5) have the property that $\ell(T_D) \geq \ell(T) \geq \ell(S_K)$.   □

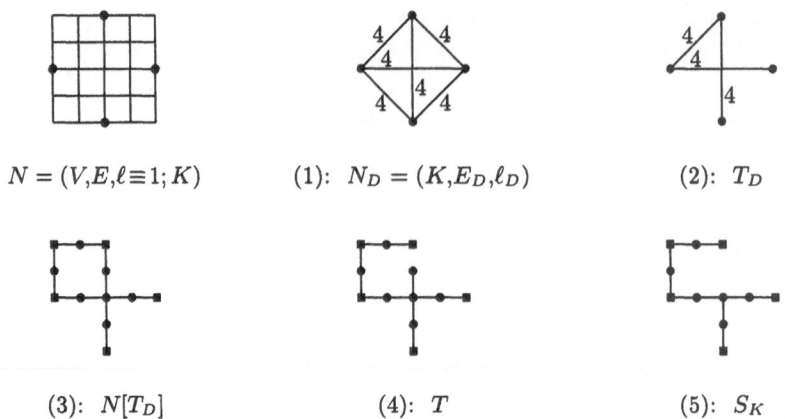

$$N = (V,E,\ell \equiv 1; K) \qquad (1): \ N_D = (K,E_D,\ell_D) \qquad (2): \ T_D$$

$$(3): \ N[T_D] \qquad\qquad (4): \ T \qquad\qquad (5): \ S_K$$

Figure 6.2 Illustration of the various step of the MST-algorithm

A straightforward analysis of the MST-algorithm shows that the most time consuming step in the algorithm is the computation of the distance network N_D. This network can be computed by k shortest path computations which each require at most $\mathcal{O}(n \log n + m)$ time. We omit a detailed discussion of these computational aspects here. Instead we will now modify the algorithm to further improve its time complexity.

6.2   Improving the time complexity

As we have seen, the running time of the MST-algorithm is dominated by the computation of the distance network N_D. But N_D is likely to contain a lot more information than we actually need. The important for the algorithm is how a terminal t is connected to those terminals which are "close" to t. Those terminals which are "far away" do not really matter.

A variant of the MST-algorithm, proposed by Mehlhorn in 1988, is based on this observation. It replaces the computation of the distance network N_D by the (faster) computation of a sparser network N_D^*. This network N_D^* is defined in such a way that every minimum spanning tree of N_D^* is also a minimum spanning tree of the distance network N_D. If we thus replace in Algorithm 6.3 the complete distance network N_D by this new network N_D^*, then the bound on the length $\ell(S_K)$ of the obtained Steiner tree from Theorem 6.4 remains valid.

In order to define the network N_D^*, we borrow a well-known notion from computational geometry. Given n points $x_1, \ldots, x_n$ in the Euclidean plane, the so-called *Voronoi region* of a point x_i consists of all those points in the Euclidean

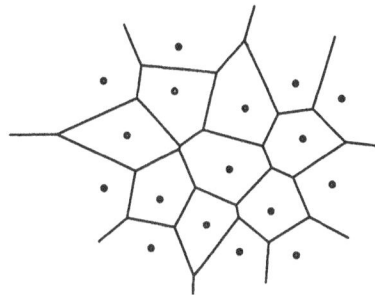

Figure 6.3 Voronoi regions of the Euclidean plane

plane which are closer to x_i than to any other point x_j, $j \neq i$. See Figure 6.3 for an example. In the graph case we proceed similarly. Namely, given a network $N = (V,E,\ell)$ and a terminal set K the *Voronoi region* $\Psi(s)$ of a terminal $s \in K$ consists of all vertices in V which are at least as close to s as to any other terminal, i.e., all vertices $v \in V$ for which $p(s,v) \leq p(t,v)$ for all $t \in K$. In fact, for our purposes this definition is slightly inadequate. If a vertex v has the *same* minimum distance to two or more different terminals, then according to the above definition it would belong to the Voronoi regions of all these terminals. We would, however, prefer every vertex to belong to *exactly one* set $\Psi(s)$. In the following we will therefore assume that in case of such ties the vertex v is (arbitrarily) assigned to exactly one of these closest terminals. That is, we assume that the sets $(\Psi(s))_{s \in K}$ do form a partition of V. We will later show that such a partition can actually be found quickly.

The network N_D^* is defined with respect to these Voronoi regions:

$$
\begin{aligned}
N_D^* &= (K, E_D^*, \ell_D^*) \quad \text{where} \\
\{x,y\} \in E_D^* &\Leftrightarrow \exists \{u,v\} \in E \quad \text{such that} \quad u \in \Psi(x) \text{ and } v \in \Psi(y), \\
\ell_D^*(x,y) &= \min\{p(x,u) + \ell(u,v) + p(v,y) \mid u \in \Psi(x), v \in \Psi(y), \{u,v\} \in E\}.
\end{aligned}
$$

The following lemma shows that this network N_D^* has the required property.

Lemma 6.5 *Every minimum spanning tree of N_D^* is also a minimum spanning tree of N_D.*

Proof. Let T_D be a minimum spanning tree in N_D which minimizes the number of edges $|E(T_D) \setminus E_D^*|$ which do not belong to E_D^*. If there is more than tree satisfying $E(T_D) \subseteq E_D^*$ we choose amongst those one which minimizes the length $\ell_D^*(T_D)$ of the tree T_D with respect to the length function in N_D^*.

We claim that any tree T_D defined in this way is in fact a minimum spanning tree in N_D^* for which $\ell_D(T_D) = \ell_D^*(T_D)$. Assume not. Then there exist two terminals s and t such that $\{s,t\} \in E(T_D)$ and one of the following two conditions holds:

(1) $\{s,t\} \in E(T_D) \setminus E_D^*$.

(2) $\ell_D^*(s,t) > \ell_D(s,t)$.

Let $\pi : V \to K$ be the function which assigns to each vertex v the terminal in whose Voronoi region v lies, i.e., $v \in \Psi(\pi(v))$. Furthermore let $s = v_0, v_1, \ldots, v_r = t$ be a shortest s-t path in N.

Removing the edge $\{s,t\}$ from T_D breaks the tree into two components, one containing s and the other containing t. As $\pi(v_0) = s$ and $\pi(v_r) = t$ and s and t lie in different components of $T_D \setminus \{s,t\}$, there has to exist an index i such that $\pi(v_{i-1})$ and $\pi(v_i)$ are contained in different components of $T_D \setminus \{s,t\}$ as well. Note that, by construction, $\{\pi(v_{i-1}), \pi(v_i)\}$ is an edge of N_D^* and satisfies

$$
\begin{aligned}
\ell_D(\pi(v_{i-1}), \pi(v_i)) &\leq \ell_D^*(\pi(v_{i-1}), \pi(v_i)) \\
&\leq p(\pi(v_{i-1}), v_{i-1}) + \ell(v_{i-1}, v_i) + p(v_i, \pi(v_i)) \\
&\leq p(s, v_{i-1}) + \ell(v_{i-1}, v_i) + p(v_i, t) \\
&= \ell_D(s,t) \\
(\ &< \ell_D^*(s,t) \quad \text{in case (2) holds.})
\end{aligned}
$$

The first three inequalities follow immediately from the definition of ℓ_D^* and the Voronoi regions $\Psi(v)$. The equality in the fourth line holds because $v_0, \ldots, v_r$ is a shortest s-t path.

If we replace the edge $\{s,t\}$ by the edge $\{\pi(v_{i-1}), \pi(v_i)\}$, we obtain another spanning tree in N_D (recall that by choice of i, $\pi(v_{i-1})$ and $\pi(v_i)$ are contained in different components of $T_D \setminus \{s,t\}$). Let us call this tree $\widetilde{T}_D$. Then

$$
\ell_D(\widetilde{T}_D) = \ell_D(T_D) - \ell_D(s,t) + \ell_D(\pi(v_{i-1}), \pi(v_i)) \leq \ell_D(T_D).
$$

As T_D was chosen to be a minimum spanning tree this implies that $\widetilde{T}_D$ is a minimum spanning tree in N_D as well. If (1) holds then $\widetilde{T}_D$ contains more edges from E_D^* than T_D, contradicting the choice of T_D. So (1) cannot hold and we therefore have that $E(T_D) \subseteq E_D^*$. Now assume (2) holds. Then

$$
\ell_D^*(\widetilde{T}_D) = \ell_D^*(T_D) - \ell_D^*(s,t) + \ell_D^*(\pi(v_{i-1}), \pi(v_i)) < \ell_D^*(T_D),
$$

again contradicting the choice of T_D. So T_D is in fact a minimum spanning tree in N_D^* such that $\ell_D(T_D) = \ell_D^*(T_D)$.

To conclude the proof of Lemma 6.5 we still have to show that *any* minimum spanning tree of N_D^* is also a minimum spanning tree in N_D. To see this, let T_D^* be an arbitrary minimum spanning tree in N_D^*. Clearly, T_D^* is also a spanning tree in N_D. As $\ell_D \leq \ell_D^*$, we also deduce that $\ell_D(T_D^*) \leq \ell_D^*(T_D^*)$. If we now recall that we already know that the lengths of minimum spanning trees in N_D and N_D^* are equal, this immediately implies that T_D^* is a minimum spanning tree in N_D.

□

We still have to show that the network N_D^* can be computed quickly. In order to do so, we first observe that the partition of the vertex set into the Voronoi regions $\Psi(s)$ can be obtained in $\mathcal{O}(n \log n + m)$ time. The main idea here is that this partition can be obtained by a single shortest path computation using Dijkstra's Algorithm 4.2. To see how, add a new vertex s_0 to V and connect s_0 to all terminals by an edge of length 0. Next we compute shortest paths from s_0 to all vertices in V. Clearly, as the edge lengths from s_0 to all terminals are 0, the first inner vertex s of a shortest path from s_0 to v is the terminal $\pi(v)$ to whose Voronoi region v belongs and the length of this path is exactly $p(s,v)$.

At that point we know for each vertex $v \in V$ its closest terminal $\pi(v)$ and the length $p(v,\pi(v))$ of the path from v to $\pi(v)$. Using this information we now have to compute the edge set E_D^* and the length function ℓ_D^*. Observe that each edge $\{u,v\} \in E$ generates exactly one edge in E_D^*, namely the edge $\{\pi(u),\pi(v)\}$ with a (potential) length $p(u,\pi(u)) + \ell(u,v) + p(v,\pi(v)))$. The difficulty is that some edges in E_D^* might be generated several times, and we have to detect the smallest of the corresponding length values. If we store the edges of E_D^* in an adjacency matrix, then selecting the minimum value is trivial: we run once over the edge list E, compute for each edge $\{u,v\} \in E$ in constant time the edge $\{\pi(u),\pi(v)\}$, check whether this edge is already in the network N_D^*, if not, we store it (with tentative length $p(u,\pi(u)) + \ell(u,v) + p(v,\pi(v)))$, if yes, we compare its current length with $p(u,\pi(u)) + \ell(u,v) + p(v,\pi(v)))$ and keep the smaller value. However, storing the graph as an adjacency matrix adds the term $\mathcal{O}(k^2)$ to the running time of the algorithm. In order to avoid this term, we have to use some clever trick.

First we compute in $\mathcal{O}(m)$ time for each edge $\{u,v\} \in E$ the two triples

$$(\; \pi(u), \;\; \pi(v), \;\; p(u,\pi(u)) + \ell(u,v) + p(v,\pi(v)) \;)$$
$$(\; \pi(v), \;\; \pi(u), \;\; p(u,\pi(u)) + \ell(u,v) + p(v,\pi(v)) \;).$$

We collect all theses triples in a list of length $2m$. Observe that if this list would be sorted according to the first component and within those entries sharing the same first component according to the second component, we could easily compute E_D^* and ℓ_D^* by running over the list once: we merely have to detect for each occurring pair (x,y) of first components (which will form an edge in E_D^*) the minimum value in the third component (which will be the length $\ell_D^*(x,y)$ of this edge).

It thus just remains to show that we can find this sorted order quickly. The idea is to use the so-called *bucket sort*, a sorting algorithm which is especially efficient if the items to be sorted come from a relatively small universe. In our case the "universe" is the set K of vertices of N_D^*. So we initialize an array of size $k = |K|$ and distribute the items of our list in this array according to their *second* component. Next we recombine the k list into a single list, in which the entries are now sorted according to their second component. Finally, we repeat the above process of distributing the entries into k lists and recombining these lists afterwards, this time using the first component in order to distribute the items. One easily checks that after recombining the k list into a single list, the new list has the desired order. Clearly, the total time required for this sorting procedure is just $\mathcal{O}(k+m)$. Thus, we have shown:

Lemma 6.6 *The network N_D^* can be computed in $\mathcal{O}(n \log n + m)$ steps.*   □

As a spinoff, working with the network N_D^* instead of the full distance network N_D has another advantage. Recall that computing an arbitrary minimum spanning tree T_D of the network N_D and transforming this tree back into N by replacing every edge of T_D by the corresponding shortest path yields a graph $N[T_D]$ which is not necessarily a tree. For minimum spanning trees in N_D^* one can, however, show that $N[T_D^*]$ is always a tree.

Lemma 6.7 *Let T_D^* be a minimum spanning tree of N_D^* and let $N[T_D^*]$ be the subgraph of N which is obtained by replacing every edge of T_D^* by the corresponding shortest path. Then $N[T_D^*]$ is a tree.*

Proof. First, we observe that by construction of N_D^*, $N[T_D^*]$ restricted to a Voronoi region $\Psi(s)$ does not contain a cycle. (To see this, recall that we constructed N_D^* by a single call to Dijkstra's algorithm. Hence we know from Problem 4.1 that the union of all shortest paths from s_0 to v form a tree. Therefore the union of all v-$\pi(v)$ paths, $v \in V$, form a collection of trees.) Furthermore, for two different Voronoi regions $\Psi(s)$ and $\Psi(t)$ there can exist at most one edge in $N[T_D^*]$ connecting a vertex in $\Psi(s)$ to a vertex in $\Psi(t)$ – call this edge the *crossing edge* for s and t. Hence, the subgraph of $N[T_D^*]$ induced by $\Psi(s) \cup \Psi(t)$ cannot contain a cycle. Now assume that $N[T_D^*]$ contains a cycle passing through more than two Voronoi regions, i.e., using more than two crossing edges. Then, obviously, the corresponding edges in T_D^* form a cycle - which is a contradiction.
                                                                                    □

These observations lead to the following simple algorithm.

Algorithm 6.8 (MEHLHORN'S ALGORITHM)
Input: Network $N = (V, E, \ell; K)$.
Output: Steiner tree S_M for N.
(1) Compute the distance network $N_D^* = (K, E_D^*, \ell_D^*)$.
(2) Compute a minimum spanning tree T_D^* in N_D^*.
(3) Transform T_D^* into a Steiner tree $S_M := N[T_D^*]$ of N by replacing every edge of T_D^* by the corresponding path.

Theorem 6.9 *Let $N = (V, E, \ell; K)$ be a network. Then Mehlhorn's algorithm computes in $\mathcal{O}(n \log n + m)$ steps a Steiner tree S_M for N such that*

$$\ell(S_M) \ \leq \ \left(2 - \frac{2}{k}\right) \cdot smt(N).$$

Proof. The correctness of Mehlhorn's algorithm and the bound on the length of S_M follows immediately from Lemma 6.1 in connection with Lemma 6.5 and Lemma 6.7.

By Lemma 6.6, the network N_D^* can be computed in $\mathcal{O}(n \log n + m)$ steps. By Theorem 4.5 a minimum spanning tree T_D^* in N_D^* can be obtained in $\mathcal{O}(k \log k + m)$ steps. Finally, $S_M := N[T_D^*]$ can easily be computed in $\mathcal{O}(m)$ steps, by including for each edge $\{\pi(v), \pi(u)\}$ in T_D^* the edge $\{u, v\}$ and the shortest paths from v and u to $\pi(v)$ resp. $\pi(u)$.   □

It is worthwhile to observe that the complexity of Mehlhorn's algorithm is of the same order of magnitude as that of Dijkstra's algorithm (cf. Section 4.1). Note also that for terminal sets of cardinality two Mehlhorn's algorithm always computes a shortest path connecting these two terminals.

6.3 Excursion: Machine scheduling

In 1966 Ron Graham wrote a paper entitled *"Bounds on certain multiprocessing anomalies"*. In this paper he considered approximation algorithms for scheduling problems. Actually, at that time the study of approximation algorithms was still quite uncommon. Graham's paper is therefore considered to be one of the roots of the field of approximation algorithms. This section is devoted to the problem studied by Graham.

The input of a scheduling problem consists of n jobs $J_1, \ldots, J_n$. The jobs come with processing times $p_1, \ldots, p_n$, which we will assume to be positive integers. Each job has to be assigned (scheduled) to one of m identical machines $M_1, \ldots, M_m$ in such a way that the maximum load of a machine is minimized. Here the *load* of a machine is the sum of the processing times of all jobs assigned to that machine. We assume that each machine can only work on one job at a given time (and will never be idle if there is a job which is not yet processed). The maximum load corresponds thus to the maximum completion time of all jobs. This value is also called the *makespan* of the schedule. In order to come closer to realistic models of real-world problems, which usually come with additional constraints, many different versions of scheduling problems have been considered in the literature. They range from rather common constraints like release and/or due dates, classes of precedence constraints up to an endless number of variants and special models. In this section, however, we will only consider the simplest case.

MACHINE SCHEDULING:

Given: A set of n jobs with positive integral processing times $p_1,\dots,p_n$, and a number m of machines.

Find: A schedule assigning each of the n jobs to exactly one of the m machines which minimizes the makespan.

In the rest of this section we will use C_{opt} to denote the minimum makespan of a given machine scheduling problem. We also use the letter σ to denote a schedule. Formally, a schedule consists of two functions. Namely, a function $\mu : \{1,\dots,n\} \to \{1,\dots,m\}$ that assigns to each job i to a machine $\mu(i)$, and a function $t : \{1,\dots,n\} \to \mathbb{R}^+$ that assigns to each job i a starting time $t(i)$. In order to form a proper schedule these two functions also have to satisfies the following condition: for all pairs of jobs $1 \le i < j \le n$ we have $\mu(i) \ne \mu(j)$ or $t(i) + p_i \le t(j)$ or $t(j) + p_j \le t(i)$ (i.e. no two jobs are executed simultaneously on the same machine). Observe that with this notation at hand we can write C_{opt} as

$$C_{\mathrm{opt}} = \min_{\sigma}\ \max_{1\le i \le m} [t(i) + p_i],$$

where the minimum is over all schedules $\sigma = (\mu,t)$.

Observe that the KNAPSACK problem in which the weight of each item is equal to its profit, i.e. $w_i = p_i$, and the capacity of the knapsack is $B = \frac{1}{2}\sum_{i=1}^{n} p_i$ is a special case of the MACHINE SCHEDULING problem with $m = 2$ machines. This suggests that we might be able to solve the latter problem using a dynamic programming approach as well. So let us try.

The key for a successful application of dynamic programming is a good choice for the function f, which has to be computed recursively. How should we choose this function for the machine scheduling problem? On the one hand it should be as simple as possible, on the other hand it should nevertheless capture all "relevant" properties of a schedule. So, what are the relevant properties of a schedule? As we are only interested in the makespan (but not in the precise location of a particular job), it should be plausible that these "relevant" properties are just m numbers, namely the load of each of the m machines. In fact, we can even do without the mth load, as it is equal to the total sum of all processing times minus the load of the other $m - 1$ machines. This suggests the following choice for the function f:

$$f(i,t_1,\dots,t_{m-1}) = \begin{cases} 1 & \text{if there exists an assignment } \mu : \{1,\dots,i\} \to \{1,\dots,m\} \\ & \text{for the first } i \text{ jobs such that } \sum_{s\in\mu^{-1}(j)} p_s = t_j \\ & \text{for all } 1 \le j \le m-1 \\ 0 & \text{otherwise.} \end{cases}$$

Computing this function recursively is very easy: the ith job has to be scheduled on one of the m machines. That is, $f(i,t_1,\dots,t_{m-1}) = 1$ if and only if at least one of the m values $f(i-1,t_1-p_i,t_2,\ldots,t_{m-1}),\ \ldots,\ f(i-1,t_1,\ldots,t_{m-2},t_{m-1}-p_i)$, and $f(i-1,t_1,\ldots,t_{m-2},t_{m-1})$ is equal to one. As all these values are either 0 or 1 we can also write this as

$$f(i,t_1,\ldots,t_{m-1}) \quad := \quad \max\{f(i-1,t_1,\ldots,t_{m-1}),$$
$$\max_{1\leq j\leq m-1}\{f(i-1,t_1,\ldots,t_{j-1},t_j-p_i,t_{j+1},\ldots,t_{m-1})\}\}.$$

With this observation at hand we obtain an algorithm for machine scheduling in a way which should seem familiar by now.

Algorithm 6.10 (MACHINE SCHEDULING – DYNAMIC PROGRAMMING)

Input: A set of n jobs with positive integral processing times $p_1,\ldots,p_n$, and a number m of machines.

Output: The value C_{opt} of the optimal makespan.

{ *Initialization* }
$T := \sum_{i=1}^{n} p_i$;
for all $(t_1,\ldots,t_{m-1}) \in \{0,\ldots,T\}^{m-1}$ **do** $f[0,t_1,\ldots,t_{m-1}] := 0$;
$f[0,0,\ldots,0] := 1$;
{ *Recursion* }
for $i := 1$ **to** n **do**
    for all $(t_1,\ldots,t_{m-1}) \in \{0,\ldots,T\}^{m-1}$ **do**
        $f[i,t_1,\ldots,t_{m-1}] := \max\{f[i-1,t_1,\ldots,t_{m-1}],$
        $\max\{f[i-1,t_1,\ldots,t_j-p_i,\ldots,t_{m-1}] \mid j=1,\ldots,m-1$ s.t. $t_j \geq p_i\}\}$;
{ *Determine optimal value* }
return $\min\{C \mid \exists(t_1,\ldots,t_{m-1}) \in \{0,\ldots,C\}^{m-1}$ s.t. $T-\sum_{j=1}^{m-1} t_j \leq C$
    and $f[n,t_1,\ldots,t_{m-1}]=1\}$.

Theorem 6.11 *Algorithm 6.10 correctly computes the optimal makespan C_{opt} in time $\mathcal{O}(nmT^{m-1})$, where $T = \sum_{i=1}^{n} p_i$.* □

Remark 6.12 It is not difficult to change Algorithm 6.10 so that it computes not only the *value* of the optimal makespan but also a schedule achieving this makespan. Essentially, we just have to change the definition of the function f slightly: instead of storing a 1 if a schedule exists we store the *number* of the machine to which the ith job is assigned in such a schedule. Clearly, this changes nothing in the overall complexity of the algorithm. Just the details get more messy. We leave them to the reader, cf. Problem 6.7.

What have we achieved so far? – From a practical point of view not very much. Algorithm 6.10 solves the problem optimally, but the required time is exorbitantly large, in particular for large m's. On the other hand, can we really do much better? Unfortunately, this is quite unlikely, as a seemingly much easier special case is already $\mathcal{NP}$-complete.

m-MACHINE SCHEDULING:

Instance: A set of n jobs with positive integral processing times $p_1, \ldots, p_n$, an integer number B.

Question: Does there exist a schedule σ with makespan at most B?

Theorem 6.13 *For every $m \geq 2$ the decision problem m-MACHINE SCHEDULING is $\mathcal{NP}$-complete.*

Proof. For $m = 2$ this follows from the fact that 2-MACHINE SCHEDULING is a generalization of PARTITION, cf. Problem 3.10. To see this, let $p_i = a_i$ and $B = \frac{1}{2} \sum_{i=1}^{n} a_i$ and observe that every schedule with makespan at most B corresponds to a valid partition of the values $a_1, \ldots, a_n$ and vice versa.

For all $m > 2$ the $\mathcal{NP}$-completeness of m-MACHINE SCHEDULING follows from the fact that each scheduling problem for 2 machines can be converted to an equivalent one for m-machines by adding $m - 2$ jobs with processing time B.

□

In the light of Theorem 6.13, we will therefore not try to design a better algorithm for computing an optimal schedule, but instead aim at finding one which just computes a "good" schedule. As it turns out, in doing so Algorithm 6.10 will prove to be very useful, despite its huge running time. In fact, for small values of m this algorithm is not too bad anyway. If m is fixed then T^{m-1} is just a polynomial in T. Does that mean that we have found a polynomial time algorithm for an $\mathcal{NP}$-complete problem? Of course not. The situation here is similar as for the knapsack problem in Section 5.3: the input for the 2-machine scheduling problem consists just of $\sum_{i=1}^{n} \lceil \log_2 p_i \rceil$ bits and $T = \sum_{i=1}^{n} p_i$ therefore does not depend polynomial on the input. (But of course it follows that for fixed m Algorithm 6.10 is a pseudo-polynomial time algorithm.)

In addition, we can use Algorithm 6.10 to obtain a polynomial time approximation algorithm for all m. The key word here is *scaling*. This is a technique which has proved to be very useful in many contexts. Let us see how it can be applied to the scheduling problem. The main trick is to modify the instance before Algorithm 6.10 is called. This modification is done in such a way that the T-value of the modified problem is bounded by a polynomial in n. Of course, if we modify the instance and, hence, loose some information about the input, we cannot expect that an optimal schedule for the modified problem is still an optimal schedule for the original problem. But, if we are careful enough, it might still be "close" to the optimal schedule. This is what we will do.

Algorithm 6.14 (MACHINE SCHEDULING II – SCALING)

Input: A set of n jobs with positive integral processing times $p_1, \ldots, p_n$, a number m of machines, and an integer $k \in \mathbb{N}$.

Output: A schedule σ with makespan $C \leq (1 + \frac{1}{k})C_{\mathrm{opt}}$.

{ *Compute scaled problem* }

$p_{\max} := \max\{p_i \mid 1 \leq i \leq n\}$;

for $i := 1$ **to** n **do** $\quad p'_i := \lceil kp_i n / p_{\max} \rceil$;

{ *Solve scaled problem* }

Use Algorithm 6.10 (modified according to Problem 6.7) with respect to the processing times p'_i to compute the schedule σ.

Theorem 6.15 *For every $k \geq 1$ Algorithm 6.14 computes in time $\mathcal{O}(m \cdot k^{m-1} \cdot n^{2m-1})$ a schedule σ such that its makespan C satisfies $C \leq (1 + \frac{1}{k})C_{\mathrm{opt}}$.*

Proof. Let $T' := \sum_{i=1}^{n} p'_i$. Then $T' = \sum_{i=1}^{n} \lceil \frac{kp_i n}{p_{\max}} \rceil \leq n \cdot kn$. The complexity of Algorithm 6.14 is thus $\mathcal{O}(n \cdot m \cdot (kn^2)^{m-1}) = \mathcal{O}(m \cdot k^{m-1} \cdot n^{2m-1})$, as claimed. Now consider the makespan. Here we use the following notation. C denotes the makespan of the schedule σ with respect to the processing times p_i, C' the makespan of the same schedule with respect to the scaled processing times p'_i. Similarly, C_{opt} denotes the makespan of an optimal schedule with respect to the processing times p_i, while C'_{opt} denotes the makespan of an optimal schedule with respect to the scaled processing times p'_i. As $p_i \leq \frac{p_{\max}}{kn} p'_i$ we know that $C \leq \frac{p_{\max}}{kn} \cdot C'$. On the other hand, $C' = C'_{\mathrm{opt}}$ (as we used Algorithm 6.10 with respect to the processing times p'_i to compute the schedule σ). Furthermore, an optimal schedule with respect to the processing times p_i is a schedule with makespan at most $\frac{kn}{p_{\max}} C_{\mathrm{opt}} + n$ with respect to the scaled processing times p'_i. That is, we have $C'_{\mathrm{opt}} \leq \frac{kn}{p_{\max}} C_{\mathrm{opt}} + n$. Combining all these inequalities we get:

$$C \leq \frac{p_{\max}}{kn} \cdot C' = \frac{p_{\max}}{kn} \cdot C'_{\mathrm{opt}} \leq \frac{p_{\max}}{kn} \cdot \left(\frac{kn}{p_{\max}} C_{\mathrm{opt}} + n \right) \leq \left(1 + \frac{1}{k} \right) \cdot C_{\mathrm{opt}},$$

where the last inequality comes from the fact that $C_{\mathrm{opt}} \geq p_{\max}$. $\qquad \square$

Observe that Theorem 6.15 implies that for fixed m Algorithm 6.14 is a polynomial time algorithm, even if k depends on n, say for example $k = n$. Clearly, for such large k the schedule obtained is then really close to an optimal schedule. And for, say, $m = 2$ the complexity would then be just $\mathcal{O}(n^4)$, which is not bad in light of the the the fact that finding an optimal schedule is $\mathcal{NP}$-hard. But what happens for larger m? Then the running time of Algorithm 6.14 is clearly not acceptable. In the rest of this section we will therefore design approximation algorithms which have a polynomial running time for all values of m.

We start with a very simple algorithm, called *List Scheduling*. In fact, this is an algorithm from the seminal paper of Graham mentioned in the beginning of this section.

Algorithm 6.16 (MACHINE SCHEDULING III — LIST SCHEDULING)
Input: A set of n jobs with positive integral processing times $p_1, \ldots, p_n$,
            a number m of machines.
Output: A schedule σ with makespan $C \leq (2 - \frac{1}{m})C_{\text{opt}}$.

Consider the jobs one-by-one, assign each job to that machine which (at that time) has the least load.

Theorem 6.17 *Algorithm 6.16 computes in time* $\mathcal{O}(m \cdot n)$ *a schedule* σ *such that its makespan* C *satisfies* $C \leq (2 - \frac{1}{m})C_{\text{opt}}$.

Proof.   At termination of the algorithm, the load of at least one machine is C. Consider the last job which was assigned to this machine. Assume it was the ith job. Then at the moment when it was assigned the load of its machine was $C - p_i$. By the rule of the list scheduling algorithm this implies that at that time the load of *all* machines was at least $C - p_i$. Otherwise it would have been assigned to another machine. This implies that after the ith job has been scheduled the total load of all m machines is at least $m(C - p_i) + p_i$. This shows that

$$m(C - p_i) + p_i \leq \sum_{j=1}^{n} p_j.$$

On the other hand, we clearly also have

$$\sum_{j=1}^{n} p_j \leq mC_{\text{opt}},$$

as in any schedule with makespan C_{opt} the total load of all scheduled jobs can be at most mC_{opt}. Combining both inequalities we get

$$m(C - p_i) + p_i \leq mC_{\text{opt}} \qquad \Longleftrightarrow \qquad C \leq C_{\text{opt}} + (1 - \frac{1}{m})p_i.$$

Observing that $C_{\text{opt}} \geq p_{\max} \geq p_i$ completes the proof of the theorem.   □

It is easy to see that the factor $2 - \frac{1}{m}$ is best possible. Consider, e.g., the following problem: $n = m(m - 1) + 1$ and $p_1 = \cdots = p_{n-1} = 1$, $p_n = m$. Then $C_{\text{opt}} = m$, but list scheduling computes a schedule with makespan $2m - 1$. On the other hand, for this instance it is quite obvious how we could improve the performance of the algorithm: we should consider the jobs in a different order! In fact, one can show that sorting the jobs before applying list scheduling improves the worst case performance of list scheduling quite a bit. We leave the proof of the following theorem to the reader (Problem 6.9).

Theorem 6.18 *If list scheduling is applied to jobs whose processing times are sorted so that $p_1 \geq \cdots \geq p_n$, then the makespan C of the computed schedule satisfies $C \leq (\frac{4}{3} - \frac{1}{3m})C_{\text{opt}}$.* □

For this modified version of list scheduling one can also find instances (try!) which show that the factor $\frac{4}{3} - \frac{1}{3m}$ is best possible. It is therefore plausible to ask whether one can do much better. As it turns out, one can. Before we show how, we consider a slightly easier problem. Namely, we consider the problem of constructing a schedule with makespan not much larger than a *given* value C (which may, but need not, be equal to the optimal makespan C_{opt}).

> **Algorithm 6.19** (MACHINE SCHEDULING IV — FEASIBILITY)
> **Input:** A set of n jobs with positive integral processing times $p_1, \ldots, p_n$, a number m of machines, an integer $k \in \mathbb{N}$, and a value $C > 0$.
> **Output:** A schedule σ with makespan at most $(1 + \frac{1}{k})C$ or FAILURE.
>
> { *Compute rounded problem* }
> **for** $i := 1$ **to** n **do**
>   **if** $p_i < \frac{C}{k}$ **then**
> $p_i' := 0$
>   **else**
> $p_i' := (1 + \frac{1}{k})^t \cdot \frac{C}{k}$, where $t \in \mathbb{N}_0$ is chosen such that $\frac{p_i}{1+\frac{1}{k}} < p_i' \leq p_i$;
>
> { *Solve rounded problem* }
> Use complete enumeration (see proof of Lemma 6.20) to solve the scheduling problem with processing times p_i' (considering only the jobs with processing times $p_i' > 0$).
> If no schedule with makespan $\leq C$ exists return FAILURE.
>
> { *Schedule jobs with $p_i' = 0$* }
> Consider the jobs with $p_i' = 0$ one-by-one, assigning each job to a machine which (at that time) has load at most C.
> If that is not possible for some job return FAILURE.

Lemma 6.20 *For all $k \in \mathbb{N}$ and $C \geq C_{\text{opt}}$ Algorithm 6.19 computes in time $\mathcal{O}(n + (m+1)^{(k+1)\lceil 2k \ln k \rceil + 2})$ a schedule with makespan at most $(1 + \frac{1}{k})C$.*

Proof. Consider an arbitrary value $C \geq C_{\text{opt}}$. We claim that the rounded processing times p_i' take on only $s := \lceil 2k \ln k \rceil + 1$ different positive values. Namely,

$$\frac{C}{k}, \quad (1 + \frac{1}{k}) \cdot \frac{C}{k}, \quad (1 + \frac{1}{k})^2 \cdot \frac{C}{k}, \quad \ldots, \quad (1 + \frac{1}{k})^{s-1} \cdot \frac{C}{k}.$$

To see this, it suffices to observe that $1 + x > e^{\frac{1}{2}x}$ for all $0 < x < 1$ and that hence $(1 + \frac{1}{k})^{s-1} \cdot \frac{C}{k} > e^{\frac{s-1}{2k}} \cdot \frac{C}{k} \geq C \geq C_{\text{opt}} \geq p_{\max}$.

Let us, for the moment disregard all jobs with (rounded) processing time zero. We claim that we can find in time $\mathcal{O}((m+1)^{(k+1)^s})$ a schedule for the remaining jobs with makespan at most C (with respect to the rounded processing times p_i'). To see this, we first observe that in any such schedule at most k jobs can be assigned to the same machine, as all nonzero processing times p_i' are greater than or equal to C/k. We can thus represent the assignment for a single machine by a vector $\rho = (\rho_1, \ldots, \rho_s)$, where ρ_i denotes the number of jobs with size $(1+\frac{1}{k})^{i-1}\frac{C}{k}$ that are scheduled on this machine. Of course, such a vector ρ represents a valid assignment if and only if it satisfies

$$\sum_{i=1}^{s} \rho_i \cdot (1+\frac{1}{k})^{i-1}\frac{C}{k} \leq C. \tag{6.1}$$

As each ρ_i is an integer of size at most k, we can easily compute all valid vectors $\rho^1, \ldots, \rho^z \in \{0, \ldots, k\}^s$ that satisfy equation (6.1) in time $s \cdot (k+1)^s$ by complete enumeration.

Let n_j denote the number of jobs with (rounded) processing time $(1+\frac{1}{k})^{j-1} \cdot \frac{C}{k}$, for $j = 1, \ldots, s$. A schedule for the rounded problem corresponds to an integral vector $x \in \{0, \ldots, m\}^z$, so that the ith component x_i specifies how many machines have an assignment according to the scheme ρ^i. And, vice versa, each vector $x \in \{0, \ldots, m\}^z$ corresponds to a valid schedule iff

$$\sum_{i=1}^{z} x_i = m \quad \text{and} \quad \sum_{i=1}^{z} \rho_j^i x_i = n_j \quad \text{for all } j = 1, \ldots, s \tag{6.2}$$

As $z \leq (k+1)^s$ we can, in time $\mathcal{O}((m+1)^{(k+1)^s} \cdot (k+1)^s \cdot (s+1))$, enumerate all vectors $x \in \{0, \ldots, m\}^z$, and check for each vector whether it satisfies (6.2). Observe that, as $p_i' \leq p_i$ for all jobs and as $C \geq C_{\text{opt}}$, such a schedule does exist. What can we say about its makespan with respect to the original processing times? As $p_i \leq (1+\frac{1}{k})p_i'$, this makespan is at most $(1+\frac{1}{k})C$, which is exactly what we want.

Finally, let us consider the "small' jobs, i.e., the jobs with rounded processing time zero or, equivalently, those with processing time $p_i < \frac{C}{k}$. Clearly, scheduling such a job on a machine with (current) load at most C will keep the makespan below $(1+\frac{1}{k})C$. Also, the algorithm can only fail to assign such a job, if *all* machines have a load larger than C. But this would imply that $mC < \sum_{i=1}^{n} p_i \leq mC_{\text{opt}}$, contradicting our assumption that $C \geq C_{\text{opt}}$. $\square$

Remark 6.21 In the proof of Lemma 6.20 we actually grossly overestimated the running time required for finding a schedule with respect to the rounded processing times p_i'. Recall that we just needed to find a vector $x \in \{0, \ldots, m\}^z$ that satisfies (6.2). Observe also that all equalities in (6.2) are linear in the x_i's. That is, we are just looking for a nonnegative integer solution to this linear programming problem. It is an important observation that in this linear programming problem the number of variables as well as the number of constraints depend only on k and not on n or m. We may thus apply a result of Lenstra who showed that such an integer linear programming problem can be solved in time polynomial in the size of the input. As in our case the size of the coefficients in equation (6.2) depend only logarithmically on n and m, this part of the algorithm can actually be achieved in $\mathcal{O}(n)$ time. This reduces the overall complexity of the algorithm to $\mathcal{O}(n)$. Note, however, that the constant hidden in the big O depends exponentially on k.

Having Algorithm 6.19 at hand, it is now quite straightforward to design a good approximation algorithm. We just apply Algorithm 6.19 repeatedly within a binary search procedure.

> **Algorithm 6.22** (MACHINE SCHEDULING V — BINARY SEARCH)
> **Input:** A set of n jobs with positive integral processing times $p_1, \ldots, p_n$, a number m of machines, and an integer $k \in \mathbb{N}$.
> **Output:** A schedule σ with makespan $C \leq (1 + \frac{1}{k})^2 C_{\mathrm{opt}}$.
>
> $L := p_{\max}$;
> $U := n \cdot p_{\max}$;
> **repeat**
> $C := (U + L)/2$;
>   **if** Algorithm 6.19 finds a schedule with makespan $\leq (1 + \frac{1}{k})C$ **then**
> $U := C$;
>   **else**
> $L := C$;
> **until** $U - L \leq p_{\max}/k$;
> Use Algorithm 6.19 to find a schedule with makespan $\leq (1 + \frac{1}{k})U$;

Note that if at termination of the algorithm $U < n \cdot p_{\max}$ then the last line is not really necessary, as the desired schedule has been computed earlier already.

Theorem 6.23 *For all $k \in \mathbb{N}$ Algorithm 6.22 computes in time $\mathcal{O}(n \log n)$ a schedule with makespan at most $(1 + \frac{1}{k})^2 C_{\mathrm{opt}}$. (Note that the constant hidden in the big O notation depends on k but not on n or m.)*

Proof. We first observe that the last call to Algorithm 6.19 will never return FAILURE. For $U < n \cdot p_{\max}$ this is obvious from the way the algorithm works. For $U = n \cdot p_{\max} \geq C_{\mathrm{opt}}$, Lemma 6.20 guarantees that Algorithm 6.19 will find a schedule with makespan at most $(1 + \frac{1}{k})U$.

What can we say about the makespan of the final schedule? Clearly, because of Lemma 6.20 we know that $L \leq C_{\mathrm{opt}}$ at all times. That is, at termination of the algorithm we have $U \leq L + \frac{p_{\max}}{k} \leq (1 + \frac{1}{k})C_{\mathrm{opt}}$. According to Lemma 6.20 the makespan C of the schedule computed in the last call to Algorithm 6.19 thus satisfies

$$C \leq (1 + \tfrac{1}{k})U \leq (1 + \tfrac{1}{k})^2 C_{\mathrm{opt}},$$

as claimed. The bound on the complexity follows from Remark 6.21 together with the observation that the binary search stops after at most $\log nk = \log n + \mathcal{O}(1)$ iterations. □

With a bit more care in the initialization of L and U in Algorithm 6.22, one can show that in fact a constant number of iterations of the binary search procedure suffice to obtain a schedule with makespan at most $(1 + \frac{1}{k})C_{\text{opt}}$. We leave the details to the reader, cf. Problem 6.10.

Problems

6.1   Use the MST-ALGORITHM to compute a Steiner tree for the following net-
      work (the terminals are the vertices denoted by a square):

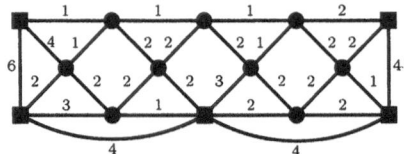

      Compare its length with that of a Steiner minimum tree.

6.2   Modify step (3) of the MST-ALGORITHM as follows:

      (3') Compute the set S of all vertices which are contained in the shortest
            paths corresponding to the edges in T_D. Let $N[T_D]$ denote the subnet-
            work induced by the set S.

      Let S'_K denote the Steiner tree which is computed by the modified algo-
      rithm, while S_K denotes the Steiner tree computed by the original algo-
      rithm. Is it true that $\ell(S'_K) \leq \ell(S_K)$? Prove it or give a counterexample.

6.3   Consider the following algorithm:

      Algorithm 6.24 (TAKAHASHI-MATSUYAMA ALGORITHM)
      Input: Network $N = (V, E, \ell; K)$.
      Output: Steiner tree S_K for N.
      Choose $v \in K$ arbitrarily and let $S_K := \emptyset$, $X := \{v\}$, $L := K \setminus \{v\}$.
      while $L \neq \emptyset$ **do**
            Compute for all $w \in L$ a shortest path P_w from w to some vertex in X.
            Compute w_0 such that $\ell(P_{w_0}) \leq \ell(P_w)$ for all $w \in L$.
            Let $L := L \setminus \{w_0\}$, $X := X \cup V(P_{w_0})$, and $S_K := S_K \cup E(P_{w_0})$.

      Show that the TAKAHASHI-MATSUYAMA ALGORITHM computes a Steiner
      tree S_k that satisfies $\ell(S_k) \leq (2 - \frac{2}{k})\ell(S_{\text{opt}})$, where S_{opt} is a Steiner minimum
      tree in N. Show also that the algorithm can be implemented in such a way
      that its running time is bounded by $O(k \cdot (n \log n + m))$.

6.4   For a network $N = (V, E, \ell; K)$ let S_{MST} denote the Steiner tree returned by the MST-algorithm, while S_{TM} denotes the Steiner tree returned by the Takahashi-Matsuyama Algorithm. Give examples for networks such that
a) $\ell(S_{MST}) < \ell(S_{TM})$, and
b) $\ell(S_{TM}) < \ell(S_{MST})$.

6.5   Show that, assuming $\mathcal{P} \neq \mathcal{NP}$, there cannot exist a constant $B \in \mathbb{N}$ and an algorithm $\mathcal{A}$ that returns for every network $N = (V, E, \ell; K)$ in polynomial time a Steiner tree $S_{\mathcal{A}}$ such that $\ell(S_{\mathcal{A}}) \leq \ell(S_{\text{opt}}) + B$, where S_{opt} is a Steiner minimum tree in N.

6.6   a) Is the problem 2-MACHINE SCHEDULING strongly $\mathcal{NP}$-complete? Prove it or give a pseudo-polynomial time algorithm.
b) Show that the decision version of MACHINE SCHEDULING in which the number of machines is part of the input is strongly $\mathcal{NP}$-complete. (Hint: Use Problem 5.11.)

6.7   Modify Algorithm 6.10 such that it returns not only the value of the optimal makespan C_{opt}, but also a schedule which achieves this makespan.

6.8   Consider an input to machine scheduling for which there exists an optimal schedule with the property that to every machine at most two jobs are assigned. Show that in this case list scheduling always yields an optimal schedule, provided the jobs are sorted in such a way that the processing times satisfy $p_1 \geq p_2 \geq \cdots \geq p_n$. Does a similar result hold for any instance containing at most $2m$ jobs? Prove it or give a counterexample.

6.9   Prove Theorem 6.18.

6.10 Show that Algorithm 6.22 can be modified so that a schedule with makespan $(1 + \frac{1}{k})C_{\text{opt}}$ is obtained in time $\mathcal{O}(n)$.

6.11 Consider the following variant of machine scheduling. The input consists of n jobs with positive integral processing times $p_1, \ldots, p_n$ and weights $w_1, \ldots, w_n$, and a number m of machines. The aim is to find a schedule for which $\sum_{i=1}^{n} w_i t(i)$ is minimized.
Show that this problem is solvable in time $\mathcal{O}(n \log n)$ for $m = 1$, and is $\mathcal{NP}$-complete for all $m \geq 2$. (Hint for $m = 2$: Find a reduction from PARTITION, cf. Problem 3.10.)

Notes

Algorithm 6.3 was suggested several times in the literature, see e.g. Choukhmane (1978), Kou, Markowsky, and Berman (1981), Plesník (1981), and Iwainsky, Canuto, Taraszow and Villa (1986). The improvement of the running time presented in Section 6.2 is due to Mehlhorn (1988), the observation of Lemma 6.7 is from Floren (1991). The algorithm in Problem 6.3 is due to Takahashi and Matsuyama (1980). There also exist many other approximation algorithms which guarantee the same bound as in Theorem 6.4. Overviews can be found in Chapter II.4 of the book by Hwang, Richards, Winter (1992) and in Widmayer (1986a, 1986b).

Voronoi regions are named after the Russian mathematician G. Voronoi. Fast algorithms for computing Voronoi regions in the Euclidean plane can be found in the text book by Preparata and Shamos (1985), the standard reference for an introduction to computational geometry.

The seminal paper of Ronald Graham on machine scheduling mentioned in the introduction of Section 6.3 is Graham (1966). Algorithm 6.10 is based on ideas from Rothkopf (1966), see also Błazewicz, Ecker, Pesch, Schmidt, and Węglarz (1996). The idea of using scaling in order to turn a pseudopolynomial time algorithm into a polynomial time approximation algorithm is due to Ibarra and Kim (1975). Sahni (1976) presents a slightly different approach. Woeginger (1999) gives a general framework for transforming pseudopolynomial time algorithms based on dynamic programming into polynomial time approximation schemes.

Theorem 6.18 is due to Graham (1969). Algorithms 6.19 and 6.22 are from Hochbaum and Shmoys (1987), our presentation also used ideas from Hochbaum (1997). The reference to the result of Lenstra mentioned in Remark 6.21 is Lenstra (1983). For additional background on linear and integer programming we recommend Schrijver (1986).

For a complete overview on the many different models and aspects of machine scheduling and additional pointers to the literature we refer the reader to the book by Błazewicz, Ecker, Pesch, Schmidt and Węglarz (1996) and the surveys by Lawler, Lenstra, Rinnooy Kan, and Shmoys (1993) and Hall (1997).

References to general aspects of approximation algorithms can be found in the note sections of the next two chapters.

7

More on Approximation Algorithms

In this Chapter we will extend the idea which led to approximation algorithms with performance ratio 2 in order to obtain approximation algorithms with a better performance ratio for the Steiner tree problem. Recall that the main reason for the value 2 of the performance ratio of the algorithm studied so far was Lemma 6.1. It shows that the maximum ratio between the length of a minimum spanning in the complete distance network and the length of a Steiner minimum tree is at most two. Recall also that the complete distance network is just a graph on the terminal set K in which the length of each edge is equal to the length of the shortest path between the corresponding terminals in the original network N. Now observe that a shortest path between two terminals can also be viewed as a Steiner minimum tree for these two terminals. This observation motivates the following generalization: instead of just using Steiner minimum trees for all pairs of vertices we could also try to find Steiner minimum trees for all 3 or, say, r element subsets of K. This would lead to a weighted "hyper"graph instead of the complete distance network.

As we will show in the next section, minimum spanning trees in these hypergraphs can indeed be used to construct short Steiner trees. Unfortunately, we will also show in this section that finding these minimum spanning trees is $\mathcal{NP}$-hard as well. In Section 7.2 we will therefore design a suitable greedy approach that, while not guaranteed to find a minimum spanning tree, will at least find a spanning tree that can be used to compute a good Steiner tree in the original network.

7.1   Minimum spanning trees in hypergraphs

A *hypergraph* $H = (V,F)$ is a generalization of the notion of a graph. In a hypergraph the set F of (hyper)edges is an arbitrary family of subsets of V and not just a family of 2-element subsets as in the definition of a graph. Many notions and results of graph theory generalize to hypergraphs. Here we restrict our considerations to paths, cycles, trees and forests. A *u-v-path* (of length $l \geq 2$) in H is a sequence $u = x_1, e_1, \ldots , x_l, e_l, x_{l+1} = v$ of vertices and edges such that the x_i are distinct vertices, the e_i are distinct edges, $x_1 \in e_1$, $x_{l+1} \in e_l$, and $x_i \in e_{i-1} \cap e_i$ for all $i = 2, \ldots , l$. A *cycle* (of length $l \geq 2$) in H is a path $x_1, e_1, \ldots , x_{l-1}, e_{l-1}, x_l$ with the additional property that x_1 and x_l are contained in an edge e_l different from all other e_i's. Note that, contrary to graphs, hypergraphs may contain cycles of length two. The following figure shows a cycle of length four and a cycle of length two.

A hypergraph H is *connected* if for every pair u,v of vertices there is a path from u to v. A hypergraph H is a *tree* if and only if H is a connected graph that contains no cycle. A *spanning tree* of a hypergraph H is a subhypergraph T of H that is a tree and satisfies $V(T) = V(H)$. Observe that a cycle of length two is a connected hypergraph that does not contain a spanning tree. That is, again contrary to graphs, not every connected hypergraph contains a spanning tree.

An *r-uniform* hypergraph is a hypergraph with all its edges having cardinality exactly r. That is, graphs are 2-uniform hypergraphs. An *r-bounded* hypergraph is a hypergraph with all its edges having cardinality at most r.

A *weighted* hypergraph $H = (V,F,w)$ is a hypergraph together with a function $w : F \to \mathbb{R}$. A minimum spanning tree in a weighted hypergraph $H = (V,F,w)$, denoted by $mst(H)$, is a spanning tree in H whose total weight is minimum among all spanning trees. If H contains no spanning tree we set $mst(H)$ to infinity.

We now return to the Steiner tree problem. Let $N = (V,E,\ell; K)$ be a network. With respect to this network we define for every $r \geq 2$ a weighted r-bounded hypergraph $H_r(N) = (K,E_r,\ell_r)$ on the vertex set K as follows. The edge set E_r consists of all subsets of K of cardinality at most r, and the weight $\ell_r(e)$ of an edge $e \in E_r$ is the length of a Steiner minimum tree for e in the network N. The hypergraph $H_r(N)$ is called the *r-distance hypergraph* of N. Notice that $H_2(N)$ is just the usual distance network $N_D = (K,E_D,\ell_D)$ of N. For simplicity, we denote by $mst_r(N) := mst(H_r(N))$ the length of a minimum spanning tree in the r-distance hypergraph $H_r(N)$.

Lemma 7.1 *Let $N = (V,E,\ell; K)$ be a network with terminal set K and let $H_r(N)$ be as defined above. Then*

$$mst_r(N) \geq smt(N) \qquad (7.1)$$

with equality if there exists a Steiner minimum tree T for K such that all full components of T contain at most r vertices.

Proof. Let T be a minimum spanning tree in $H_r(N)$. For every edge e in T choose a Steiner minimum tree T_e for e in N. Then the fact that T is a spanning tree in $H_r(N)$ implies that $S = \bigcup_{e \in T} T_e$ is a connected subgraph of (V,E) which contains all vertices in K. This shows (7.1).

Assume now that there exists a Steiner minimum tree T for K such that all of its full components contain at most r vertices. Then the edges corresponding to the sets of terminals in the full components of T form a spanning tree in $H_r(N)$ whose length is equal to that of T. ◻

Lemma 7.1 shows that $mst_r(N)$ is at least as large as $smt(N)$. As we will see shortly, $mst_r(N)$ can in fact also be bounded from above by $smt(N)$ times an appropriate constant. Before we do so, we show that r-distance hypergraphs can be computed in polynomial time.

Lemma 7.2 *Let r be a fixed constant. Then the r-distance hypergraph $H_r(N)$ of a network $N = (V,E,l; K)$ can be computed in time $\mathcal{O}(n^2 \log n + n\,m + k^{r+1}n^2)$. Conversely, if T is a spanning tree of $H_r(N)$, then a Steiner tree S for K in N satisfying $l(S) \leq \ell_r(T)$ can be computed in time $\mathcal{O}(n^3)$.*

Proof. In order to construct the hypergraphs $H_r(N)$ we have to compute a Steiner minimum tree for each subset of K of size at most r. There are $\sum_{i=2}^{r} \binom{k}{i} \leq \sum_{i=2}^{r} k^i/i! \leq k^{r+1}$ such subsets. For each of these subsets we compute a Steiner minimum tree with Algorithm 5.6 of Dreyfus and Wagner. Recall from the proof of Theorem 5.7 that the initialization phase of the Dreyfus and Wagner algorithm (the computation of the shortest paths between all pairs of vertices) requires time $\mathcal{O}(n^2 \log n + n\,m)$. The remaining steps of the Dreyfus Wagner algorithm require time $\mathcal{O}(2^r n^2 + 3^r n)$ for a terminal set of size at most r. As the initialization step has to be performed only once, we can therefore compute the Steiner minimum trees for each subset of K of size at most r in time $\mathcal{O}(n^2 \log n + n\,m + k^{r+1}n^2)$. (Recall that r is assumed to be a fixed constant!)

To see the second part, assume that a spanning tree T of $H_r(N)$ with edges $T_1, \ldots, T_s$ is given. In order to construct the desired Steiner tree, we first compute for every $T_i \subseteq K$ a Steiner minimum tree S_i in N. Note that, by definition of the function ℓ_r, these Steiner trees have lengths $\ell(S_i) = \ell_r(T_i)$. (Preferably, these trees are already stored when $H_r(N)$ is constructed. Otherwise, this can be done in time $\mathcal{O}(n^2 \log n + n\,m + s \cdot n^2) = \mathcal{O}(n^3)$ in the same way as outlined above.) Now consider the subnetwork $S' = \bigcup_{i=1}^s S_i$ of N. By construction, S' is a connected subnetwork of N which covers K and satisfies $\ell(S') \leq \ell_r(T)$. We can therefore use the Jarník-Prim Algorithm 4.4 to find in $\mathcal{O}(n \log n + m)$ time a minimum spanning tree in this subnetwork, and subsequently a Steiner tree S in N such that $\ell(S) \leq \ell_r(T)$.   □

Lemma 7.2 implies that if we could show for some fixed r that (1) minimum spanning trees in $H_r(N)$ can be determined in polynomial time and (2) minimum spanning trees in $H_r(N)$ have the property that they are not much larger than Steiner minimum trees, then this would imply a polynomial time approximation algorithm for the Steiner problem with good performance ratio. We now make these ideas precise.

Let ρ_r denote the least upper bound of the ratio $mst_r(N)/smt(N)$ for all networks $N = (V, E, \ell; K)$. The values ρ_r are called *Steiner ratios*. Lemma 7.1 shows that for every r we have $\rho_r \geq 1$. From Lemma 6.1 it follows that ρ_2 is at most 2. In Example 6.2 we have seen that in fact $\rho_2 = 2$. On the other hand, for larger values of r the value ρ_r can be much smaller. In fact, it tends to one for r tending to infinity. Before we show this (Theorem 7.4), we present a lower bound for ρ_r.

Example 7.3 NETWORKS N WHERE $mst_r(N)$ IS LARGE COMPARED TO $smt(N)$. Consider the complete binary tree T_h of height h. This is the tree on $2^h - 1$ vertices containing exactly one vertex of degree 2 (the root) and all other vertices having either degree 3 (the branching vertices) or degree 1 (the leaves). Being complete means that T_h has 2^{h-1} leaves and the paths from the root to each leaf have lengths exactly $h - 1$. A vertex v is said to belong to the ith level of T_h if the path from the root to v has length $i - 1$. Thus, the root itself is the only vertex on the first level of T_h and all leaves are on the hth level. We assume now h to be an odd number. In order to obtain a network from T_h, let the set $L(T_h)$ of leaves of T_h be the set of terminals. The weight of an edge connecting a vertex from level i to a vertex from level $i + 1$ is given by 2^{h-1-i}. As T_h is a tree, the length $smt(T_h)$ of a Steiner minimum tree for $L(T_h)$ is just the total length of all edges. Thus,

$$smt(T_h) = \sum_{i=1}^{h-1} 2^i \cdot 2^{h-1-i} = (h-1) \cdot 2^{h-1}.$$

For convenience, we label the vertices of T_h with 0-1 vectors of length h as follows: the root gets label $(0 \dots 01) \in 2^h$. The vertices on the $(i+1)$st level get labels $(0 \dots 01x)$ where $x \in 2^i$. We assume T_h to be embedded into the plane and the vertices on the same level to be labeled increasingly with respect to the lexicographic order on the labels. Compare T_5 as an example:

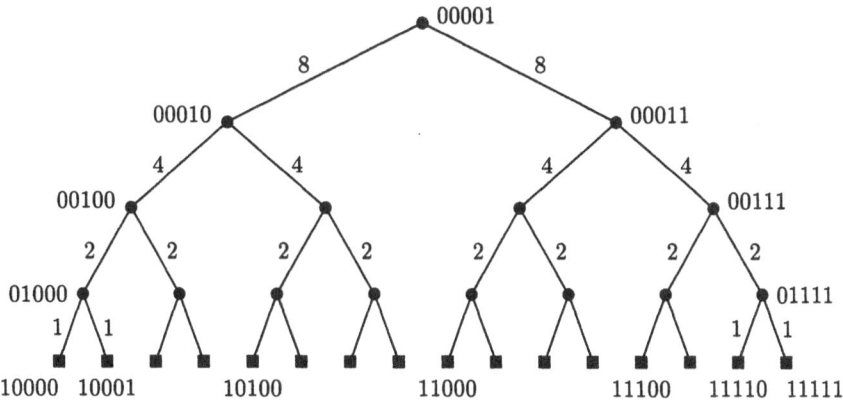

Let $\mathcal{H}$ be the 4-bounded hypergraph on the leaves of T_h which is associated with the tree T_h. In $\mathcal{H}$ we define a minimum spanning tree by considering some suitable sets $\mathcal{X}_i$ of hyperedges. A set $\mathcal{X}_i$ contains a hyperedge for each of the 2^{i-1} vertices on level i. For each such vertex consider the subtree rooted at this vertex: it contains exactly four vertices at level $i+2$. For each of these four vertices consider the leftmost leaf of the subtree rooted at this vertex. These four vertices will form a hyperedge. Formally, the sets $\mathcal{X}_i$ are defined as follows:

$$\mathcal{X}_1 \quad = \quad \{\{(1000\ldots0), (1010\ldots0), (1100\ldots0), (1110\ldots0)\}\}$$

$$\vdots$$

$$\mathcal{X}_i \quad = \quad \{\{(1x000^{h-i-2}), (1x010^{h-i-2}), (1x100^{h-i-2}), (1x110^{h-i-2})\} : x \in 2^{i-1}\}$$

$$\vdots$$

$$\mathcal{X}_{h-2} \quad = \quad \{\{(1x00), (1x01), (1x10), (1x11)\} : x \in 2^{h-3}\}$$

It is not difficult to prove (cf. Problem 7.1) that $\mathcal{H}^* := \bigcup_{j=1}^{\frac{h-1}{2}} \mathcal{X}_{2j-1}$ is a minimum spanning tree in $\mathcal{H}$. In order to compute its length we first observe that each edge F in $\mathcal{X}_i$ has length

$$w(F) = 2 \cdot 2^{h-1-i} + 4 \cdot \sum_{j=0}^{h-2-i} 2^j = 2^{h-i} + 4 \cdot (2^{h-1-i} - 1) = 3 \cdot 2^{h-i} - 4.$$

Hence, we obtain

$$w(\mathcal{H}^*) \quad = \quad \sum_{j=1}^{(h-1)/2} 2^{(2j-1)-1} \cdot (3 \cdot 2^{h-(2j-1)} - 4)$$

$$= \quad \frac{3}{2}(h-1)2^{h-1} - \sum_{j=1}^{(h-1)/2} 4^j = \frac{3}{2}(h-1)2^{h-1} - \frac{2}{3}2^h + \frac{4}{3}.$$

Thus, $w(\mathcal{H}^*)/smt(T_h) = 3/2 \cdot (1 - \mathcal{O}(1/h))$ and therefore $\rho_4 \geq 3/2$, considering the case that h tends to infinity.

It is easy to extend Example 7.3 to show that $\rho_{2^s} \geq (s+1)/s$. With some more work, Example 7.3 can also be modified in order to show that for every $r \geq 2$ with $r = 2^s + t$ so that $0 \leq t < 2^s$ it is true that

$$\rho_r \geq \frac{(s+1)2^s + t}{s2^s + t}. \tag{7.2}$$

We leave the proof of this result to the reader, cf. Problem 7.2 and Problem 7.3. Instead, we show that these examples actually do exhibit the worst cases.

Theorem 7.4 *Let* $r \geq 2$ *be an integer such that* $r = 2^s + t$ *and* $0 \leq t < 2^s$. *Then*

$$\rho_r = \frac{(s+1)2^s + t}{s2^s + t}.$$

Example 7.5 VALUES OF ρ_r FOR SMALL r.

r	2	3	4	5	6	7	8	16	32	64
ρ_r	2	1.67	1.5	1.44	1.4	1.36	1.33	1.25	1.2	1.17

Before we start proving Theorem 7.4, we state a technical proposition. Its proof is a straightforward induction. We leave it to the reader, cf. Problem 7.4.

Proposition 7.6 *For every complete binary tree there exists a one-to-one mapping, say* g, *from the internal vertices (i.e., all vertices which are not leaves) to the leaves such that for any internal vertex* x *the value* $g(x)$ *is a descendant of* x *and the paths from* x *to* $g(x)$ *are pairwise edge-disjoint.* □

Proof of Theorem 7.4. Let $N = (V, E, \ell; K)$ be a network and let $r \geq 2$ be an integer such that $r = 2^s + t$ where $0 \leq t < 2^s$. In the light of equation (7.2) it obviously suffices to construct a spanning tree T in $H_r(N)$ such that

$$\frac{\ell_r(T)}{smt(N)} \leq \frac{(s+1)2^s + t}{s2^s + t}. \tag{7.3}$$

Let S be a Steiner minimum tree in N for K. We may assume without loss of generality that S is a full Steiner tree, as otherwise we could handle each component separately. That is, the vertices in K are exactly the leaves in S. Furthermore, by duplicating vertices and adding edges of length 0, we may also assume that S is a complete binary tree with root, say, $r(S)$ and height h. In order to find a spanning tree in $H_r(N)$ which is sufficiently small to satisfy (7.3), we first define an appropriate labeling on the vertices of S. We label the internal vertices of S inductively with sets of size 2^s chosen from $\{1, 2, \ldots, s\,2^s + t\}$ (cf. Figure 7.1 for an illustration):

- Label the root $r(S)$ of S with the set $\{1,2,\ldots,2^s\}$.

- For $2 \le i \le s$, label all vertices on the ith level with set $\{(i-1)2^s+1,\ldots,i2^s\}$.

- For $i > s$ we construct the label sets as follows. Let w be a vertex on the ith level. Then w has a unique ancestor v on level $(i-s)$. Assume that w is the jth descendent on level i (in some arbitrary but fixed ordering) of vertex v. Assume further that the label set of v is $\{\xi_1,\ldots,\xi_{2^s}\}$. Then we label vertex w with the set $\{\xi_1,\ldots,\xi_{j-1},\xi_1^*,\ldots,\xi_t^*,\xi_{j+t},\ldots,\xi_{2^s}\}$ where $\xi_1^*,\ldots,\xi_t^*$ are those numbers which are not used in label sets of any of the s immediate predecessors of w. (The addition in the index is modulo 2^s. That is, for $j = 2^s$ we label w with the set $\{\xi_2^*,\ldots,\xi_t^*,\xi_t,\ldots,\xi_{2^s-1},\xi_1^*\}$.)

This labeling has three important properties which are easily verified:

(P1) The sets of labels on s consecutive vertices on a descending path in S are pairwise disjoint.

(P2) The sets of labels on $s+1$ consecutive vertices on a descending path in S contain all numbers from $\{1,\ldots,s2^s + t\}$.

(P3) For each vertex v of S the following is true. If v is a vertex on level i and j is some number contained in the label set of v then the subtree rooted at v contains exactly $2^s - t$ vertices on level $i + s$ and $2t$ vertices on level $i + s + 1$ which contain j in its label set.

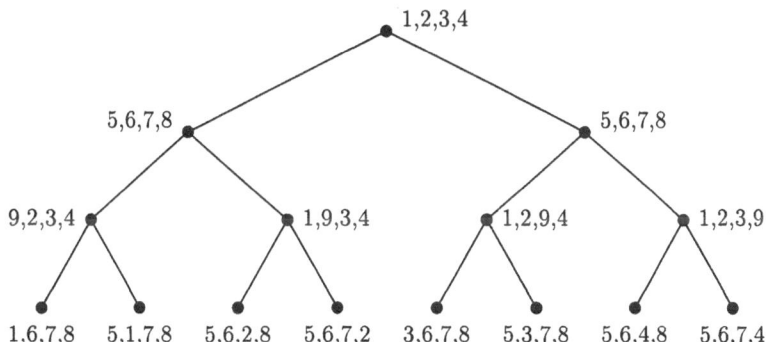

Figure 7.1 Illustration of the labeling defined in the proof of Theorem 7.4. Here $s = 2$, $t = 1$, $r = 2^2 + 1 = 5$ and S is the complete binary tree of height 4. The label sets are subsets of size 4 of the set $\{1,\ldots,9\}$.

For each number $j \in \{1,\ldots,s2^s + t\}$ we will now build a family $\mathcal{X}_j$ of subtrees of S, where each subtree contains at most r leaves. Observe that each such subtree will thus uniquely identify an edge in $H_r(N)$. Furthermore, each family $\mathcal{X}_j$ will in fact define a spanning tree in $H_r(N)$.

Fix some $j \in \{1,\ldots,s2^s + t\}$. Then the roots of the subtrees contained in $\mathcal{X}_j$ will be $r(S)$ and each vertex of S which contains j in its label set. For each such root vertex v we first connect v to all vertices below v that (i) contain j

in its label set and (ii) are at most $s + 1$ levels below v; we call these vertices intermediate leaves. (Observe that property (P2) implies that if $v = r(S)$ and j is not in the label set of $r(S)$ then there are at most $2^s \leq r$ such intermediate leaves. If $v \neq r(S)$ then (P1) and (P2) imply that all intermediate leaves are either s or $s + 1$ levels below v. Furthermore, property (P3) implies that there are at most $2^s - t + 2t = r$ intermediate leaves.) Each intermediate leaf w is then connected to the leaf $g(w)$, where g is the function from Proposition 7.6. If the subtree rooted at v contains, besides v, no other vertex with j in its label set, then we connect v to all leaves (which in this case are also considered to be intermediate leaves) in its subtree. (Note that property (P2) implies that the subtree can contain at most $2^s \leq r$ leaves.)

Recall that, by assumption, every terminal t is a leaf in S. Observe also that each leaf is contained in at least one of the subtrees of $\mathcal{X}_j$ (consider the first ancestor, say x, containing j in its label). Furthermore, one easily checks that each edge of S is contained in at least one subtree of $\mathcal{X}_j$. Hence it follows that if we take for each subtree of $\mathcal{X}_j$ the edge in $H_r(N)$ corresponding to the (at most r) leaves of this subtree, then these edges form a connected subhypergraph of $H_r(N)$ that contains all vertices. Observe that this also implies that we can find a spanning tree $\mathcal{T}_j$ in $H_r(N)$ such that the length of $\mathcal{T}_j$ is bounded by the sum of the lengths of all subtrees in $\mathcal{X}_j$.

We now bound this sum. First consider the paths from the root vertices of the subtrees in $\mathcal{X}_j$ to the intermediate leaves. As each edge of S is contained in exactly one such path, the sum of the lengths of these paths is equal to $\ell(S) = smt(N)$. Next consider the paths from intermediate leaves v to the leaves $g(v)$. Let L_j denote the sum of length of these path for all subtrees in $\mathcal{X}_j$. Since each internal vertex v in S appears in exactly 2^s trees as intermediate leaf (once for every number in its label set), the length of the path from v to $g(v)$ is counted exactly 2^s times in the sum $L_1 + L_2 + \cdots + L_{s2^s+t}$. Since these paths are disjoint for different intermediate leaves, it follows that

$$L_1 + L_2 + \cdots + L_{s2^s+t} \leq 2^s \cdot smt(N) \,.$$

Hence, there exists j such that

$$L_j \leq \frac{2^s}{s2^s + t} \cdot smt(N) \,.$$

Observe that for this j the length of the spanning tree $\mathcal{T}_j$ is thus bounded by

$$\ell_r(\mathcal{T}_j) = \sum_{T \in \mathcal{X}_j} \ell(T) = smt(N) + L_j$$

$$\leq \frac{(s+1)2^s + t}{s2^s + t} \cdot smt(N),$$

which concludes the proof of the theorem.

$\square$

The second question we started with is thus answered quite satisfactory: the length of a minimum spanning tree in $H_r(N)$ tends to the length of a Steiner minimum tree in N if r is getting large.

In order obtain an approximation algorithm with performance ratio less than 2, it remains to show that one can find minimum spanning trees in a given r-bounded hypergraph quickly. Unfortunately, this turns out to be difficult.

Theorem 7.7 *Finding a minimum spanning tree in a weighted r-bounded hypergraph is $\mathcal{NP}$-hard for every $r \geq 4$.*

Proof. From Theorem 3.11 we know that for every $r \geq 4$ the R-RESTRICTED STEINER PROBLEM is $\mathcal{NP}$-complete. As a polynomial time algorithm for finding a minimum spanning tree in the r-distance hypergraph $H_r(N)$ of N solves the R-RESTRICTED STEINER PROBLEM in the network N as well (cf. Lemma 7.1), the existence of such an algorithm therefore implies that $\mathcal{P} = \mathcal{NP}$. □

Actually, the proof of Theorem 7.7 shows slightly more than stated in the theorem.

Corollary 7.8 *Finding a minimum spanning tree in a r-distance hypergraph is $\mathcal{NP}$-hard for every $r \geq 4$.* □

Using the $\mathcal{NP}$-completeness of TRIPARTITE MATCHING it also follows that even finding a minimum spanning tree in an unweighted 4-uniform hypergraph is difficult.

Theorem 7.9 *Deciding whether an unweighted 4-uniform hypergraph contains a spanning tree is $\mathcal{NP}$-complete.*

Proof. By Theorem 3.14, TRIPARTITE MATCHING is $\mathcal{NP}$-complete. Let an instance from TRIPARTITE MATCHING be given, i.e., let X, Y, and Z be three pairwise disjoint sets each containing n elements and let $S \subseteq X \times Y \times Z$. Consider the unweighted hypergraph $H = (V, F)$ with vertex set $V = \{a\} \cup X \cup Y \cup Z$, where the vertex a is not contained in $X \cup Y \cup Z$, and edge set $F = \{\{a\} \cup e \mid e \in S\}$. Obviously, H is a 4-uniform hypergraph such that there exists a perfect matching $M \subseteq S$ if and only if $\{\{a\} \cup m \mid m \in M\}$ forms a spanning tree in H. □

For $k = 3$ deciding whether an *unweighted* 3-uniform hypergraph contains a spanning tree is known to be in $\mathcal{P}$. The status of the problem of finding a minimum spanning tree in a *weighted* 3-uniform hypergraph is still open. I.e., a polynomial time algorithm is not known, but so far it was also not shown that this problem is $\mathcal{NP}$-hard.

We close this section with the remark that there are also no polynomial time approximation algorithms known for the minimum spanning subgraph problem and the minimum spanning tree problem in r-bounded hypergraphs which yield a better performance ratio than 2 for the Steiner problem. Presently the best known polynomial time approximation algorithm for the minimum spanning subgraph problem in r-bounded hypergraphs has performance ratio H_{r-1}. (Here H_n denotes the nth harmonic number. It has size about $H_n = (1+o(1))\log n$.) For $r = 3$ this algorithm reduces to a 3/2-approximation algorithm for the minimum spanning subgraph problem in 3-bounded hypergraphs. Together with the fact that $\rho_3 = 5/3$, this yields only an 5/2-approximation algorithm for the Steiner problem.

7.2   Improving the performance ratio I

In the previous section we have seen that the most straightforward way to obtain better performance ratios for the Steiner tree problem, namely, to just solve the minimum spanning tree problem in r-bounded hypergraph, fails: this problem turns out to be $\mathcal{NP}$-hard even in the special case of r-distance hypergraphs.

In this section we will, however, see that one can nevertheless use spanning trees in r-distance hypergraphs to obtain polynomial time approximation algorithms for the Steiner problem with performance ratio $(1+\ln 2)\rho_r$. Thus, from the fact that the Steiner ratio ρ_r tends to 1 for r tending to infinity (Theorem 7.4), we get polynomial time approximation algorithms for the Steiner problem which have performance ratio arbitrarily close to $1 + \ln 2 \approx 1.69$.

Let $N = (V,E,\ell;K)$ be a network for which we want to solve the Steiner problem. The aim is to find a short spanning tree in the corresponding r-distance hypergraph $H_r(N)$ and then to convert it into a Steiner tree in N. The strategy to find a spanning tree in $H_r(N)$ is to select step by step edges greedily. Of course, we cannot just choose the shortest edge in every step in $H_r(N)$ — this would lead to selecting only 2-element edges and thus to finding a minimum spanning tree in $H_2(N)$. Hence, we have to invent a measure on the edges indicating the use of choosing a particular edge. Roughly speaking, an edge is allowed to be relatively long if its contribution to the resulting Steiner tree is sufficiently large. How can we measure this contribution?

Let $H_2(N) = (K,E_2,\ell_2)$. For $X \subseteq E_2$ we denote by $H_2(N;X)$ the network which arises from $H_2(N)$ by replacing the lengths of all edges in X by zero. Obviously, $H_2(N;\emptyset) = H_2(N)$. Similarly, we denote for a set $Y \subset K$ by $mst_2(N;Y)$ the length of a minimum spanning tree in $H_2(N;\binom{Y}{2})$. In particular, $mst_2(N;\emptyset) = mst_2(N)$ is the length of a minimum spanning tree in $H_2(N)$. If $Y_1,\ldots,Y_j$ is a family of edges in $H_r(N) = (K,E_r,\ell_r)$, we use $H_2(N;Y_1,\cdots,Y_j)$, resp. $mst_2(N;Y_1,\ldots,Y_j)$ for $H_2(N,Y)$, resp. $mst_2(N;Y)$ where $Y = \{\{v,w\} \mid v,w \in Y_i \text{ for some } i = 1,\ldots,j\}$.

As a measure for how much an edge $Y_i \in E_r$ contributes to the resulting Steiner tree in N, we use the effect that shrinking the set Y_i has on the length of the minimum spanning tree in $H_2(N; Y_1, \cdots, Y_{i-1})$. More precisely, we use

$$mst_2(N; Y_1, \ldots, Y_{i-1}) - mst_2(N; Y_1, \ldots, Y_{i-1}, Y_i).$$

The larger this quantity is, the larger is the use of choosing this edge.

Now we describe an algorithm due to Zelikovsky for computing Steiner trees which is based on this measure. In fact, we describe a family of algorithms $\mathcal{A}_r$, one for each integer $r \geq 2$.

Algorithm 7.10 (ZELIKOVSKY'S ALGORITHM $\mathcal{A}_r$)
Input: A network $N = (V, E, \ell; K)$.
Output: A Steiner tree T_Z for K.
Compute $H_r(N) = (K, E_r, \ell_r)$;
Compute $H_2(N) = (K, E_2, \ell_2)$;
$i := 0$;
while $Y_1 \cup \cdots \cup Y_i$ is not a connected spanning subgraph of $H_r(N)$ **do**
begin
    Choose $Y_{i+1} \in E_r$ that minimizes
        $f_i(Y_{i+1}) := \ell_r(Y_{i+1}) / [mst_2(N; Y_1, \ldots, Y_i) - mst_2(N; Y_1, \ldots, Y_i, Y_{i+1})]$;
    $i := i + 1$;
end;
Compute from $Y_1 \cup \cdots \cup Y_i$ a Steiner tree T_Z for K in N;
return T_Z.

Before proving that this strategy actually leads to an approximation algorithm with a good performance ratio, we first show that Zelikovsky's algorithm is indeed a polynomial time algorithm.

Proposition 7.11 *Let* $N = (V, E, \ell; K)$ *be a network and* $r \geq 2$. *Then Zelikovsky's algorithm* $\mathcal{A}_r$ *terminates in time* $\mathcal{O}\left(n^2 \log n + nm + k^{r+1}n^2\right)$.

Proof. From Lemma 7.2 it follows that the preprocessing step, i.e., computing $H_r(N)$ and $H_2(N)$ and the postprocessing step, i.e., computing a Steiner tree in N from a spanning tree in $H_r(N)$ can be done in time $\mathcal{O}(n^2 \log n + nm + k^{r+1}n^2)$.

The while-loop will be executed at most $k - 1$ times. For each iteration the time is bounded by $\mathcal{O}(k^r(n \log n + m))$ as E_r contains at most k^r edges and computing a spanning tree can be done in time $\mathcal{O}(n \log n + m)$. Since $m \leq n^2$, the statement of the proposition follows.   □

We start the analysis of Zelikovsky's algorithm by proving a key fact about the function $mst_2(N; X)$.

Lemma 7.12 *Let $H_2(N) = (K, E_2, \ell_2)$ be the distance network of N and let $X, Y \subseteq E_2$ and $z \in E_2$. Then*

$$mst_2(N; X \cup Y) - mst_2(N; X \cup Y \cup \{z\}) \le mst_2(N; X) - mst_2(N; X \cup \{z\}).$$

Proof. Let T be a minimum spanning tree in $H_2(N; X)$ and let $T(Y)$ be the minimum spanning tree in $H_2(N; X \cup Y)$ which is obtained by successively inserting the edges of $Y \backslash E(T)$ into T and removing in every step the longest edge in the arising cycle (cf. Problem 7.5). Let $z = \{z_0, z_1\}$. Observe that $mst_2(N; X) - mst_2(N; X \cup \{z\})$ is equal to the length of the longest edge in the path connecting z_0 with z_1, in T.

Now let $c \ge 0$ be some constant. We denote by T_c, $T_c(Y)$ respectively, the subgraph of T, $T(Y)$ respectively, which is induced by all edges e satisfying $\ell_2(e) \le c$. Note that T_c and $T_c(Y)$ may by disconnected. From Problems 7.5 and 7.6 we deduce that

$$
\begin{aligned}
&mst_2(N; X) - mst_2(N; X \cup \{z\}) \\
= \quad &\min\{c \ge 0 \mid z_0 \text{ and } z_1 \text{ are in the same component of } T_c \} \\
\ge \quad &\min\{c \ge 0 \mid z_0 \text{ and } z_1 \text{ are in the same component of } T_c(Y) \} \\
= \quad &mst_2(N; X \cup Y) - mst_2(N; X \cup Y \cup \{z\}).
\end{aligned}
$$

□

Corollary 7.13 *Let $H_2(N) = (K, E_2, \ell_2)$ be the distance network of N and let $X, Y, Z \subseteq E_2$. Then*

$$mst_2(N; X \cup Y) - mst_2(N; X \cup Y \cup Z) \le mst_2(N; X) - mst_2(N; X \cup Z).$$

Proof. We prove the corollary by induction on $|Z|$. For $|Z| = 1$ we apply Lemma 7.12. So assume that the assertion is true for all $X, Y \subseteq E_2$ and all $Z' \subseteq E_2$ with $|Z'| \le t$ for some $t \ge 1$. Let $Z = Z' \cup \{z\}$ be of cardinality $t + 1$. Then from the inductive hypothesis it follows that

$$mst_2(N; X \cup Y) - mst_2(N; X \cup Y \cup Z') \le mst_2(N; X) - mst_2(N; X \cup Z')$$

and from Lemma 7.12 we infer that

$$
\begin{aligned}
&mst_2(N; X \cup Y \cup Z') - mst_2(N; X \cup Y \cup Z' \cup \{z\}) \\
\le \quad &mst_2(N; X \cup Z') - mst_2(N; X \cup Z' \cup \{z\}).
\end{aligned}
$$

Adding these two inequalities proves Corollary 7.13.   □

Next, we remind the reader on a useful fact about sequences of integers.

Proposition 7.14 *Let $a_i \geq 0$, $b_i > 0$ be integers for $i = 1, \ldots, n$. Then*

$$\frac{\sum_{i=1}^{n} a_i}{\sum_{i=1}^{n} b_i} \geq \min_{1 \leq j \leq n} \frac{a_j}{b_j}.$$

Proof.

$$\left(\min_{1 \leq j \leq n} \frac{a_j}{b_j} \right) \cdot \left(\sum_{i=1}^{n} b_i \right) = \sum_{i=1}^{n} \left(b_i \cdot \min_{j} \frac{a_j}{b_j} \right) \leq \sum_{i=1}^{n} b_i \cdot \frac{a_i}{b_i} = \sum_{i=1}^{n} a_i.$$

$\square$

Now we are ready to prove that Zelikovsky's algorithm computes a Steiner tree whose length is bounded by a suitable multiple of the length of a minimum spanning tree in the associated r-bounded hypergraph.

Theorem 7.15 *Let $N = (V, E, \ell; K)$ be a network and $r \geq 2$. Then Zelikovsky's algorithm $\mathcal{A}_r$ computes in polynomial time a Steiner tree T_Z for K such that*

$$\ell(T_Z) \leq (1 + \ln 2) \cdot mst_r(N).$$

Proof.   Let $T^* = Y_1^* \cup Y_2^* \cup \cdots \cup Y_s^*$ be a minimum spanning tree in $H_r(N)$, i.e., $\ell(T^*) = mst_r(N)$, and let $Y_1, \ldots, Y_t$ be the edges chosen by algorithm $\mathcal{A}_r$. Note that as long as $Y_1 \cup \ldots \cup Y_i$ is not a connected spanning subgraph of $H_r(N)$, there always exists an edge $e \in E_2$ (e.g., every edge in a minimum spanning tree in $H_2(N; Y_1, \ldots, Y_i)$ with nonzero length will do) such that $l_2(e) \leq mst_2(N; Y_1, \ldots, Y_i) - mst_2(N; Y_1, \ldots, Y_i, e)$. Hence, for every $i = 1, \ldots, t - 1$ we have that

$$f_i(Y_{i+1}) \leq 1. \tag{7.1}$$

Zelikovsky's algorithm selects at the $(i + 1)$st step the hyperedge $Y_{i+1} \in E_r$ which minimizes f_i. Therefore we have

$$f_i(Y_{i+1}) \leq \min_{j} f_i(Y_j^*)$$

$$= \min_{j} \frac{\ell_r(Y_j^*)}{mst_2(N; Y_1, \ldots, Y_i) - mst_2(N; Y_1, \ldots, Y_i, Y_j^*)}$$

and hence, applying Proposition 7.14, we deduce that

$$f_i(Y_{i+1}) \leq \frac{\sum_j \ell_r(Y_j^*)}{\sum_j (mst_2(N; Y_1, \ldots, Y_i) - mst_2(N; Y_1, \ldots, Y_i, Y_j^*))}.$$

The numerator of the right hand side of this inequality is equal to $mst_r(N)$. Using Corollary 7.13 we observe that the denominator is greater or equal than

$$\sum_j (mst_2(N; Y_1, \ldots, Y_i, Y_1^*, \ldots, Y_{j-1}^*) - mst_2(N; Y_1, \ldots, Y_i, Y_1^*, \ldots, Y_{j-1}^*, Y_j^*)).$$

This is a telescoping sum in which all but the first and the last term cancel. The last term is $mst_2(N; Y_1, \ldots, Y_i, Y_1^*, \ldots, Y_s^*) = 0$ since $T^* = Y_1^* \cup \cdots \cup Y_s^*$ is a minimum spanning tree in $H_r(N)$. Thus, we get

$$f_i(Y_{i+1}) \leq \frac{mst_r(N)}{mst_2(N; Y_1, \ldots, Y_i)}. \tag{7.2}$$

In order to simplify notation, we set $m_i := mst_2(N; Y_1, \ldots, Y_i)$, for $i = 0, \ldots, t$. Combining the definition of f_i and (7.1), (7.2) we obtain:

$$\ell(T_Z) = \sum_{i=1}^{t} \ell_r(Y_i) = \sum_{i=1}^{t} f_{i-1}(Y_i)(m_{i-1} - m_i)$$

$$\leq \sum_{i=1}^{t} \min\left\{1, \frac{mst_r(N)}{m_{i-1}}\right\} (m_{i-1} - m_i).$$

The sequence $m_0, \ldots, m_t$ is monotone decreasing with $m_0 = mst_2(N)$ and $m_t = 0$. Therefore, we can replace the sum by an integral:

$$\sum_{i=1}^{t} \min\left\{1, \frac{mst_r(N)}{m_{i-1}}\right\} (m_{i-1} - m_i) \leq \int_{m_t}^{m_0} \min\left\{1, \frac{mst_r(N)}{x}\right\} dx$$

$$= \int_0^{mst_r(N)} 1\, dx + mst_r(N) \int_{mst_r(N)}^{mst_2(N)} \frac{1}{x} dx$$

$$= mst_r(N) + mst_r(N) \cdot \ln\left(\frac{mst_2(N)}{mst_r(N)}\right)$$

$$= mst_r(N) \left(1 + \ln\left(\frac{mst_2(N)}{mst_r(N)}\right)\right).$$

Since $mst_r(N) \geq smt(N)$ (cf. Lemma 7.1) and $mst_2(N)/smt(N) \leq \rho_2 = 2$ this gives $\ell(T_Z) \leq (1 + \ln 2)mst_r(N)$. □

Having Theorem 7.15 at hand, it is now a straightforward exercise to obtain approximation algorithms with performance ratio arbitrarily close to $1 + \ln 2$.

Corollary 7.16 *Let* $N = (V, E, \ell; K)$ *be a network. Then for every* $\epsilon > 0$ *there exists* $r \geq 2$ *such that algorithm* $\mathcal{A}_r$ *computes in polynomial time a Steiner tree* T_Z *for* K *in* N *satisfying*

$$\ell(T_Z) \leq (1 + \epsilon)(1 + \ln 2) \cdot smt(N).$$

Proof. Just choose r sufficiently large so that

$$\rho_r = \limsup_N \frac{mst_r(N)}{smt(N)} \leq 1 + \epsilon.$$

Then Theorem 7.15 implies that Zelikovsky's algorithm $\mathcal{A}_r$ computes in polynomial time a Steiner tree T_Z in N such that $\ell(T_Z) \leq (1+\epsilon)(1+\ln 2) \cdot smt(N)$.

□

Though the performance ratio of the approximation algorithms from Corollary 7.16 is considerably better than what we have seen before (namely $1 + \ln 2 \approx 1.693$ versus 2), we have to pay dearly with running time for this achievement. Zelikovsky's algorithm $\mathcal{A}_3$ has only a performance ratio of 2.83, i.e., much worse than 2, which is obtained by the MST-algorithm, and 49 is the smallest integer such that the performance ratio of $\mathcal{A}_{49}$ is less than 2. On the other hand, $\mathcal{A}_{49}$ has running time $\mathcal{O}\left(n^2 \log n + nm + k^{50} n^2\right)$.

7.3 Excursion: The complexity of optimization problems

In the previous sections and chapters we studied various approximation algorithms. In doing so, we implicitly observed that the fact that an optimization problem is $\mathcal{NP}$-hard tells us very little about the potential existence of good approximation algorithms. For example, we had to undergo considerable effort to achieve a performance ratio of $1 + \ln 2$ for the MINIMUM STEINER PROBLEM IN GRAPHS . For MACHINE SCHEDULING on the other hand, Theorem 6.23 implies that there exists a polynomial time approximation algorithm with performance ratio $1 + \epsilon$ for every arbitrarily small $\epsilon > 0$. The aim of this section is to introduce some notions which allow us to impose a structure on the class $\mathcal{NPO}$ which is based on the approximability of the problems in $\mathcal{NPO}$.

Recall from Chapter 3 that an optimization problem Π is formally specified by a four-tuple $\langle \mathcal{I}, \text{Sol}, \text{val}, \text{goal}\rangle$, where $\mathcal{I}$ denotes the set of instances and $\text{Sol}(I)$ denotes for every instance the set of legal solutions. For every solution $x \in \text{Sol}(I)$ the (integer) function $\text{val}(I,x)$ measures the quality of the solution, and goal specifies whether we are considering a minimization or a maximization problem. The aim of an optimization problem is to find for a given instance I a solution $x_{\text{opt}} \in \text{Sol}(I)$ such that $\text{val}(I,x_{\text{opt}}) = \text{opt}(I)$, where

$$\text{opt}(I) = \begin{cases} \min\{\text{val}(I,x) \mid x \in \text{Sol}(I)\} & \text{if goal} = \text{MIN}, \\ \max\{\text{val}(I,x) \mid x \in \text{Sol}(I)\} & \text{if goal} = \text{MAX}. \end{cases}$$

A (polynomial time) *approximation algorithm* for an optimization problem $\Pi = \langle \mathcal{I},\text{Sol},\text{val},\text{goal}\rangle$ is an algorithm $\mathcal{A}$ which computes for each instance $I \in \mathcal{I}$ (in polynomial time) a solution $x_{\mathcal{A}} \in \text{Sol}(I)$. We will use the notation $\mathcal{A}(I) := \text{val}(I,x_{\mathcal{A}})$ to denote the value of the objective function for the solution obtained by algorithm $\mathcal{A}$. As the name indicates, the value $\mathcal{A}(I)$ of the solution $x_{\mathcal{A}}$ returned by the algorithm should, in some suitable sense to be defined later, approximate the optimum value $\text{opt}(I)$.

There are several possibilities to measure the "quality" of an approximation algorithm. In the previous chapters we considered the ratio $\mathcal{A}(I)/\text{opt}(I)$ and tried to design the algorithm $\mathcal{A}(I)$ in such a way that this ratio is as small as possible. (Note that so far we have only considered minimization problems!) Before we continue, let us justify why this ratio is a reasonable measure. We do that by first exploring some alternatives.

A measure which immediately comes to mind is the *absolute error*.

Definition 7.17 *Let* $\Pi = \langle \mathcal{I},\text{Sol},\text{val},\text{goal}\rangle$ *be an optimization problem. An approximation algorithm with absolute error* B *is an algorithm* $\mathcal{A}$ *such that*

$$|\mathcal{A}(I) - \text{opt}(I)| \leq B \qquad \textit{for all } I \in \mathcal{I}.$$

Unfortunately, approximation algorithms with a small absolute error are very rare. The reason is that for many problems a polynomial time approximation algorithm with small absolute error can be used to design a polynomial time algorithm which always finds an *optimum* solution. We explain this considering an example.

Lemma 7.18 *Assume there exists a constant* B *and a polynomial time approximation algorithm* $\mathcal{A}$ *for the* MINIMUM STEINER PROBLEM IN NETWORKS *with absolute error* B. *Then* $\mathcal{P} = \mathcal{NP}$.

Proof.   The proof of this result is actually very simple. Given a graph $G = (V,E)$ and a terminal set $K \subseteq V$ we will show that algorithm $\mathcal{A}$ can be used to construct a Steiner minimum tree for K in G. As STEINER PROBLEM IN GRAPHS is $\mathcal{NP}$-complete, this then implies that $\mathcal{P} = \mathcal{NP}$.

The trick is to apply algorithm $\mathcal{A}$ not to the graph $G = (V,E)$ directly, but instead to the network $N = (V,E,\ell; K)$, where $\ell(e) := B + 1$ for all edges $e \in E$. Clearly, every Steiner tree for K in the network N is also a Steiner tree for K in the graph G and vice versa. Note also that the definition of the function ℓ implies that $\ell(T)$ is a multiple of $B + 1$ for every Steiner tree T and that

$$smt(N) = (B + 1) \cdot smt(G).$$

Now observe what happens if we apply algorithm $\mathcal{A}$ to the network N. By assumption, we know that $\mathcal{A}(N) \leq smt(N) + B$ which is strictly less than $(B + 1) \cdot (smt(G) + 1)$. That is, the Steiner tree returned by the algorithm $\mathcal{A}$ contains fewer than $smt(G) + 1$ edges – which implies that the Steiner tree is in fact a Steiner minimum tree for K in G. □

Remark 7.19 A construction similar to the one in the proof of Lemma 7.18 works for many $\mathcal{NP}$-complete optimization problems. On the other hand, there do exist $\mathcal{NP}$-complete problems which have polynomial time approximation algorithms with absolute error equal to 1. In fact, it is not difficult to construct such an example: we just have to find an $\mathcal{NP}$-complete problem, where the task is to decide whether there exists a solution of value 2, and ensure that there always exists a solution of value 3 which can be found easily. See Problem 7.7 for an example.

Another measure for the quality of an approximation algorithm is the so-called *relative error*.

Definition 7.20 *Let* $\Pi = \langle \mathcal{I}, Sol, val, goal \rangle$ *be an optimization problem. An approximation algorithm with relative error* ρ *is an algorithm* $\mathcal{A}$ *such that*

$$\frac{|\mathcal{A}(I) - \mathrm{opt}(I)|}{\mathrm{opt}(I)} \leq \rho \qquad \text{for all } I \in \mathcal{I}.$$

At first sight, the relative error might look like a reasonable measure. There is, however, one significant drawback: its behavior is quite different for minimization and maximization problems. For a maximization problem *every* approximation algorithm is an approximation algorithm whose worst case relative error is bounded by 1. On the other hand one easily constructs minimization problems and associated approximation algorithms with unbounded relative error. As our ultimate aim (which we approach in Chapter 9) is to define reductions between optimization problems which are stable with respect to the approximability of the problems under considerations, the relative error would be difficult to handle. This is the reason for using the performance ratio as a measure of the quality of approximation algorithms.

Definition 7.21 *Let* $\Pi = \langle \mathcal{I}, Sol, val, goal \rangle$ *be an optimization problem. An approximation algorithm with performance ratio* $\rho \geq 1$ *is an algorithm* $\mathcal{A}$ *such that*

$$\frac{1}{\rho} \leq \frac{\mathcal{A}(I)}{\mathrm{opt}(I)} \leq \rho \qquad \text{for all } I \in \mathcal{I}.$$

An alert reader might already be wondering why we used *absolute* constants B and ρ in the Definitions 7.17, 7.20 and 7.21. In fact, this is not really necessary. One could also consider the case that, for example in Definition 7.21, the ratio $\mathcal{A}(I)/\mathrm{opt}(I)$ is bounded by $\rho(|I|)$, where ρ is some slowly growing function. In this book, however, we will not consider this case and instead concentrate on the case when ρ is close to one.

Definition 7.22 *An algorithm $\mathcal{A}$ is said to be a* polynomial time approximation scheme *for an optimization problem* $\Pi = \langle \mathcal{I}, \mathrm{Sol}, \mathrm{val}, \mathrm{goal} \rangle$ *if $\mathcal{A}$ returns for every instance $I \in \mathcal{I}$ and every rational $\epsilon > 0$ a solution $x_{\mathcal{A}} \in \mathrm{Sol}(I)$ such that $\mathcal{A}(I,\epsilon) := \mathrm{val}(I,x_{\mathcal{A}})$ satisfies*

$$\frac{1}{1+\epsilon} \leq \frac{\mathcal{A}(I,\epsilon)}{\mathrm{opt}(I)} \leq 1+\epsilon.$$

For every fixed $\epsilon > 0$ the running time of $\mathcal{A}$ has to be polynomially bounded in $|I|$. If the running time is in fact bounded by a polynomial in $|I|$ and $1/\epsilon$ then $\mathcal{A}$ is called a fully polynomial time approximation scheme.

We will see examples of polynomial time approximation schemes shortly. Before we do so, we subdivide the set $\mathcal{NPO}$ into classes of problems according to the existence of certain approximation algorithms.

Definition 7.23 $\mathcal{PO}$ *is the set of all optimization problems in $\mathcal{NPO}$ which can be solved optimally in polynomial time.*

Definition 7.24 $\mathcal{FPTAS}$ *is the set of all optimization problems in $\mathcal{NPO}$ which admit a fully polynomial time approximation scheme.*

Definition 7.25 $\mathcal{PTAS}$ *is the set of all optimization problems in $\mathcal{NPO}$ which admit a polynomial time approximation scheme.*

Definition 7.26 $\mathcal{APX}$ *is the set of all optimization problems in $\mathcal{NPO}$ which admit a polynomial time approximation algorithm with performance ratio ρ for some $\rho \geq 1$.*

From these definitions it is immediately clear that

$$\mathcal{PO} \subseteq \mathcal{FPTAS} \subseteq \mathcal{PTAS} \subseteq \mathcal{APX} \subseteq \mathcal{NPO}.$$

In the remainder of this chapter we will see that in fact, unless $\mathcal{P} = \mathcal{NP}$ all these inclusions are strict. We start by exhibiting an $\mathcal{NP}$-hard optimization problem that belongs to the class $\mathcal{FPTAS}$.

Theorem 7.27 *If* $\mathcal{P} \neq \mathcal{NP}$, *then the problem* MAXIMUM KNAPSACK *belongs to the class* $\mathcal{FPTAS} \setminus \mathcal{PO}$.

Proof. That MAXIMUM KNAPSACK does not belong to the class $\mathcal{PO}$ follows immediately from the fact that the underlying decision problem KNAPSACK is $\mathcal{NP}$-complete (cf. Section 5.3). So it remains to show that it belongs to the class $\mathcal{FPTAS}$. This is what we will do now.

Let I be an instance of MAXIMUM KNAPSACK with weights $w_1, \ldots, w_n$, profits $p_1, \ldots, p_n$ and capacity B. Note that we may assume without loss of generality that all weights w_i are less or equal to B. According to Theorem 5.11 Algorithm 5.10 finds an *optimum* packing of the knapsack in time $\mathcal{O}(n^2 \cdot p_{\max})$, where $p_{\max} = \max_i p_i$ denotes the maximum profit of an item.

Consider an arbitrary value $\epsilon > 0$. In fact, in the following we will assume that $\epsilon \leq 1/2$. This we may do without loss of generality, as otherwise we could just take $\epsilon = 1/2$ instead of the given value for ϵ. We denote by I_ϵ the instance with scaled profits $p_i' := \lfloor p_i / K_\epsilon \rfloor$, where $K_\epsilon = (\epsilon \cdot p_{\max})/n$. Note that in the instance I_ϵ the maximum profit is at most $p_{\max}/K_\epsilon = n/\epsilon$. That is, the running time of Algorithm 5.10 with respect to the instances I_ϵ is polynomially bounded in n and $1/\epsilon$.

What can we say about the packings obtained by Algorithm 5.10? Clearly, they are feasible but not necessarily optimum packings with respect to the original instances I. However, we would expect that they are at least "close" to an optimum solution. This is indeed the case. Let KNAPSACK PACKING(I_ϵ) denote the packing returned by Algorithm 5.10 when applied to instance I_ϵ. Then

$$
\begin{aligned}
\mathrm{val}(I, \text{KNAPSACK PACKING}(I_\epsilon)) \; &\geq \; K_\epsilon \cdot \mathrm{val}(I_\epsilon, \text{KNAPSACK PACKING}(I_\epsilon)) \\
&= \; K_\epsilon \cdot \mathrm{opt}(I_\epsilon) \geq \mathrm{opt}(I) - K_\epsilon n \\
&= \; \mathrm{opt}(I) - \epsilon \cdot p_{\max} \geq \mathrm{opt}(I)(1 - \epsilon).
\end{aligned}
$$

The inequality in the first line follows from the fact that $p_i \geq K_\epsilon \cdot p_i'$. Similarly, the inequality in the second line, follows from $p_i' \geq p_i/K_\epsilon - 1$ together with the observation that there are at most n items in the knapsack. As $1 - \epsilon \geq 1/(1 + 2\epsilon)$ for all $\epsilon \leq \frac{1}{2}$ this concludes the proof of Theorem 7.27. $\qquad\square$

Our next aim is to find candidates for the class $\mathcal{PTAS} \setminus \mathcal{FPTAS}$. Here the following theorem will be useful.

Theorem 7.28 *Let* $\Pi = \langle \mathcal{I}, \mathrm{Sol}, \mathrm{val}, \mathrm{goal} \rangle$ *be an optimization problem in* $\mathcal{NPO}$. *Assume that there exists a polynomial* p *such that* $\mathrm{opt}(I) \leq p(|I|)$ *for all instances* $I \in \mathcal{I}$. *Then* $\Pi \notin \mathcal{FPTAS} \setminus \mathcal{PO}$, *unless* $\mathcal{P} = \mathcal{NP}$.

Proof. Assume $\Pi \in \mathcal{FPTAS}$. Then there exists an approximation scheme $\mathcal{A}$ which returns, for given $I \in \mathcal{I}$ and rational $\epsilon > 0$, in time polynomial in $|I|$ and $1/\epsilon$ a solution $\mathcal{A}(I, \epsilon) \in \mathrm{Sol}(I)$ with performance ratio at most $1 + \epsilon$.

Apply the algorithm with $\epsilon = 1/(p(|I|) + 1)$. Then $1/\epsilon$ is bounded by a polynomial in $|I|$ and according to the assumption on $\mathcal{A}$ we have

$$\frac{1}{1 + \frac{1}{p(|I|)+1}} \leq \frac{\text{val}(I,\mathcal{A}(I,\epsilon))}{\text{opt}(I)} \leq 1 + \frac{1}{p(|I|) + 1}.$$

As $1 - \frac{1}{p(|I|)+1} \leq \frac{1}{1+\frac{1}{p(|I|)+1}}$ this implies that

$$|\text{val}(I,\mathcal{A}(I,\epsilon)) - \text{opt}(I)| \leq \frac{|\text{opt}(I)|}{p(|I|) + 1} < 1,$$

where the last inequality follows from the assumption that $|\text{opt}(I)| \leq p(|I|)$ for all instances I. As $\text{val}(I,\mathcal{A}(I,\epsilon))$ and $\text{opt}(I)$ are both integers this therefore implies that $\text{val}(I,\mathcal{A}(I,\epsilon)) = \text{opt}(I)$. That is, we have constructed an algorithm which solves Π optimally and whose running time is bounded by a polynomial in $|I|$. But this can only be if $\Pi \in \mathcal{PO}$ or $\mathcal{P} = \mathcal{NP}$.   □

A similar result holds for problems for which the optimum is bounded by a polynomial in $|I|$ and $Max(I)$ and the underlying decision problem is strongly $\mathcal{NP}$-complete (recall page 83 for the definition of these notions).

Theorem 7.29 *Let* $\Pi = \langle \mathcal{I}, \text{Sol}, \text{val}, \text{goal} \rangle$ *be an optimization problem in* $\mathcal{NPO}$ *such that the underlying decision problem* Π_D *is strongly* $\mathcal{NP}$-complete *and such that there exists a polynomial* p *so that* $\text{opt}(I) \leq p(|I|, Max(I))$ *for all instances* $I \in \mathcal{I}$. *Then* $\Pi \notin \mathcal{FPTAS}$, *unless* $\mathcal{P} = \mathcal{NP}$.

Proof. We proceed similarly as in the proof of Theorem 7.28. That is, we assume that there exists a polynomial time approximation scheme $\mathcal{A}$ for Π and apply it for a suitably defined value $\epsilon > 0$. The only difference lies in the choice of that value. We now set $\epsilon = 1/(p(|I|, Max(I)) + 1)$. In exactly the same way as in the proof of Theorem 7.28 we deduce that this choice of ϵ implies that $\mathcal{A}$ has to solve Π optimally. What can we say about the running time? A priori it need not be polynomially bounded. However, if we restrict Π to instances I that satisfy $Max(I) \leq q(|I|)$ for some polynomial q, then the running time is in fact polynomially bounded. The assumption that the underlying decision problem Π_D is strongly $\mathcal{NP}$-complete thus implies $\mathcal{P} = \mathcal{NP}$.   □

As a consequence of Theorem 7.29 we can now show that, unless $\mathcal{P} = \mathcal{NP}$, also the inclusion of $\mathcal{FPTAS}$ in the class $\mathcal{PTAS}$ is strict.

Corollary 7.30 *If* $\mathcal{P} \neq \mathcal{NP}$, *then the problem* MACHINE SCHEDULING *belongs to the class* $\mathcal{PTAS} \setminus \mathcal{FPTAS}$.

Proof. The makespan of every instance of MACHINE SCHEDULING is clearly bounded by $\sum_{i=1}^{n} p_i$, which is obviously polynomial in the length of the input times the maximum number occurring in the input. As the problem is also strongly $\mathcal{NP}$-complete (cf. Problem 6.6), we deduce from Theorem 7.29 that MACHINE SCHEDULING does not belong to the class $\mathcal{FPTAS}$. On the other hand, in Theorem 6.23 we already exhibited a polynomial time approximation scheme (although we didn't call it so at that time). This implies that MACHINE SCHEDULING does indeed belong to the class $\mathcal{PTAS}$.   □

Another consequence of Theorem 7.28 is that the Steiner tree problem is difficult.

Corollary 7.31 *The* MINIMUM STEINER PROBLEM IN GRAPHS *does not admit a fully polynomial time approximation scheme, unless* $\mathcal{P} = \mathcal{NP}$.   □

In Chapter 9 we will see that, unless $\mathcal{P} = \mathcal{NP}$, MINIMUM STEINER PROBLEM IN GRAPHS does not belong to the class $\mathcal{PTAS}$ as well. At this stage, however, the proof is still too difficult to explain and we instead consider another candidate for the class $\mathcal{APX} \setminus \mathcal{PTAS}$: we will look at the problem MINIMUM BIN PACKING, which is a close relative of the MACHINE SCHEDULING problem. In both problems we are given a set of items. But while in a machine scheduling problem we want to assign them to a given set of machines in such a way that the maximum load of a machine is minimized, the bin packing problem requires the assignment of the items to an a priori arbitrary number of "machines" (which will be called "bins" in this context) such that the load in each bin is at most some given value.

MINIMUM BIN PACKING:

Given: An integer B and n items of positive integral size $a_1, \ldots, a_n \leq B$.
Find: A partition of the items into *bins*, as few as possible, such that in each bin the total size of the items is at most B.

The fact that the bin packing problem does not belong to the class $\mathcal{PTAS}$ is not difficult to see.

Theorem 7.32 *If* $\mathcal{P} \neq \mathcal{NP}$, *then there cannot exist a polynomial time approximation algorithm for* MINIMUM BIN PACKING *with performance ratio* $3/2 - \epsilon$, *for every* $\epsilon > 0$.

Proof. Assume there exists an approximation algorithm $\mathcal{A}$ with performance ratio $3/2 - \epsilon$, for some $\epsilon > 0$. We will show that this algorithm can be used to design a polynomial time algorithm for the $\mathcal{NP}$-complete problem PARTITION (cf. Problem 3.10). Given an instance $a_1, \ldots, a_n$ of PARTITION, consider the instance $a_1, \ldots, a_n$ and $B := \frac{1}{2} \sum_{i=1}^{n} a_i$ of MINIMUM BIN PACKING. One easily checks (cf.

Problem 7.7) that the optimal solution of this instance is either 2 or 3 — and is 2 exactly if there exists a partition of the a_i into two sets with equal sum. Observe also that $(3/2 - \epsilon) \cdot 2 < 3$. The approximation algorithm $\mathcal{A}$ therefore has to find a solution which uses only two bins whenever such a solution exists. That is, the algorithm $\mathcal{A}$ indeed decides the problem PARTITION in polynomial time, which can only be the case if $\mathcal{P} = \mathcal{NP}$.   □

Our next step is to design a polynomial time approximation algorithm with performance ratio 3/2 for MINIMUM BIN PACKING. As the bin packing problem is related to MACHINE SCHEDULING, it seems like a good idea to try some of the approaches from Section 6.3. Actually, the following straightforward modification of the list scheduling Algorithm 6.16 works.

Algorithm 7.33 (BIN PACKING – FIRST FIT DECREASING)
Input: n items with positive integral sizes $a_1,\ldots,a_n$, an integer capacity B.
Output: A partition of the items in bins such that in each bin the total size
            of the items is at most B.
(1) Sort the items such that $a_1 \geq \cdots \geq a_n$.
(2) Consider the items in this order and assign each item to the first bin which can take it without violating the capacity constraint. If this is not possible, open a new bin.

Theorem 7.34 *Algorithm 7.33 is a polynomial time approximation algorithm for* MINIMUM BIN PACKING *with performance ratio 3/2.*

Proof.   Let $a_1 \geq \cdots \geq a_n$ and B be an instance of MINIMUM BIN PACKING. Let N_{opt} denote the number of bins in an optimum packing, and let N denote the number of bins used by Algorithm 7.33. We have to show that $N \leq \frac{3}{2}N_{\text{opt}}$.

Let ℓ be the number of items with size greater than $\frac{1}{2}B$. Any algorithm will have to place those items in separate bins. Hence, it follows in particular that $N_{\text{opt}} \geq \ell$.

Consider an arbitrary value j such that $\max\{\ell, \frac{1}{3}(2N - 1)\} \leq j \leq N - 1$ and let ξ denote the maximum space in any of the first j bins at termination of Algorithm 7.33. Clearly, the size of the items in each of the bins $j + 1,\ldots,N$ is larger than ξ, as otherwise the algorithm would have placed them in one of the first j bins. As $\xi < \frac{1}{2}B$ this implies furthermore that the bins with number $j + 1,\ldots,N - 1$ all contain at least two items. As the last bin contains at least one item, we can thus conclude that the total size of all items is at least

$$
\begin{aligned}
j \cdot (B - \xi) + [2(N - j - 1) + 1] \cdot \xi \ &= \ jB - (3j - 2N + 1) \cdot \xi \\
&> \ jB - \tfrac{1}{2}B \cdot (3j - 2N + 1) \qquad (7.3) \\
&= \ NB - \tfrac{1}{2}B \cdot (j + 1),
\end{aligned}
$$

where the inequality follows from $\xi < \frac{1}{2}B$ and the fact that the term in the bracket is nonnegative by choice of j. As the total size of all items constitutes a natural lower bound for $N_{\mathrm{opt}} \cdot B$, we deduce from (7.3) that $N_{\mathrm{opt}} > N - \frac{1}{2}(j+1)$ or, equivalently, $N_{\mathrm{opt}} \geq \lceil N - \frac{1}{2}j \rceil$.

We now distinguish two cases based on the divisibility of N. In both cases we either use the fact that $N_{\mathrm{opt}} \geq \ell$, if ℓ is sufficiently large, or use $N_{\mathrm{opt}} \geq \lceil N - \frac{1}{2}j \rceil$ for an appropriate value j.

First assume that $N = 3x+r$ for some $r \in \{0,1\}$. If $\ell \geq 2x+1$ then $N \leq \frac{3}{2}\ell \leq \frac{3}{2}N_{\mathrm{opt}}$. Otherwise, we let $j = 2x + r$ and observe that $N = \frac{3}{2}(N - \frac{1}{2}j) + \frac{1}{4}r$. For $r = 0$ we have $N_{\mathrm{opt}} \geq \lceil N - \frac{1}{2}j \rceil = N - \frac{1}{2}j$ and we therefore immediately deduce that $N \leq \frac{3}{2}N_{\mathrm{opt}}$. For $r = 1$ we have $\lceil N - \frac{1}{2}j \rceil = N - \frac{1}{2}j + \frac{1}{2}$ and we therefore also deduce in this case that $N \leq \frac{3}{2}N_{\mathrm{opt}}$. Now assume $N = 3x + 2$. If $\ell \geq 2x + 2$ then $N \leq \frac{3}{2}\ell \leq \frac{3}{2}N_{\mathrm{opt}}$. Otherwise, we let $j = 2x+1$ and observe $N = \frac{3}{2}(N - \frac{1}{2}(j+1)) + \frac{1}{2}$. As $N - \frac{1}{2}(j + 1)$ is an integer, the fact that $N_{\mathrm{opt}} > N - \frac{1}{2}(j + 1)$ again implies that $N \leq \frac{3}{2}N_{\mathrm{opt}}$.   □

Combining Theorem 7.32 and 7.34 we immediately obtain that the bin packing problem separates the classes $\mathcal{APX}$ and $\mathcal{PTAS}$.

Corollary 7.35 *If $\mathcal{P} \neq \mathcal{NP}$, then* MINIMUM BIN PACKING *belongs to the class $\mathcal{APX} \setminus \mathcal{PTAS}$.*   □

It is important to note that the fact that the bin packing problem is difficult to approximate relies heavily on the use of small items. Indeed, if we restrict the problem to instances in which the sizes of the items are such that every bin can contain at most 4 items, then the problem becomes much easier.

> MINIMUM BALANCED BIN PACKING:
>
> *Given:* n items of size $a_1, \ldots, a_n \in \mathbb{N}$ and a capacity $B \in \mathbb{N}$ such that $\frac{1}{5}B < a_i < \frac{1}{3}B$ for all $i = 1, \ldots, n$.
> *Find:* Partition of the items into bins, as few as possible, such that in each bin the total size of the items is at most B.

Theorem 7.36 MINIMUM BALANCED BIN PACKING *belongs to the class $\mathcal{PTAS}$.*

Proof. See Problem 7.10.   □

We close this section by exhibiting a member of the class $\mathcal{NPO} \setminus \mathcal{APX}$.

> MINIMUM TRAVELLING SALESMAN:
>
> *Given:* A complete network $N = (V, \binom{V}{2}, \ell)$.
> *Find:* A Hamilton cycle H such that $\ell(H)$ is as small as possible.

Theorem 7.37 *If* $P \neq NP$, *then* MINIMUM TRAVELLING SALESMAN *belongs to the class* $NPO \setminus APX$.

Proof. Obviously, MINIMUM TRAVELLING SALESMAN is a member of NPO. It thus remains to show that the existence of a polynomial time approximation algorithm with performance ratio ρ for some $\rho \geq 1$ implies $P = NP$. Assume A is such an algorithm. We use it to design a polynomial time algorithm that decides whether a graph $G = (V,E)$ contains a Hamilton cycle. As we know from Problem 3.9 that this problem is NP-complete, this then concludes the proof of the theorem.

Consider an arbitrary graph $G = (V,E)$. We apply algorithm A to the network $N = (V, \binom{V}{2}, \ell)$, where

$$\ell(e) := \begin{cases} 1, & \text{if } e \in E \\ n^2, & \text{otherwise.} \end{cases}$$

Clearly, if G contains a Hamilton cycle, then the length of a minimum travelling salesman tour in N is exactly $n = |V|$. A therefore has to return a tour of length at most $\rho \cdot n$. If on the other hand G does not contain a Hamilton cycle, then every salesman tour in N has to contain at least one edge e of length $\ell(e) = n^2$. As $n^2 \gg \rho \cdot n$, for n sufficiently large, algorithm A can thus indeed be used to decide whether the graph G contains a Hamilton cycle.   □

Problems

7.1   Let T_h be the complete binary tree of height h with its leaves $L(T_h)$ being terminals and a weight function w as defined in Example 7.3. Show that the subhypergraph $\mathcal{X}^*$ defined in Example 7.3 is a minimum spanning tree for the 4-bounded hypergraph $\mathcal{H}$.

7.2   Show that $\rho_3 = \frac{5}{3}$. That is, exhibit a sequence of networks N_i such that

$$\lim_{i \to \infty} \frac{mst_3(N_i)}{smt(N_i)} = \frac{5}{3} .$$

(Hint: Consider as in Example 7.3 a complete binary tree with its leaves being the terminals.)

7.3   Combine Example 7.3 and Problem 7.2 in order to show that for every $r \geq 2$ with $r = 2^s + t$ so that $0 \leq t \leq 2^s$ it is true that

$$\rho_r \geq \frac{(s+1)2^s + t}{s2^s + t} .$$

7.4   Prove Proposition 7.6.

7.5   Let $N = (V,E,\ell)$ be a network and T be a minimum spanning tree in N. For $X \subseteq E$ let $N(X)$ denote the network where the lengths of the edges in X are reduced to zero. Show that there exists a minimum spanning tree in $N(X)$ which is obtained by successively inserting the edges of $X \backslash E(T)$ into T and then removing in every step the longest edge in the cycle created by the new edge.

7.6   Let $G = (V,E)$ be a graph and $\ell_1, \ell_2 : E \to \mathbb{R}$ be two length functions such that $\ell_1 \leq \ell_2$. Let T_1 and T_2 be minimum spanning trees with respect to ℓ_1 respectively ℓ_2. For an edge $e = \{x,y\}$ in E let $c_i(e)$ denote the *longest* edge in the cycle induced by adding e to T_i. If $e \in E(T_i)$ we set $c_i(e) = \ell_i(e)$. Show that $c_1(e) \leq c_2(e)$ for all $e \in E$.

7.7   Exhibit a suitable optimization variant of the problem PARTITION (cf. Problem 3.10) that is $\mathcal{N}\mathcal{P}$-hard, but for that nevertheless a polynomial time approximation algorithm with absolute error 1 exists.

7.8   Let $\Pi = \langle \mathcal{I}, \text{Sol}, \text{val}, \text{goal} \rangle$ be an optimization problem contained in $\mathcal{N}\mathcal{P}\mathcal{O}$.
a) Show that the relative error of an approximation algorithm with performance ratio $1 + \epsilon$ is bounded by ϵ.
b) Let $0 < \epsilon \leq 1$. Show that the performance ratio of an approximation algorithm with relative error $\frac{1}{2}\epsilon$ is bounded by $1 + \epsilon$.

7.9   Prove that if Algorithm 7.33 returns a solution such that every bin contains at most two items, then this solution is in fact an optimum solution.

7.10 Prove Theorem 7.36. (Hint: Imitate the approach from Theorem 6.23.)

7.11 Consider the restriction of the travelling salesman problem to instances for which the length function ℓ satisfies the triangle inequality $\ell(u,v) \leq \ell(u,x) + \ell(x,v)$ for all $u,x,v \in V$. This is the so-called MINIMUM METRIC TRAVELLING SALESMAN problem.
a) Use the ideas from the proof of Lemma 6.1 to show that this problem belongs to the class $\mathcal{A}\mathcal{P}\mathcal{X}$.
b) Modify the algorithm from part a) to an approximation algorithm with performance ratio $3/2$. (Hint: Solve a matching problem in a suitably defined graph.)

Notes

The connection between minimum spanning subgraphs in r-bounded hypergraphs and Steiner minimum trees in networks was studied by many people. All recent approximation algorithms for solving the Steiner tree problem are based on this approach. The fact that $\rho_2 = 2$ was proven by several authors around 1980, cf. Chapter 6. Although it was believed since then that $\rho_3 < 2$, it took more than a decade before Zelikovsky (1993) proved that

$\rho_3 \leq 5/3$. This result was complemented by Du (1995) who showed $\rho_3 \geq 5/3$ concluding that $\rho_3 = 5/3$. Zelikovsky used the upper bound he had obtained on ρ_3 to design the first polynomial time approximation algorithm for the Steiner problem with performance ratio less than 2. The exact Steiner ratio ρ_r for all $r \geq 2$ (see Theorem 7.4 and Problem 7.3) was obtained by Borchers and Du (1997). A polynomial time approximation algorithm for the minimum spanning subgraph problem in r-bounded hypergraphs with performance ratio H_{r-1} was given by Wolsey (1982). In fact, he showed that the greedy algorithm for the more general problem of finding minimum spanning subsets in r-polymatroids has a performance ratio of $H_r \approx \ln r$, which is best possible for large r.

The performance of Zelikovsky's algorithm (Algorithm 7.10) was proved in Zelikovsky (1996). The presentation given in Chapter 7.2 is from Gröpl, Hougardy, Nierhoff, and Prömel (2001). Using a different objective function for choosing the hyperedges at every step, Robins and Zelikovsky (2000) were able to prove that basically the same Greedy approach as used in Zelikovsky's algorithm can be used to find a Steiner tree whose length is at most a factor $1 + \frac{\ln 3}{2} + \epsilon \approx 1.55 + \epsilon$ times the length of a Steiner minimum tree. Thereby, $\epsilon = \epsilon(r) > 0$ can be made arbitrarily small, depending on the r-bounded hypergraph $H_r(N)$ that is used in the algorithm. Presently, the best known lower bound for the performance ratio of a polynomial time approximation algorithm for the Steiner tree problem (assuming that $\mathcal{P} \neq \mathcal{NP}$) is rather close to 1, viz. 1.0025, cf. Berman and Karpinski (1999).

The first approximation algorithms were introduced at the end of the 60s. The paper by Garey and Johnson (1976) gives a nice survey of these early results. The basic terminology in the context of approximation algorithms was coined in the 70s. The two textbooks by Horowitz and Sahni (1978) and Garey and Johnson (1979) are the first comprehensive introduction to this field and served as invaluable guides for many years. Two excellent references from a modern viewpoint are Ausiello et al (1999) and Hochbaum (1997).

The algorithm from Theorem 7.27 is due to Ibarra and Kim (1975). Korte and Schrader (1981) exhibited a first candidate for the class $\mathcal{PTAS} \setminus \mathcal{FPTAS}$. Theorem 7.32 is due to Garey and Johnson (1979). The bound of $3/2$ for Algorithm 7.33 was proven by Simchi-Levi (1994). Note that Theorem 7.32 implies that this value is best possible. However, better results can be achieved if we state the bounds in the form $\mathcal{A}(I) \leq a \cdot \text{opt}(I) + b$. Here the use of the constant b makes it possible to improve the bounds on a. In particular, Johnson (1973) and Baker (1983) have shown that Algorithm 7.33 satisfies $\mathcal{A}(I) \leq \frac{11}{9} \text{opt}(I) + 4$. Theorem 7.36 is due to Sahni and Gonzales (1976). A very strong and surprising improvement for the general bin packing problem is due to Karmarkar and Karp (1982). They showed that there exists an *asymptotic* fully polynomial approximation scheme. Meaning that bin packing can be approximated within $1 + \epsilon$ in time polynomial in $|I|$ and $1/\epsilon$, if we restrict the input to instances with a sufficiently large optimum value large. For details and many more results related to bin packing we refer the reader to the surveys Garey, Johnson (1981) and Coffman, Garey, Johnson (1997).

For the MINIMUM METRIC TRAVELLING SALESMAN defined in Problem 7.11 the modification outlined in part b) is due to Christofides (1976). It is still the best known approximation algorithm for this problem.

8

Randomness Helps

So far, all our algorithms were *deterministic* algorithms. Deterministic algorithms always return the same result for the same input, independently of how often they are called. While this seems like a sensible feature that an algorithm should have, we will see in this chapter that in some cases one gets much better results if the algorithm is allowed to 'gamble'. Such *randomized* algorithms, as they are called, will no longer always return the same result for the same input. Depending on the type of the algorithm, they might even sometimes return wrong results. But if this happens only with small probability, these algorithms nevertheless can turn out to be very useful. In fact, as we will see in the last section of this chapter, without the use of randomized algorithms modern cryptography (and subsequently electronic commerce) would be impossible.

8.1 Probabilistic complexity classes

In this section we will introduce several complexity classes which are based on randomized algorithms. Before we do so we still have to define precisely what a randomized algorithm is. We start with an example.

Example 8.1 Assume p_i and q_j are polynomials in n variables $x_1, \ldots, x_n$. How efficiently can we decide whether

$$\prod_{i=1}^{k} p_i(x_1, \ldots, x_n) = \prod_{j=1}^{l} q_j(x_1, \ldots, x_n)?$$

Clearly, there is an immediate algorithm for solving this problem. Just expand both polynomials and compare the coefficients of their monomials. A moments thought, however, brings the realization that this algorithm is usually not a polynomial time algorithm. (Consider for example the case that we have $\sqrt{n}$ polynomials of degree $\sqrt{n}$. Then the input size is $\mathcal{O}(n)$, but expanding the polynomials usually leads to superpolynomially many different monomials.) Can we do better? Yes, at least in some sense. Here is how: choose random numbers $a_1, \ldots, a_n \in \{1, \ldots, d\}$, for a suitable constant d. Check (in linear time) whether the product of the polynomials p_i evaluate at $(a_1, \ldots, a_n)$ to the same number as the product of the polynomials q_j. If this is not the case we know that the two polynomials are definitely not identical. On the other hand, if the numbers are equal, then they could be identical or could be different. As, however, every polynomial that is not identically zero has only a finite number of zeros, we know that, if d is chosen appropriately (see Problem 8.2) then the probability that $(a_1, \ldots, a_n)$ is a zero of the difference of the two products is at most $\frac{1}{2}$. If we repeat this test therefore, say, 40 times, then the probability that all differences are zero even though the two products are not identical is at most 2^{-40}, a number whose decimal representation starts with 12 zeros.

Informally speaking, a randomized algorithm is an algorithm which may toss an unbiased coin after every step in the computation and base its next step on the outcome 'head' or 'tail' of the coin. Formally, we say that a *randomized algorithm* for a RAM has (read-only) access to an additional memory cell R, which contains a string of zeros and ones. The output of the algorithm for a particular instance I may now depend on the contents of the register R at the beginning of the execution of the algorithm. Hence, it makes sense to speak of the *probability* that the algorithm gives a particular answer. For a precise calculation of this probability we assume that the length of the 0-1 string in memory cell R is exactly $T_{\mathcal{A}}(|I|)$ and that all such strings are equally likely. (Note that the assumption that a randomized algorithm may toss an unbiased coin *once* after each step, or equivalently read *one* random bit, implies that for input I the algorithm can use at most $T_{\mathcal{A}}(|I|)$ random bits.)

With this formalization at hand we are now able to introduce new complexity classes involving randomized algorithms. A decision problem $\Pi = \langle \mathcal{I}, \text{Sol} \rangle$ is said to belong to some probabilistic complexity class (to be named later) if there exists a randomized polynomial time algorithm $\mathcal{A}$ such that

$$\Pr\left[\mathcal{A}(I) \text{ accepts}\right] = \begin{cases} \text{``high''} & \text{if } \text{Sol}(I) \neq \emptyset, \\ \text{``low''} & \text{if } \text{Sol}(I) = \emptyset. \end{cases}$$

There are several suitable substitutions for "high" and "low". As a first try we let "high" mean larger than 1/2 and "low" mean at most 1/2. This leads to the class $\mathcal{PP}$ A decision problem $\Pi = \langle \mathcal{I}, \text{Sol} \rangle$ is said to belong to complexity class $\mathcal{PP}$ (*probabilistic polynomial time*) if and only if there exists a randomized polynomial time algorithm $\mathcal{A}$ such that

$$\Pr\left[\mathcal{A}(I) \text{ accepts}\right] \begin{cases} > \frac{1}{2} & \text{if } \mathrm{Sol}(I) \neq \emptyset, \\ \leq \frac{1}{2} & \text{if } \mathrm{Sol}(I) = \emptyset. \end{cases}$$

Example 8.2 As an example for a problem in $\mathcal{PP}$ consider the problem $\frac{1}{2}$-SAT: given a Boolean formula F in conjunctive normal form, decide whether more than half of all truth assignments are satisfying assignments for F. A randomized algorithm for showing that $\frac{1}{2}$-SAT $\in \mathcal{PP}$ is easily designed. The algorithm just chooses an assignment uniformly at random and then checks whether it is a satisfying assignments for F. If so then it accepts, otherwise it rejects.

For problems in $\mathcal{PP}$ the difference between the acceptance probabilities for instances I with empty resp. nonempty solution set $\mathrm{Sol}(I)$ can be exponentially small. This means in particular that in order to reduce the error probability to "small" values like *e.g.* 10^{-10} an exponential number of repetitions of the algorithm is necessary. This can be avoided by bounding the acceptance probabilities away from one half. This leads to the class $\mathcal{BPP}$: a decision problem $\Pi = \langle I, \mathrm{Sol}\rangle$ is said to belong to complexity class $\mathcal{BPP}$ if and only if there exists a randomized polynomial time algorithm $\mathcal{A}$ and a constant $\epsilon > 0$ such that

$$\Pr\left[\mathcal{A}(I) \text{ accepts}\right] \begin{cases} \geq \frac{1}{2} + \epsilon & \text{if } \mathrm{Sol}(I) \neq \emptyset, \\ \leq \frac{1}{2} - \epsilon & \text{if } \mathrm{Sol}(I) = \emptyset. \end{cases}$$

Note that the definition of the class $\mathcal{BPP}$ only requires the existence of some constant $\epsilon > 0$. One can, however, show that this constant can in fact be chosen arbitrarily close to one half.

Lemma 8.3 *If* $\Pi = \langle I, \mathrm{Sol}\rangle$ *is a problem in* $\mathcal{BPP}$ *then there exists for every polynomial* q *a randomized polynomial time algorithm* $\mathcal{A}'$ *such that*

$$\Pr\left[\mathcal{A}'(I) \text{ accepts}\right] \begin{cases} \geq 1 - 2^{-q(|I|)} & \text{if } \mathrm{Sol}(I) \neq \emptyset, \\ \leq 2^{-q(|I|)} & \text{if } \mathrm{Sol}(I) = \emptyset. \end{cases}$$

Proof. Let $\epsilon > 0$ be the constant and $\mathcal{A}$ be the randomized algorithm showing that $\Pi \in \mathcal{BPP}$. We construct the algorithm $\mathcal{A}'$ as follows:

$t := \lceil -2q(|I|)/\log_2(1 - 4\epsilon^2)\rceil;\ s := 0;$
repeat t **times**:
    Call algorithm $\mathcal{A}(I)$;
    if $\mathcal{A}(I) = $ ACCEPTS **then** $s := s + 1$;
if $s \geq t/2$ **then return** ACCEPT **else return** REJECT.

In order to bound the acceptance probability of algorithm $\mathcal{A}'$ let us consider an arbitrary but fixed instance I and let

$$p := \Pr[\mathcal{A}(I) = \text{ACCEPT}].$$

If $\text{Sol}(I) \neq \emptyset$, then $p \geq \frac{1}{2} + \epsilon$ and hence in particular $p^i(1-p)^{t-i} \leq p^{t/2}(1-p)^{t/2} = (p - p^2)^{t/2} \leq (\frac{1}{4} - \epsilon^2)^{t/2}$ for all $i \leq t/2$. Therefore

$$\Pr[\mathcal{A}'(I) \text{ rejects}] = \sum_{i=0}^{\lceil t/2 \rceil - 1} \binom{t}{i} p^i (1-p)^{t-i}$$

$$\leq \sum_{i=0}^{t/2} \binom{t}{i} (\frac{1}{4} - \epsilon^2)^{t/2} \leq (\frac{1}{4} - \epsilon^2)^{t/2} \cdot \sum_{i=0}^{t} \binom{t}{i}$$

$$= (\frac{1}{4} - \epsilon^2)^{t/2} \cdot 2^t = (1 - 4\epsilon^2)^{t/2} \leq 2^{-q(|I|)},$$

by choice of t. The case $\text{Sol}(I) = \emptyset$ follows similarly.   □

It can be shown that the class $\mathcal{BPP}$ is still rather large. For example, it is rather unlikely that it is a subset of $\mathcal{NP}$. In fact, according to present knowledge it could even contain all of $\mathcal{NP}$, even though this is considered to be unlikely, compare Problem 8.1. In the light of Lemma 8.3 one might wonder, however, whether it is at all possible to further restrict the class $\mathcal{BPP}$ without dropping all the way into $\mathcal{P}$. In fact, this is indeed possible and Example 8.1 also indicates how. Namely, we allow only one-sided errors. A decision problem $\Pi = \langle \mathcal{I}, \text{Sol} \rangle$ is said to belong to complexity class $\mathcal{RP}$ if and only if there exists a randomized polynomial time algorithm $\mathcal{A}$ such that

$$\Pr[\mathcal{A}(I) \text{ accepts}] \begin{cases} \geq \frac{1}{2} & \text{if } \text{Sol}(I) \neq \emptyset, \\ = 0 & \text{if } \text{Sol}(I) = \emptyset. \end{cases}$$

Clearly, by iterating the algorithm one can achieve arbitrarily low error probabilities.

Lemma 8.4 *If $\Pi = \langle \mathcal{I}, \text{Sol} \rangle$ is a problem in $\mathcal{RP}$ then there exists for every polynomial q a randomized polynomial time algorithm $\mathcal{A}'$ such that*

$$\Pr[\mathcal{A}'(I) \text{ accepts}] \begin{cases} \geq 1 - 2^{-q(|I|)} & \text{if } \text{Sol}(I) \neq \emptyset, \\ = 0 & \text{if } \text{Sol}(I) = \emptyset. \end{cases}$$

Proof. Let $\mathcal{A}$ be an algorithm which shows that $\Pi \in \mathcal{RP}$. Construct the algorithm $\mathcal{A}'$ as follows. $\mathcal{A}'$ calls $\mathcal{A}$ for the same input $q(|I|)$ times, each times using fresh random bits. If at least one of these calls accepts, then $\mathcal{A}'$ accepts, otherwise $\mathcal{A}'$ rejects. Using the acceptance probabilities of algorithm $\mathcal{A}$ we easily deduce that $\mathcal{A}'$ satisfies the claimed error probabilities.   □

Algorithms corresponding to problems in $\mathcal{RP}$ are the type of randomized algorithms which occur most often. More precisely, if one speaks of finding a randomized algorithm for some problem, then one usually tacitly assumes that the algorithm should satisfy the requirements in the definition of the class $\mathcal{RP}$. Namely, it should run in polynomial time and should have only a one sided error. Note that if we consider optimization problems then the latter property comes for free, as the value of a (feasible) solution can never be smaller than that of the optimum solution.

Recall that we already met such a kind of randomized algorithm in Example 8.1. Another example (namely, an algorithm for deciding whether a given integer is prime) will be the main topic of Section 8.4. In fact, these are examples where the randomized algorithm is the only polynomial time algorithm which is known for the problem. It is worth pointing out, however, that randomized algorithms are sometimes also used even though a deterministic polynomial time algorithm is known for the problem. This is due to the fact that randomized algorithms are often much simpler to implement than their deterministic counterparts. We show such an approach for the problem of finding a perfect matching in a bipartite graph.

Example 8.5 (FINDING PERFECT MATCHINGS) Let $G = (A \uplus B, E)$ be a bipartite graph. Assume $A = \{a_1, \ldots, a_n\}$ and $B = \{b_1, \ldots, b_n\}$. We define an $n \times n$ matrix $M = (m_{ij})$, where $m_{ij} = x_{ij}$ if $\{a_i, b_j\} \in E$, and $m_{ij} = 0$ otherwise. One easily observes that the determinant $\det(M)$ of the matrix M is a polynomial in n^2 variables x_{ij}, in which every variable has degree at most one. It should not be difficult for the reader to convince himself that this polynomial is identically zero if and only if G does not contain a perfect matching. With the help of Example 8.1 and Problem 8.2 it is therefore easy to design a randomized algorithm for *testing* whether G contains a perfect matching. But how can we *find* one? Actually, this is not difficult at all. We just temporarily remove an arbitrary edge and check whether the graph still has a perfect matching. If so, then we make the deletion of the edge permanent. If not, then we know that this edge belongs to the perfect matching. So we reinsert it and (permanently) remove all edges incident to it. We then repeat this process with another edge and continue in this way until we have found a perfect matching. In Problem 8.5 you are asked to fill in the details of this algorithm. We note that this approach can be generalized to finding a maximum matching in arbitrary graphs and networks, cf. Problem 8.11 and notes.

The asymmetry in the definition of the class $\mathcal{RP}$ calls for another definition. Switching the side with zero error probability leads to the class $co\mathcal{RP}$: a decision problem $\Pi = \langle \mathcal{I}, \text{Sol} \rangle$ is said to belong to complexity class $co\mathcal{RP}$ if and only if there exists a randomized polynomial time algorithm $\mathcal{A}$ such that

$$\Pr\left[\mathcal{A}(I) \text{ accepts}\right] \begin{cases} = 1 & \text{if } \mathrm{Sol}(I) \neq \emptyset, \\ \leq \frac{1}{2} & \text{if } \mathrm{Sol}(I) = \emptyset. \end{cases}$$

Finally, we let

$$\mathcal{ZPP} := \mathcal{RP} \cap \mathrm{co}\mathcal{RP}.$$

In Problem 8.3 the reader will be asked to show that the class $\mathcal{ZPP}$ can also be defined as containing exactly those problems for which a randomized algorithm exists that *never* gives a false answer, but instead it may refuse an answer with a certain (small) probability.

Theorem 8.6 *The following inclusions are all valid:*

$$P \subseteq \mathcal{ZPP} \quad \begin{matrix} \mathcal{RP} & \subseteq & \mathcal{NP} \\ \subseteq & \subseteq & \subseteq \\ & \mathcal{BPP} & \subseteq & \mathcal{PP} \\ \subseteq & \subseteq & \subseteq \\ \mathrm{co}\mathcal{RP} & \subseteq & \mathrm{co}\mathcal{NP} \end{matrix}$$

Proof. First we remark that the inclusions between probabilistic complexity classes follow immediately from their definitions. The relation $P \subseteq \mathcal{ZPP}$ is also obvious. To see the inclusion $\mathcal{RP} \subseteq \mathcal{NP}$ we choose as the $\mathcal{NP}$-algorithm the $\mathcal{RP}$-algorithm and interpret the sequence of random bits as the "proof" required by an $\mathcal{NP}$-machine. This works as, by definition, for every $\mathcal{RP}$-algorithm there exists a random sequence $\tau \in \{0,1\}^*$ for which the algorithm accepts if and only if $\mathrm{Sol}(I) \neq \emptyset$. The inclusion $\mathrm{co}\mathcal{RP} \subseteq \mathrm{co}\mathcal{NP}$ follows similarly. To see that $\mathcal{NP} \subseteq \mathcal{PP}$ consider the following algorithm for SAT: toss a coin, if head comes up we accept, if tails comes up we choose an assignment at random and accept if it is a satisfying assignment and reject otherwise. The inclusion $\mathrm{co}\mathcal{NP} \subseteq \mathcal{PP}$ follows again similarly.   □

It is believed that all inclusions of Theorem 8.6 are strict. At present, however, this has not been shown for any of the inclusions. We also point out that even though the class $\mathcal{ZPP}$ is believed to be different from P, for practical purposes the fact that a problem belongs to the class $\mathcal{ZPP}$ is often as helpful as the fact that it belongs to the class P, see Problem 8.4 for details. In the last section of this chapter we will illustrate this for one of the most prominent members of the class $\mathcal{ZPP}$: the problem to decide whether a given integer is a prime number. Before we come to that we first consider two applications of randomized methods to the Steiner problem. In the next section we show how the use of 'randomness' allows a quite different approach to approximation algorithms for the Steiner problem than the ones we have seen so far.

8.2 Improving the performance ratio II

In Chapter 7 we saw that algorithms for finding a minimum spanning tree in r-bounded hypergraphs would immediately imply good approximation algorithms for the Steiner tree problem. Unfortunately, we also saw that finding a minimum spanning tree is $\mathcal{NP}$-complete for all $r \geq 4$. In this section we will see that the situation is slightly better for the case $r = 3$. Namely, we will show that there exists a *randomized* algorithm for the case that the edge weights are polynomially bounded in the number of vertices. More precisely, we will show:

Theorem 8.7 *There exists a randomized algorithm that finds in weighted 3-bounded hypergraphs $H = (V,F,w)$ with probability at least $1/2$ a minimum spanning tree (if a spanning tree exists at all). The running time of the algorithm is bounded by a polynomial in n and $w_{\max} = \max\{w(f) \mid f \in F\}$.*

Note that the algorithm of this theorem is in general *not* a polynomial time algorithm, as then the running time would have to be a polynomial in $\log w_{\max}$ and not in $w_{\max}$. Recall from Section 5.3 that an algorithm with this dependence on the weights of the input is called a pseudopolynomial time algorithm. Using a scaling technique similar to the one which we saw in Section 6.3 we can derive from the algorithm in Theorem 8.7 a (randomized) approximation scheme.

Corollary 8.8 *For every $\epsilon > 0$ there exists a randomized polynomial time algorithm that finds in every weighted 3-bounded hypergraph H with probability at least $1/2$ a spanning tree T such that $w(T) \leq (1+\epsilon)mst(H)$, provided H contains a spanning tree.*

Proof. Let $H = (V,F,w)$ be a weighted 3-bounded hypergraph on n vertices and let $w_{\max} := \max\{w(f) \mid f \in F\}$. For a given $\epsilon > 0$ we set $t := \epsilon \cdot w_{\max}/n$ and define a new hypergraph $H' = (V,F,w')$ with the same vertex and edge set and weight function w' given by

$$w'(f) := \left\lceil \frac{w(f)}{t} \right\rceil \qquad \text{for all } f \in F.$$

Observe that, by construction,

$$mst(H') \leq \frac{1}{t}mst(H) + n - 1$$

and that $w'_{\max} := \max\{w'(f) \mid f \in F\} = \mathcal{O}(\frac{n}{\epsilon})$. Hence, the algorithm of Theorem 8.7 finds with probability at least $1/2$ in time polynomially bounded in n (recall that $\epsilon > 0$ is a fixed constant!) a minimum spanning tree T in H'. If T is indeed a minimum spanning tree in H' we can bound its weight with respect to the weight function $w(\cdot)$ as follows:

$$w(T) \le t \cdot w'(T) = t \cdot mst(H') \le mst(H) + tn \le mst(H) + \epsilon w_{\max}.$$

That is, if we could guarantee that $mst(H) \ge w_{\max}$ we would be home. Unfortunately, this is not true in general. But another trick helps here.

Let $w_1 < \cdots < w_s$ be the different weights in H sorted in increasing order. For every $1 \le i \le s$ we define a hypergraph $H_i = (V, F_i, w)$ by deleting all edges from H which have weight larger than w_i. That is, we let $F_i := \{f \in F \mid w(f) \le w_i\}$. Furthermore, let T_{opt} be an (arbitrary) minimum spanning tree in H and let i_0 be defined such that w_{i_0} is the weight of an edge of maximum weight in T_{opt}. Note that this implies that H_{i_0} has a spanning tree. Then, clearly, $w_{i_0} \le mst(H_{i_0}) = mst(H)$. That is, if we use the scaling technique outlined above to compute a minimum spanning tree successively in all hypergraphs H_i, for $1 \le i \le s$, and return of the at most s spanning trees one with minimum weight, this tree T will be, with probability at least $\frac{1}{2}$, a spanning tree in H such that $w(T) \le (1 + \epsilon)mst(H)$. The running time of this modified algorithm is, of course, still bounded by a polynomial in n.   □

For the Steiner tree problem we combine Theorem 8.7 and Corollary 8.8 with Lemma 7.1 and Theorem 7.4 to immediately obtain the following results:

Corollary 8.9 *There exists a randomized polynomial time approximation algorithm with performance ratio 5/3 for the* STEINER PROBLEM IN GRAPHS.   □

Corollary 8.10 *For all $\epsilon > 0$ there exists a randomized polynomial time approximation algorithm with performance ratio $5/3 + \epsilon$ for the* STEINER PROBLEM IN NETWORKS.   □

Note that a performance ratio of $5/3 \approx 1.67$ is slightly better than that of $1 + \ln 2 \approx 1.69$ from Corollary 7.16. On the other hand we pay a little for this improvement by obtaining only a randomized instead of a deterministic algorithm. In return, the running time of the algorithms of Corollaries 8.9 and 8.10 are much better than those hidden in the $(1 + \ln 2)$-approximation scheme of Corollary 7.16. Nevertheless they are still much to high to be of practical importance and we therefore don't bother to estimate them precisely.

In Section 3.2 we saw that the Steiner tree problem remains $\mathcal{NP}$-hard even if we restrict the input to graphs for which there exists a Steiner minimum tree such that all full components contain at most 4 terminals. The above results imply that this situation changes if we further restrict the instances to those for which there exists a Steiner minimum tree such that all full components contain at most 3 terminals.

Corollary 8.11 *There exists a randomized polynomial time approximation algorithm which solves the* MINIMUM 3-RESTRICTED STEINER PROBLEM IN GRAPHS.
                                  □

Corollary 8.12 *There exists a randomized pseudopolynomial time approxima-tion algorithm which solves the* MINIMUM 3-RESTRICTED STEINER PROBLEM IN NETWORKS. □

The rest of this section is devoted to the proof of Theorem 8.7. As we will see, our approach here is completely different from that in Chapter 7. Actually, most of the time we will not work in graphs or hypergraphs at all. Instead we proceed similarly as in Example 8.5. That is, we transform the problem into an algebraic setting and find the minimum spanning tree by computing determi-nants of appropriately defined matrices. We start by introducing the concept of a pfaffian.

Let $A = (a_{ij})$ be a skew-symmetric matrix (i.e., $A^T = -A$) of size $2n \times 2n$ and let $\mathcal{P}$ be the set of all partitions of $\{1, \ldots, 2n\}$ into pairs. For an element $p = \{\{i_1, i_2\}, \ldots, \{i_{2n-1}, i_{2n}\}\}$ of $\mathcal{P}$ we denote by $\sigma(p)$ the sign of the permutation

$$\begin{pmatrix} 1 & 2 & \cdots & 2n-1 & 2n \\ i_1 & i_2 & \cdots & i_{2n-1} & i_{2n} \end{pmatrix}$$

and by $\rho(p)$ the product

$$\rho(p) := \prod_{j=1}^{n} a_{i_{2j-1} i_{2j}}.$$

One easily verifies, see Problem 8.7, that $\sigma(p) \cdot \rho(p)$ is independent of the order of the pairs and the order within the pairs of p. Therefore

$$\mathrm{pf}(A) = \sum_{p \in \mathcal{P}} \sigma(p) \cdot \rho(p)$$

is well defined. It is called the *pfaffian* of A. A well-known result from linear algebra, which we quote without proof, says:

Lemma 8.13 *If A is a skew-symmetric matrix A of size $2n \times 2n$ and B an arbitrary $2n \times 2n$ matrix, then*

$$\det(A) = [\mathrm{pf}(A)]^2 \qquad and \qquad \mathrm{pf}(BAB^T) = \det(B) \cdot \mathrm{pf}(A).$$

For our application within the spanning tree problem we will construct a matrix A from a suitably defined set of column vectors. These column vec-tors will come in pairs a_i, b_i. For every such pair we consider the rank 2 matrix $a_i b_i^T - b_i a_i^T$ and let A denote a weighted sum of these rank 2 matrices. In the following lemma we deduce some properties of such matrices A. We use the no-tation $(a_{i_1} | b_{i_1} \cdots a_{i_n} | b_{i_n})$ to denote a matrix whose first column vector is a_{i_1}, the second b_{i_1}, and so on.

Lemma 8.14 *Let $m \geq n$ and $a_1, b_1, \ldots, a_m, b_m$ be vectors in $\mathbb{R}^{2n}$. Furthermore, let $x_1, \ldots, x_m$ be m variables. Then the $2n \times 2n$ matrix*

$$A = \sum_{i=1}^{m} x_i(a_i b_i^T - b_i a_i^T)$$

is skew-symmetric and satisfies

$$\mathrm{pf}(A) = \sum_{1 \leq i_1 < \cdots < i_n \leq m} x_{i_1} \cdots \cdots x_{i_n} \cdot \det(a_{i_1}|b_{i_1} \cdots |a_{i_n}|b_{i_n}).$$

Proof. We first consider the case $n = m$. If the a_i's and b_i's are appropriate unit vectors, namely, $a_i = e_{2i-1}$ and $b_i = e_{2i}$ for all $i = 1, \ldots, n$, where e_j denotes the j-th unit vector, then $e_{2i-1}e_{2i}^T - e_{2i}e_{2i-1}^T$ is a $2n \times 2n$ matrix with only two nonzero elements, namely "+1" in row $2i - 1$ and column $2i$ and "−1" in row $2i$ and column $2i - 1$. Hence,

$$\mathrm{pf}(A) \;=\; \mathrm{pf} \begin{pmatrix} 0 & x_1 & & & & & \\ -x_1 & 0 & & & & & \\ & & 0 & x_2 & & & \\ & & -x_2 & 0 & & & \\ & & & & \ddots & & \\ & & & & & 0 & x_n \\ & & & & & -x_n & 0 \end{pmatrix}$$

$$= \; x_1 \cdots \cdots x_n.$$

For arbitrary vectors a_i and b_i let $B = (a_1|b_1|\cdots|a_n|b_n)$. Then $a_i = Be_{2i-1}$ and $b_i = Be_{2i}$ and

$$a_i b_i^T - b_i a_i^T = B(e_{2i-1}e_{2i}^T - e_{2i}e_{2i-1}^T)B^T.$$

Therefore, using Lemma 8.13

$$\begin{aligned} \mathrm{pf}(A) \;&=\; \mathrm{pf}(B(\sum_{i=1}^{n} x_i(e_{2i-1}e_{2i}^T - e_{2i}e_{2i-1}^T))B^T) \\ &=\; \det(B) \cdot \mathrm{pf}(\sum_{i=1}^{n} x_i(e_{2i-1}e_{2i}^T - e_{2i}e_{2i-1}^T)) \\ &=\; \det(B) \cdot x_1 \cdots \cdots x_n. \end{aligned}$$

For the general case $m \geq n$ we observe that every element of the matrix A is a homogeneous linear form in the variables $x_1, \ldots, x_m$. (More precisely, the (j,k)th element of A is $\sum_{i=1}^{m} x_i[(a_i)_j(b_i)_k - (b_i)_j(a_i)_k]$.) From the definition of the pfaffian it therefore follows that $\mathrm{pf}(A)$ is a homogeneous polynomial of (total) degree n in $x_1, \ldots, x_m$.

We claim that pf(A) can't contain a monomial in which less than n of the x_i's have a positive exponent. Indeed, assume it would. Then set all other x_i's to zero. This does not affect the value of this special monomial. That is, pf(A) still contains at least one nonzero monomial. On the other hand the matrix A is now singular, that is $\det(A) = [\mathrm{pf}(A)]^2$ has to evaluate to the zero-polynomial, yielding the desired contradiction.

Finally, let $1 \leq i_1 < \cdots < i_n \leq m$ be arbitrary, but fixed. We obtain the coefficient of the monomial $x_{i_1} \cdot \ldots \cdot x_{i_n}$ in pf(A) by setting $x_j = 0$ for all $j \notin \{i_1, \ldots, i_n\}$. Thus, the special case $n = m$ treated above implies that this coefficient is $\det(a_{i_1}|b_{i_1}|\ldots|a_{i_n}|b_{i_n})$. □

We now show how to apply Lemma 8.14 to hypergraphs. Let $H = (V, F)$ be a 3-uniform hypergraph on $2n+1$ vertices. For every edge $f = \{i,j,k\}$ in F we pick one vertex arbitrarily, say i, and let $e_f = \{i,j\}$ and $\tilde{e}_f = \{i,k\}$: Let $G = (V, E)$ be a (multi-)graph on the same vertex set as H and with edge set $E = \{e_f, \tilde{e}_f \mid f \in F\}$. The proof of the following easy observation is left as an exercise.

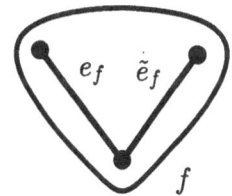

Proposition 8.15 *A set $\{f_1, \ldots, f_n\} \subseteq F$ forms a spanning tree in H if and only if $\{e_{f_1}, \tilde{e}_{f_1}, \ldots, e_{f_n}, \tilde{e}_{f_n}\}$ forms a spanning tree in G.* □

That is, the problem of finding a minimum spanning tree in a 3-uniform hypergraph is equivalent to the problem of finding a minimum spanning tree in a (multi-)graph, where the edges are "paired", i.e., either both or none are in the tree.

For every pair of edges e_f, $\tilde{e}_f$ we define two $2n$-dimensional vectors a_f and b_f as follows:

$$(a_f)_i = \begin{cases} 1 & \text{if } e_f = \{i,j\} \text{ with } j < i, \\ -1 & \text{if } e_f = \{i,j\} \text{ with } j > i, \\ 0 & \text{otherwise,} \end{cases}$$

and

$$(b_f)_i = \begin{cases} 1 & \text{if } i \in \tilde{e}_f = \{i,j\} \text{ with } j < i, \\ -1 & \text{if } \tilde{e}_f = \{i,j\} \text{ with } j > i, \\ 0 & \text{otherwise.} \end{cases}$$

(Note that the a_f's and b_f's are essentially the incidence vectors of e_f and $\tilde{e}_f$, except that the $(2n+1)$st component has been cut off and that we oriented every edge from the smaller to the large vertex.) The following fact resembles a well-known property of the incidence matrix of a graph. The proof is again left as an exercise to the reader.

Proposition 8.16 *Let $f_1, \ldots, f_n$ be n edges in F. Then*

$$|\det(a_{f_1}|b_{f_1}|\ldots|a_{f_n}|b_{f_n})| = \begin{cases} 1 & \text{if } f_1, \ldots, f_n \text{ form a spanning tree in } H, \\ 0 & \text{otherwise.} \end{cases}$$

□

Finally, let A denote the $2n \times 2n$ matrix

$$A = \sum_{f \in F} 2^{w(f)}(a_f b_f^T - b_f a_f^T). \tag{8.1}$$

Clearly, this matrix is skew-symmetric, and we will now show how Lemma 8.14 can be used to design an efficient algorithm for finding a minimum spanning tree in a hypergraph.

Assume H is a 3-uniform hypergraph on $2n + 1$ vertices and $w : F \to \mathbb{N}_0$ is a weight-function such that H has a *unique* spanning tree of minimum weight. Then Lemma 8.14 can be used to compute not only the weight of the minimum spanning tree but also the minimum spanning tree itself.

Lemma 8.17 *Assume that H is a 3-uniform hypergraph on $2n + 1$ vertices and $w : F \to \mathbb{N}_0$ is a weight-function such that H has a unique spanning tree T_0 of minimum weight, say, w_0. Then the matrix A defined in (8.1) satisfies $\det(A) \neq 0$ and 2^{2w_0} is the highest power of 2 that divides $\det(A)$. Moreover, if we let $A_f = A - 2^{w(f)}(a_f b_f^T - b_f a_f^T)$ for all $f \in F$, then*

$$f \in T_0 \qquad \text{if and only if} \qquad \frac{\det(A_f)}{2^{2w_0}} \text{ is even.}$$

Proof. Combining Lemma 8.13, 8.14, and Proposition 8.16 we deduce that

$$\det(A) = [\text{pf}(A)]^2 = \left[\sum_T 2^{w(T)} \cdot \delta_T\right]^2,$$

where the sum is over all spanning trees T of H and $\delta_T \in \{-1, +1\}$ for all such trees. Hence,

$$\det(A) = \sum_{i \geq 0} c_i 2^{2w_0 + i} \qquad \text{with } |c_0| = 1 \text{ and appropriate } c_1, c_2, \ldots \in \mathbb{Z}.$$

The first part of the theorem follows. For the second part just observe that A_f is the matrix corresponding to the hypergraph H from which the edge f is removed. As the weight of a minimum spanning tree in this new hypergraph is either still w_0, if $f \notin T_0$, or larger than w_0, the reasoning above therefore implies that

$$\det(A_f) = \begin{cases} 2^{2w_0} + c_f \cdot 2^{2w_0 + 1} & \text{if } f \notin T_0, \\ c_f \cdot 2^{2w_0 + 2} & \text{if } f \in T_0, \end{cases}$$

for appropriate constants $c_f \in \mathbb{Z}$.

□

Clearly, Lemma 8.17 implies a (deterministic) algorithm for finding a minimum spanning tree, whenever this tree is unique. The running time of this algorithm is bounded by $m+1$ times the time required to compute the determinant of the matrices A resp. A_f. Clearly, the determinant of a $2n \times 2n$ matrix can an be obtained with $\mathcal{O}(n^3)$ operations by Gaussian elimination. However, we have to be a bit careful here. The matrices A and A_f contain entries of size $2^{w_{\max}}$, which is of size exponential in the input. As outlined in Chapter 2, this violates our tacit assumption of "good behaviour" (see page 28). We therefore have to count the number of *bit operations* and not just the number of operations per se. Taking this into account we deduce that the running time also dependes on $w_{\max}$. On the other hand, it is not too difficult to show that the numbers generated within the Gaussian elimination procedure stay polynomially bounded in the size of the original matrix A. The total running time of the algorithm is thus bounded by appropriate polynomial in n and $w_{\max}$.

In order to apply this approach to general 3-bounded hypergraphs we need a way to transform any such hypergraph into a related one in which the minimum spanning tree is indeed unique. To achieve this, we will use randomization. We start with a lemma.

Lemma 8.18 *Let S be a finite set with n elements and let $A_1 \ldots ,A_t \subseteq S$ be arbitrary subsets of S. For every element $s \in S$ we choose uniformly and independently at random an integer $r(s)$ from the interval $[1,2n]$ and define the weight $w(f)$ of a set A_i by $w(A_i) := \sum_{s \in A_i} r(s)$. Then*

$$\Pr\left[\textit{There exists a unique set of minimum weight}\right] \geq \frac{1}{2}.$$

Proof. We call an element $s \in S$ *ambiguous* if and only if there exist two sets of minimum weight, one that contains s and one that does not. Obviously, it suffices to bound the probability that there exists such an ambiguous element. To achieve this, we assume without loss of generality that every element is contained in some set A_i but not in all.

Choose $s_0 \in S$ arbitrarily and suppose for the moment that all values $r(s)$ are fixed, except the one for the element s_0. Define a threshold $\alpha(s_0)$ for s_0 as follows. If $r(s_0) \leq \alpha(s_0)$ then s_0 is contained in some set of minimum weight, while if $r(s_0) > \alpha(s_0)$ then s_0 is contained in no minimum weight set. Observe that the vertex s_0 is ambiguous if and only if $r(s_0) = \alpha(s_0)$. Hence,

$$\Pr\left[s_0 \text{ is ambiguous}\right] \leq \frac{1}{2n}.$$

Since there are exactly n elements, we conclude

$$\Pr\left[\text{There exists an ambiguous element}\right] \leq \sum_{s \in S} \Pr\left[s \text{ is ambiguous}\right] \leq \frac{n}{2n} = \frac{1}{2}.$$

$\square$

Corollary 8.19 *Let $H = (V,F,w)$ be a weighted 3-uniform hypergraph on $2n + 1$ vertices containing at least one spanning tree. For every edge $f \in F$ choose uniformly and independently at random an integer $r(f)$ from the interval $[1, 2\binom{2n+1}{3}]$ and define a weight function $w' : F \to \mathbb{N}$ by:*

$$w'(f) := 3n^4 \cdot w(f) + r(f).$$

Then

$$\Pr\left[\text{There exists a unique minimum spanning tree with respect to } w'\right] \geq \frac{1}{2}.$$

Moreover, every minimum spanning tree with respect to w' is also a minimum spanning tree with respect to w.

Proof. Observe that the weight of a spanning tree in H with respect to the weight function r is at most $n \cdot 2\binom{2n+1}{3} \leq \frac{8}{3}n^4$. Notice furthermore that after multiplying the weight function w with $3n^4$ the weight of a spanning tree in H is a multiple of $3n^4$. By adding the values $r(f)$ to all edges we further increase the weight of a spanning tree by at most $\frac{8}{3}n^4 < 3n^4$. That is, the weight function w' preserves the order of the spanning trees of H with different weights. It just disturbs the order of those spanning trees in H which have the same weight "a little" – in fact, just enough to reach uniqueness with respect to w'. To see this we apply Lemma 8.18 as follows: the set S consists of all *edges* of H and the sets A_i correspond to the minimum spanning trees of H. Lemma 8.18 implies that with probability at least $1/2$ there exists a unique set A_i of minimum weight, and hence a unique minimum spanning tree of H with respect to the weight function w'. $\square$

Algorithm 8.20 (MINIMUMSPANNINGTREE)
Input: A weighted 3-uniform hypergraph $H = (V,F,w)$ on $2n + 1$ vertices.
Output: A spanning tree T of H or FAILURE.
(1) Compute the weight function w' as defined in Corollary 8.19;
(2) Compute the matrix A as defined in (8.1) on page 144 with respect to the weight function w' and let w_0 be the largest integer such that 2^{2w_0} divides $\det(A)$.
    Let $T := \emptyset$;
(3) For every edge $f \in F$ do:
        Compute $\det(A_f)/2^{2w_0}$;
        if $\det(A_f)/2^{2w_0}$ is even **then** $T := T \cup \{f\}$;
(4) **if** T is a spanning tree of H **then**   **return** T
                                else   **return** FAILURE.

Combining Corollary 8.19 with the observation about computing the determinants in the paragraph following the proof of Lemma 8.17 we obtain the following result.

Theorem 8.21 MINIMUMSPANNINGTREE *is a randomized algorithm which returns a minimum spanning tree with probability at least $1/2$, provided H contains a spanning tree. The running time is polynomial in n and $w_{\max}$.* □

To complete the proof of Theorem 8.7 we only have to show that computing minimum spanning trees in 3-bounded hypergraphs can be reduced to the problem of computing a minimum spanning tree in a suitably defined 3-uniform hypergraph.

Proof of Theorem 8.7. Let $H = (V, F, w)$ be a 3-bounded weighted hypergraph on n vertices. Construct a weighted 3-uniform hypergraph $\tilde{H} = (\tilde{V}, \tilde{F}, \tilde{w})$ as follows. The vertex set $\tilde{V}$ consists of all vertices of V plus $n-1$ new vertices $z_1, \ldots, z_{n-1}$, and

$$
\begin{aligned}
\tilde{F} \; = \; & \{e \mid e \in F, |e| = 3\} \\
& \cup \{e \cup \{z_i\} \mid e \in F, |e| = 2, 1 \le i \le n-1\} \\
& \cup \{\{v, z_i, z_j\} \mid v \in V, 1 \le i < j \le n-1\}.
\end{aligned}
$$

New edges containing exactly one z-vertex will be called type I edges, while those containing two z-vertices are of type II. The weight function $\tilde{w}$ is defined as follows:

$$
\tilde{w}(f) = \begin{cases}
w(f) & \text{if } f \in F, \\
w(f) + M & \text{if } f \text{ is a type I edge, and} \\
2M & \text{if } f \text{ is a type II edge,}
\end{cases}
$$

where M is a suitable large constant, see below. We leave it as an exercise for the reader to verify that this construction has the following properties, which completes the proof of Theorem 8.7:

(i) Every spanning tree T in H gives rise to a spanning tree $\tilde{T}$ in $\tilde{H}$ of weight $\tilde{w}(\tilde{T}) = w(T) + (n-1)M$ by replacing every edge of cardinality 2 in T by a type I edge (using different z-vertices) and adding type II edges until every z-vertex is contained in exactly one edge.

(ii) If $\tilde{T}$ is a spanning tree in $\tilde{H}$ of weight $\tilde{w}(\tilde{T}) < nM$ then every z-vertex is covered by exactly one edge of $\tilde{T}$. Thus, $\tilde{T}$ corresponds to a spanning tree T of H of weight $w(T) = \tilde{w}(\tilde{T}) - (n-1)M$.

(iii) Let $M \ge n \cdot \max\{w(f) \mid f \in F\}$. Then a minimum spanning tree in $\tilde{H}$ of length less than nM corresponds to a minimum spanning tree of H, whereas the fact that the length of a minimum spanning tree in $\tilde{H}$ is at least nM means that H contains no spanning tree. Furthermore, if H is the complete hypergraph (e.g. the hypergraph $H_3(N, K)$ in the reduction from the Steiner tree problem) then choosing $M = \max\{w(f) \mid f \in F\}$ suffices to achieve this property. □

8.3   An almost always optimal algorithm

Up to now we used randomness to make an algorithm more powerful. In turn, we obtained an algorithm that gives for *every* input a good answer with a certain (high) probability. In this section we will consider a different setup. Namely, we will consider a deterministic algorithm and study its behavior for a *random* input. In this way we will obtain statements about the *average* quality of the algorithm. Note that this contrasts the strategy which we pursued so far, where we were only interested in *worst case* analysis of the complexity and the performance ratio of an algorithm.

While analyzing the average behavior of an algorithm might sound like a plausible approach, we start with a warning. One has to be very careful using average case analysis. The quality of such an analysis depends heavily on the distribution of the input. The same algorithm can have a good or a bad average performance just by changing this distribution. Hence, a statement about the average case behavior of an algorithm can be very illuminating. But it might also be quite useless, if a "wrong" distribution has been used. Unfortunately, it is usually very difficult to choose a distribution which is at the same time meaningful and analyzable.

In this section we will show that for the Steiner tree problem in graphs our first approximation algorithm, the MST-algorithm from Section 6.1, already has a rather good performance on average. More precisely, we will see that, though it has in the worst case a performance ratio 2, it solves for "almost all" inputs the Steiner tree problem "almost" optimally. As input distribution we choose the random graph model $G_{n,p}$ from Section 1.2, where the first k vertices are the terminals. We will use the notation $G_{n,p,k}$ to denote such a problem. Note that p as well as k will usually depend on n. In fact, we will restrict ourselves in this section to the case $k = \lfloor \ln n \rfloor$. This keeps the calculations slightly simpler.

Theorem 8.22 *Let $k = \lfloor \ln n \rfloor$ and let p be a function such that $p \gg \ln n / n$. Then*

$$\Pr\left[|T_{MST}(G_{n,p,k})| \le (1 + o(1)) \cdot |T_{\text{opt}}(G_{n,p,k})| \right] = 1 - o(1),$$

where T_{opt} and T_{MST} denote the Steiner minimum tree resp. the Steiner tree returned by the MST-algorithm.

The proof of Theorem 8.22 splits into two parts. First we show that a Steiner minimum tree in $G_{n,p,k}$ contains almost surely a certain number of edges. In the second part we will then show that the number of edges in the Steiner tree returned by the MST-algorithm can almost surely be bounded from above by roughly the same function.

We start by establishing an upper bound on the number of Steiner trees for k terminals containing exactly t additional vertices.

Lemma 8.23 *The number of Steiner trees for k terminals containing exactly t additional vertices is bounded from above by $(k+t-2)! \cdot (k+t)^{k-2}/(k-2)!$.*

Proof. Without loss of generality assume that the set of terminals consists of the vertices labeled $1,\ldots,k$, while the non-terminals are labeled $k+1,\ldots,k+t$. We define a mapping between the set of all such Steiner trees and the set of strings $a_1 a_2 \ldots a_{k+t-2}$ over the alphabet $\{1,\ldots,k+t\}$ as follows. Let b_1 be the leaf in the Steiner tree with minimum label. Then a_1 is defined as the vertex adjacent to b_1. Now remove b_1 and the edge $\{a_1, b_1\}$ from the tree and repeat the process. That is, let b_2 be the leaf with the minimum label in the new tree and let a_2 be its neighbor. Continue in this way until the tree consists only of a single edge. Note that during this process non-terminals might become leaves.

This mapping is clearly injective. Furthermore, by construction the generated string contains all integers corresponding to vertices which are *not* leaves and hence in particular all integers $k+1,\ldots,k+t$. An upper bound for the number of Steiner trees with t non-terminals is thus given by the number of strings of length $k+t-2$ over the alphabet $\{1,\ldots,k+t\}$ which contain each element of $\{k+1,\ldots,k+t\}$ at least once. (In fact, one can show that the number of Steiner trees is equal to this number, but we do not need this fact here.)

Let us construct these strings. There are $k+t-2$ possibilities to choose the first occurrence of $k+1$. For the first occurrence of $k+2$ there are then $k+t-3$ possibilities. And so on. Thus there are $(k+t-2)!/(k-2)!$ possibilities to choose the first occurrences of each element in $\{k+1,\ldots,k+t\}$. Finally, we choose for each of the remaining $k-2$ positions an arbitrary element from $\{1,\ldots,k+t\}$. This clearly overestimates the desired number, since some of these choices might contradict the fact that the previously chosen positions should represent the *first* occurrences of $k+i$. Nevertheless, it implies the desired upper bound. $\square$

Lemma 8.24 *Let $k = \lfloor \ln n \rfloor$ and p be a function such that $p \gg 1/n$. Then*

$$\Pr\left[|T_{\text{opt}}(G_{n,p,k})| > (k-1)\frac{\ln n}{\ln pn} \cdot \left(1 - \frac{4\ln\ln n}{\ln n}\right) \right] = 1 - o(1).$$

Proof. We proceed similarly as in Section 1.2 where we showed that for $p \gg \ln n/n$ the random graph $G_{n,p}$ is almost surely connected. Namely, we first calculate the expected number of Steiner trees for $G_{n,p,k}$ which contain at most $t_0 := (k-1) \cdot \frac{\ln n}{\ln pn} \cdot (1 - \frac{4\ln\ln n}{\ln n}) - k + 1$ non-terminals and then show that this expectation tends to zero.

Let X denote the random variable counting the number of Steiner trees in $G_{n,p,k}$ which contain at most t_0 non-terminals. Then X can be written as a sum of zero-one variables enumerating all Steiner trees of the complete graph on $V_n = \{1,\ldots,n\}$ which contain at most t_0 non-terminals. By Lemma 8.23 there exist at most $\binom{n-k}{i} \cdot (k+i-2)! \cdot (k+i)^{k-2}/(k-2)!$ such trees with exactly i non-terminals. Since the expectation of any of these zero-one variables is p^{k+i-1}, linearity of expectation implies that

$$\Pr\left[|T_{\text{opt}}(G_{n,p,k})| \le t_0 + k - 1\right] \le \text{Ex}[X]$$

$$\le \sum_{i=0}^{t_0} \binom{n-k}{i} \cdot (k+i-2)! \cdot (k+i)^{k-2} \cdot p^{k+i-1}/(k-2)!$$

$$\le \sum_{i=0}^{t_0} n^i \cdot \binom{k+i-2}{k-2} \cdot (k+i)^{k-2} \cdot p^{k+i-1}$$

$$= n^{-k+1} \sum_{i=0}^{t_0} \binom{k+i-2}{k-2} \cdot (k+i)^{k-2} \cdot (np)^{k+i-1}.$$

One easily checks that the term within the sum is maximal for $i = t_0$. Hence, using that $\binom{k+t_0-2}{k-2} \le (k+t_0-2)^{k-2}$ and that $(k+t_0-2)(k+t_0) \le (k+t_0-1)^2$ we deduce furthermore that

$$\Pr\left[|T_{\text{opt}}(G_{n,p,k})| \le t_0 + k - 1\right]$$
$$\le n^{-k+1} \cdot (t_0+1) \cdot (np)^{k+t_0-1} \cdot (k+t_0-2)^{k-2} \cdot (k+t_0)^{k-2}$$
$$\le n^{-k+1} \cdot (np)^{k+t_0-1} \cdot (k+t_0-1)^{2k-3}.$$

The definition of t_0 implies that $n^{-k+1} \cdot (np)^{k+t_0-1} = (\ln n)^{-4(k-1)}$ and that $k + t_0 - 1 \le (k-1) \ln n / \ln pn \le (\ln n)^2$. Hence, we have shown that

$$\Pr\left[|T_{\text{opt}}(G_{n,p,k})| \le t_0 + k - 1\right] \le (\ln n)^{-2}.$$

□

For the proof of the upper bound we use the second moment method introduced in Section 1.2.

Lemma 8.25 *Let $k = \lfloor \ln n \rfloor$ and p be a function such that $p \gg \ln n/n$. Then*

$$\Pr\left[|T_{MST}(G_{n,p,k})| \le (k-1) \cdot \lceil \tfrac{\ln n}{\ln pn} + 1 \rceil\right] = 1 - o(1).$$

Proof. Let $t_1 := \lceil \tfrac{\ln n}{\ln pn} + 1 \rceil$. It suffices to show that almost surely the minimum spanning tree of the complete distance graph induced by the k terminals has length at most $(k-1) \cdot t_1$. Instead of considering arbitrary spanning trees we restrict our attention to very special ones. Namely those corresponding to paths in $G_{n,p}$ which start in terminal 1, end in terminal k, and visit all other terminals in increasing order in such a way that each of the subpaths between subsequent terminals contains exactly $t_1 - 1$ additional vertices.

We will show the existence of such a subgraph by considering every pair of successive terminals separately. Clearly, if we can show that a path of length t_1 between a given pair of terminals exists with probability at least $1 - o(1/k)$, then all paths exist simultaneously with probability $1 - o(1)$. So from now on we consider without loss of generality only terminals 1 and 2. There are $(n-k) \cdot \ldots \cdot (n-k-(t_1-2)) = ((1 - \mathcal{O}(\tfrac{k+t_1}{n})) \cdot n)^{t_1-1}$ ways to choose the additional vertices (in the order in which they appear on the path). Let X be the random variable counting the number of such paths in $G_{n,p,k}$. Then

$$\mathrm{Ex}\,[X] = ((1 - \mathcal{O}(\tfrac{k+t_1}{n})) \cdot n)^{t_1-1}p^{t_1} \geq \frac{1}{1 + o(1/k)} \cdot np,$$

by the definition of t_1. From Section 1.2 we know that in case $\mathrm{Ex}\,[X]$ tends to infinity, we can bound $\Pr[X = 0]$ by using the second moment method:

$$\Pr[X = 0] \leq \frac{\mathrm{Ex}\,[X^2] - (\mathrm{Ex}\,[X])^2}{(\mathrm{Ex}\,[X])^2}.$$

That is, we need to show that $\mathrm{Ex}\,[X^2] = (1 + o(1/k))(\mathrm{Ex}\,[X])^2$. We write X as the sum of zero-one variables X_i enumerating all paths of the desired form. Then

$$\mathrm{Ex}\,[X^2] = \sum_i \mathrm{Ex}\,[X_i^2] + \sum_{i \neq j} \mathrm{Ex}\,[X_i X_j].$$

As $X_i^2 = X_i$ the first summand is equal to the expectation of X. Evaluating the second summand is more difficult. Obviously, $X_i X_j$ is also a zero-one variable. $\mathrm{Ex}\,[X_i X_j]$ is therefore equal to p^x where x is the number of edges contained in the union of the two paths represented by X_i and X_j. This number may vary between $t_1 + 2$ and $2t_1$. We would be home if we could show that only those terms matter for which the two paths are completely disjoint. To see that this is indeed the case we fix X_i and bound the number of X_j's which have exactly $r = r_1 + r_2 + r'$ edges in common with X_i. Namely, the first r_1 edges, the last r_2 edges and r' edges from the interior of the path. Assume these r' edges split into $c \leq r'$ components. Then there are at most $\binom{t_1}{2}^c$ possibilities to choose these r' edges. We construct the paths corresponding to the desired X_j's as follows. First we choose the position of the c components in the path, in at most $2^c \cdot t_1^c$ ways (for each component there are two possibilities for its orientation and at most t_1 possibilities for its starting point in X_j). Then we choose the additional $(t_1 - 1) - (r_1 + r_2 + r' + c)$ vertices, in at most $n^{(t_1-1)-(r_1+r_2+r'+c)}$ ways.

For fixed X_i and fixed r_1, r_2 and r' the sum of the corresponding terms $\mathrm{Ex}\,[X_i X_j]$ is therefore bounded from above by

$$\sum_{c \geq 0} \binom{t_1}{2}^c \cdot 2^c t_1^c \cdot n^{(t_1-1)-(r_1+r_2+r'+c)} \cdot p^{2t_1-r}$$

$$\leq \sum_{c \geq 0} (t_1^3 n^{-1})^c \cdot n^{t_1-1} p^{t_1} \cdot p^{t_1} \cdot (np)^{-r}$$

$$= (1 + o(\tfrac{1}{k}))^2 \cdot \mathrm{Ex}\,[X] \cdot p^{t_1} \cdot (np)^{-r}.$$

as $t_1^3/n = o(1/k)$ and $\mathrm{Ex}\,[X] \geq 1/(1 + o(1/k)) \cdot n^{t_1-1}p^{t_1}$. Hence

$$\sum_{i \neq j} \mathrm{Ex}\,[X_i X_j] = \sum_i \sum_{r_1,r_2,r'} (1 + o(\tfrac{1}{k}))^2 \cdot \mathrm{Ex}\,[X] \cdot p^{t_1} \cdot (np)^{-r}.$$

As $r = r_1 + r_2 + r'$ and $\sum_{i \geq 0}(np)^{-i} = 1 + o(1/k)$, as $np \gg k$, this implies

$$\sum_{i \neq j} \mathrm{Ex}\,[X_i X_j] \;=\; (1 + o(\tfrac{1}{k}))^2 \cdot \mathrm{Ex}\,[X] \cdot \big(1 + o(\tfrac{1}{k})\big)^3 \cdot \sum_i p^{t_1}$$

and therefore $\mathrm{Ex}\left[X^2\right] = (1 + o(1/k))^5 \cdot (\mathrm{Ex}\,[X])^2 = (1 + o(1/k)) \cdot (\mathrm{Ex}\,[X])^2$, as claimed. □

Proof of Theorem 8.22. Lemmas 8.24 and 8.25 imply that

$$|T_{\mathrm{opt}}(G_{n,p,k})| > (k-1)\tfrac{\ln n}{\ln np}(1 - o(1))$$

and

$$|T_{MST}(G_{n,p,k})| \leq (k-1)\tfrac{\ln n}{\ln np}(1 + o(1))$$

with probability $1 - o(1)$. The theorem follows. □

8.4 Excursion: Primality and cryptography

Prime numbers and number theory have always been a fascinating topic for mathematicians – starting with Euclid in Alexandria around 300 BC all the way up to Andrew Wiles in Cambridge, who proved Fermat's famous "Last Theorem" only a few years ago.

But prime numbers are not only interesting for mathematicians. It also turned out that the seemingly simple problem

PRIMES:

Instance: An integer $n \in \mathbb{N}$.
Question: Is n a prime number?

plays a major rôle in computer science as well. It is one of the very few decision problems which is known to belong to the class $\mathcal{ZPP}$, but nevertheless till today no deterministic polynomial time algorithm has been found.[1] This is all the more important (or annoying) in view of the fact that prime numbers play a central rôle in cryptography, a topic with increasing significance in the light of today's electronic communication systems. Here one needs prime numbers with several hundred digits. And one has to find them quickly.

---

[1] To avoid possible confusion, let us emphasize here that the input to problem PRIMES consists just of $\lceil \log_2 n \rceil$ bits of the binary encoding of n. Testing all numbers 1 up to $\sqrt{n}$ (or applying Erathostenes' sieve) is thus *not* a polynomial time algorithm.

In this section we will present a very simple and efficient *randomized* algorithm which decides whether a given number n is prime – with just a tiny probability of error. Actually, the fundamental idea of our algorithm is almost trivial. We just choose an integer $1 < a < n$ at random and check whether it "witnesses" that n is not a prime number. If it does, we answer COMPOSITE, otherwise PRIME. Clearly, if we could show that if n is a composite number then, say, at least half of all the integers in the set $\{2, \ldots, n-1\}$ are witnesses for n's compositeness, then this approach works:

- If n is prime, then there are no witnesses and the answer of our algorithm will always be PRIME.

- If on the other hand n is composite it suffices to get the answer COMPOSITE in at least one of, say, one hundred independent runs of our algorithm. Now the probability that we get the (wrong) answer PRIMES in all 100 tries is just the probability that we do *not* pick a single witness in all tries. As the probability for picking a witness in a single run of the algorithm is at least $1/2$, the probability that our algorithm foils us by answering PRIME one hundred times is at most 2^{-100}.

Our task is thus to show that every composite number has sufficiently many witnesses. In addition we need, of course, an algorithm which quickly finds out whether a given number a *is* indeed a witness.

We start with some notation. Let $\mathbb{Z}_n := \{0, 1, \ldots, n-1\}$ and

$$\mathbb{Z}_n^* := \{x \in \mathbb{Z}_n \setminus \{0\} \mid \gcd(x,n) = 1\}, \qquad \phi(n) := |\mathbb{Z}_n^*|,$$

where $\gcd(x,n)$ denotes the greatest common divisor of x and n and $\phi(n)$ is the so-called Euler function, named after the Swiss mathematician Leonard Euler whom we already met in Section 1.1. Clearly, $\mathbb{Z}_n$ is a group with respect to addition modulo n and $\mathbb{Z}_n^*$ is a group with respect to multiplication modulo n. As all numbers $a \neq 0$ in $\mathbb{Z}_n \setminus \mathbb{Z}_n^*$ are witnesses for the fact that n is composite, a first (naive) way for a primality test as sketched above would be to simply choose a number a from $\mathbb{Z}_n \setminus \{0\}$ uniformly at random and to answer PRIME if and only if $\gcd(a,n) = 1$. In order to show that this approach indeed works, we have to check two things. Can we test whether $\gcd(a,n) = 1$ efficiently? And, are there sufficiently many witnesses? While the answer to the first question is indeed *yes*, as we will shortly see, the answer to the second question is, unfortunately, *no* in general (see Problem 8.13). We therefore have to develop more sophisticated tests for witnessing n's compositeness.

Before we will do that we recall Euclid's algorithm, which computes the greatest common divisor of two numbers m and n. In fact we will state here the so-called extended version, as this will prove useful to us later on.

Euclid(m,n,x,y)

Input: Integers $m \geq n \in \mathbb{N}$.
Output: Integers $x,y \in \mathbb{Z}$ such that $\gcd(m,n) = mx + ny$.

if $n = 0$ **then**
    $x := 1;\ y := 0$
else
    Euclid$(n, m \bmod n, x', y')$;
    $x := y';\ y := x' - y' \cdot (m \operatorname{div} n)$;

Lemma 8.26 *For all $m \geq n \in \mathbb{N}$ algorithm* Euclid(m,n,x,y) *computes integers $x,y \in \mathbb{Z}$ such that $\gcd(m,n) = mx + ny$. Moreover, the running time of the algorithm is polynomial in $\log(n + m)$.*

Proof.   The correctness is easily proven by induction on the number of recursions. For the running time we have to show in addition that the number of recursions is bounded by $\mathcal{O}(\log n)$. The details are left as an exercise, cf. Problem 8.14.   □

Corollary 8.27 (Chinese Remainder Theorem) *For all $m,n \in \mathbb{N}$ such that $\gcd(m,n) = 1$ and all $a \in \mathbb{Z}_m$ and $b \in \mathbb{Z}_n$ there exists a unique $t \in \mathbb{Z}_{mn}$ such that*

$$
\begin{aligned}
t &\equiv a \pmod{m} \qquad \text{and} \\
t &\equiv b \pmod{n}.
\end{aligned}
$$

Proof.   First we observe that it suffices to show for every pair a,b the *existence* of such an element t. The *uniqueness* follows then immediately from the fact that the cardinality of $\mathbb{Z}_{mn}$ is exactly equal to the number of pairs a,b. To see the existence we apply Lemma 8.26 to observe that there exist integers $x,y \in \mathbb{Z}$ such that $mx + ny = 1 = \gcd(m,n)$. Then $t := (bmx + any) \bmod mn$ is the desired element, as one easily checks.   □

Corollary 8.28 *For all $n \in \mathbb{N}$ and all $a \in \mathbb{Z}_n^*$ the multiplicative inverse a^{-1} can be computed in time polynomial in $\log n$.*

Proof.   We apply Lemma 8.26 to compute integers x,y such that $\gcd(a,n) = 1 = ax + yn$. Without loss of generality we may assume that $x \in \mathbb{Z}_n$. Then $ax \equiv 1 \pmod{n}$. That is, $x = a^{-1}$ is the desired inverse of a.   □

Theorem 8.29 (Euler) *For all $n \in \mathbb{N}$:*

$$
a^{\phi(n)} \equiv 1 \pmod{n} \qquad \forall a \in \mathbb{Z}_n^*. \tag{8.2}
$$

Proof. Let k be the minimum integer such that $a^k \equiv 1 \pmod{n}$. Note that such an integer exists as a^k modulo n can take on only n different values and the fact that $a^l \equiv a^k \pmod{n}$ for some $l > k$ implies that $a^{l-k} \equiv 1 \pmod{n}$. Then $\{a, a^2, \ldots, a^k\}$ forms a subgroup of Z_n^* and therefore k divides $\phi(n) = |Z_n^*|$. Hence, all $a \in Z_n^*$ satisfy

$$a^{\phi(n)} \equiv (a^k)^{\frac{\phi(n)}{k}} \equiv 1^{\frac{\phi(n)}{k}} \equiv 1 \pmod{n}.$$

$\square$

Corollary 8.30 (Fermat) *For all $n \in \mathbb{N}$:*

$$n \text{ prime} \qquad \Longleftrightarrow \qquad a^{n-1} \equiv 1 \pmod{n} \quad \forall a \in Z_n \setminus \{0\}. \qquad (8.3)$$

Proof. If n is prime then $\phi(n) = n - 1$ and $Z_n^* = Z_n \setminus \{0\}$. The direction from left to right is thus just a special case of Euler's theorem. To see the reverse direction let $g < n$ be an arbitrary divisor of n. By the assumption of the right hand side of (8.3) we have that $g^{n-1} \equiv 1 \pmod{n}$. In other words, $g^{n-1} - 1 = k \cdot n = k' \cdot g$ for appropriate $k, k' \in \mathbb{N}$. Obviously, this can only hold for $g = 1$. That is, n has no non-trivial divisor, which is to say n is prime. $\square$

Fermat's theorem will be the basis of our primality tester. Before we apply it, let us add a remark about the efficiency of computing $a^k \bmod n$. Clearly, this may be calculated by performing $k - 1$ multiplications with a. This, however, would not be a polynomial time algorithm, as the size of the involved numbers would get too large. We can avoid this difficulty by repeated squaring and multiplication with a at all positions corresponding to a 1 in the binary representation of k. This is best explained via an example. Consider $a = 43$, $k = 50$ and $n = 67$. The binary representation of 50 is 110010. Hence,

$$43^{50} = (((((43)^2 \cdot 43)^2 \cdot 1)^2 \cdot 1)^2 \cdot 43)^2 \cdot 1$$

and we can therefore compute $43^{50} \bmod 67 = 15$ as follows:

$$(((((43^2 \bmod 67) \cdot 43 \bmod 67)^2 \bmod 67)^2 \bmod 67)^2 \cdot 43 \bmod 67)^2 \bmod 67.$$

Clearly, this requires just $\mathcal{O}(\log k)$ multiplications and modulo computations of numbers with at most $\mathcal{O}(\log n)$ bits.

With this fact at hand we are now ready to state our primality tester.

PrimalityTest(n)
(1) Choose an integer $a \in \{2, \ldots, n-1\}$ uniformly at random;
(2) **if** $\gcd(a, n) \neq 1$ or n even **then**
(3)   **return** COMPOSITE

(4) **else if** $a^{n-1} \not\equiv 1 \pmod{n}$ **then**
(5)   **return** COMPOSITE
(6) **else**
(7)   **for all** $d \geq 1$ s.t. $2^d \mid (n-1)$ **do**
(8)   **if** $1 < \gcd(a^{(n-1)/2^d} - 1, n) < n$ **then**
(9)   **return** COMPOSITE
(10)**else**
(11)   **return** PRIME.

Before stating our main theorem we collect some properties of this algorithm.

Lemma 8.31 *The running time of algorithm* `PrimalityTest` *is polynomial in* $\log n$.

Proof. This follows immediately from the above observation about the computation of $a^k \bmod n$ and Lemma 8.26.   □

Lemma 8.32 *Algorithm* `PrimalityTest` *always gives the correct answer if* $n \geq 3$ *is a prime number.*

Proof. If n is prime, then, clearly, the condition in line (2) will never be satisfied. Similarly, Fermat's theorem implies that the condition in line (4) will never be true. Finally, consider the condition in line (8). As there exists no integer in $2,\ldots,n-1$ that divides n, this condition will never be true. The algorithm will therefore correctly answer PRIME, regardless of the choice for a in line (1).   □

In order to show that `PrimalityTest` identifies composite numbers with sufficiently high probability, we start with a simple lemma about the multiplicative group $\mathbb{Z}_p^*$ for prime numbers p.

Lemma 8.33 *For every prime number p the multiplicative group $\mathbb{Z}_p^*$ contains a generator, i.e. an element a such that $\mathbb{Z}_p^* = \{a, a^2, \ldots, a^{p-1}\}$.*

As it is well known that the multiplicative group of every finite field is cyclic, this lemma is an immediate consequence of the fact that for prime numbers p the set $\mathbb{Z}_p$ with addition and multiplication modulo p is a field. For sake of completeness we also add a proof using just basic principles.

Proof. The *order* of an element $x \in \mathbb{Z}_p^*$ is the smallest integer $k \in \mathbb{N}$ such that $x^k \equiv 1 \pmod{p}$. Let a be an element in $\mathbb{Z}_p^*$ with maximum order, say m. We have to show that $m = p - 1$. To see this, we first show that the order of all elements in $\mathbb{Z}_p^*$ divides m. Assume not. That is, assume there exists an element $b \in \mathbb{Z}_p^*$ of order k such that $k \nmid m$. This can only be if there exists a prime number q such that

$$m = q^{e_m} \cdot m', \quad \gcd(m',q) = 1,$$
$$k = q^{e_k} \cdot k', \quad \gcd(k',q) = 1, \quad \text{and} \quad e_k > e_m.$$

Consider the element $c := a^{q^{e_m}} \cdot b^{k'}$ and let t denote its order. Then

$$1 \equiv c^{m't} \equiv a^{mt} \cdot b^{k'm't} \equiv b^{k'm't} \pmod{p}$$
$$1 \equiv c^{q^{e_k}t} \equiv a^{q^{e_m+e_k}t} \cdot b^{kt} \equiv a^{q^{e_m+e_k}t} \pmod{p}$$

As k is the order of b the first equivalence implies that $k \mid k'm't$ and hence in particular that $q^{e_k} \mid t$. Similarly, the second equivalence implies that $m \mid q^{e_m+e_k}t$ and hence that $m' \mid t$. This implies that $q \cdot m \mid t$, contradicting the fact that m was chosen as the maximum of the orders of the elements in $\mathbb{Z}_p^*$.

Now consider $S := \{a, a^2, \ldots, a^m\}$. Clearly, $|S| = m$ and every element of S is a solution of $x^m - 1 \equiv 0 \pmod{p}$. As a polynomial of degree m can have at most m different solutions, this shows that *every* solution of $x^m - 1 \equiv 0 \pmod{p}$ is contained in S. But, as the order of every element in $\mathbb{Z}_p^*$ divides m, we have $b^m \equiv 1 \pmod{p}$ for all elements $b \in \mathbb{Z}_p^*$, and hence $\mathbb{Z}_p^* \subseteq S$. This implies that $m = p - 1$, as desired.   □

Lemma 8.34 *If n is a composite number and there exists a prime p so that i) $p^2 \mid n$ or ii) $p \mid n$ and $(p-1) \nmid (n-1)$, then algorithm* PrimalityTest *will give the correct answer with probability at least $1/2$.*

Proof. Clearly, the answer will be correct whenever we happen to choose an integer $a \in \mathbb{Z}_n \setminus \mathbb{Z}_n^*$. It thus suffices to show that there are not too many a's in $\mathbb{Z}_n^*$ for which the algorithm gives a wrong answer. In fact we will show that

$$|\{a \in \mathbb{Z}_n^* \mid a^{n-1} \equiv 1 \pmod{n}\}| \leq \tfrac{1}{2}|\mathbb{Z}_n^*|. \tag{8.4}$$

Clearly, this suffices to prove the claim of the lemma.

Before we prove (8.4) we show the much weaker statement that there exists at least one $a_0 \in \mathbb{Z}_n^*$ so that $a_0^{n-1} \not\equiv 1 \pmod{n}$. Here we use our assumption on n. First assume that the prime p satisfies $p^2 \mid n$. Then $a_0 := 1 + \frac{n}{p}$ is the desired element: it is indeed relatively prime to n and $p^2 \mid n$ implies that

$$(a_0)^{n-1} \equiv (1 + \frac{n}{p}) \cdot \ldots \cdot (1 + \frac{n}{p}) \equiv 1 + (n-1) \cdot \frac{n}{p} \not\equiv 1 \pmod{n}.$$

For the second case let $r := (n-1) \bmod (p-1)$. By our assumption we have $r \neq 0$. In this case we show the existence of the element a_0 as follows. We choose an $x \in \mathbb{Z}_p^*$ such that $x^r \not\equiv 1 \pmod{p}$ (which exists according to Lemma 8.33) and apply the Chinese Remainder Theorem to find an integer $a_0 \in \mathbb{Z}_n$ such that

$$a_0 \equiv x \pmod{p} \quad \text{and}$$
$$a_0 \equiv 1 \pmod{\tfrac{n}{p}}.$$

Then $\gcd(a_0,n) = 1$. Furthermore, $(a_0)^{n-1} \equiv 1 \pmod{n}$ would imply that $(a_0)^{n-1}$ is also congruent to 1 modulo p, contradicting the fact that, according to Fermat's Theorem 8.30, $(a_0)^{n-1} \equiv x^{n-1} \equiv x^r \not\equiv 1 \pmod{p}$. That is, a_0 is the desired element in $\mathbb{Z}_n^*$.

Finally, we show (8.4). Our strategy is as follows: let $u_1, \ldots, u_k$ denote all the elements in $\mathbb{Z}_n^*$ that satisfy $u_i^{n-1} \equiv 1 \pmod{n}$. For every u_i we will construct a "partner" $w_i \in \mathbb{Z}_n^*$ such that the w_i's are pairwise different and $w_i^{n-1} \not\equiv 1 \pmod{n}$. Clearly, this implies that $k \leq \frac{1}{2}|\mathbb{Z}_n^*|$. Actually, the construction of the w_i's is very easy: we just let $w_i := a_0 u_i \bmod n$. Then $w_i - w_j \equiv a_0(u_i - u_j) \not\equiv 0 \pmod{n}$ for $i \neq j$, as a_0 is relatively prime to n and $|u_i - u_j| < n$. Furthermore, $(w_i)^{n-1} \equiv (a_0)^{n-1} \cdot (u_i)^{n-1} \pmod{n}$, and as $(a_0)^{n-1}$ is not congruent to one, but $(u_i)^{n-1}$ is, this implies that $(w_i)^{n-1} \not\equiv 1 \pmod{n}$, as desired.   □

Lemma 8.33 implies that algorithm `PrimalityTest` gives the correct answer with probability at least $1/2$ for "many" composite numbers. Perhaps even for all? While this might seem plausible at first sight, it is unfortunately not true. In fact, one can show that there do exist infinitely many integers so that *all* integers n $a \in \mathbb{Z}_n^*$ satisfy $a^{n-1} \equiv 1 \pmod{n}$. These are the so-called *Carmichael numbers*. In fact, the reader is invited to verify that we need lines (7)-(9) of the algorithm only for these Carmichael numbers numbers. (But, of course, the algorithm doesn't know whether n is a Carmichael number and therefore has to run through the lines (7)-(9) for all n).

Lemma 8.35 *If $n = p_1 \cdot \ldots \cdot p_r$ is an odd composite number such that the p_i's are pairwise different and $(p_i - 1) \mid (n - 1)$ for all $1 \leq i \leq r$, then algorithm* `PrimalityTest` *will give the correct answer with probability at least $1/2$.*

Proof. We will show that at least half of all a's in $\mathbb{Z}_n^*$ will satisfy the condition in line (8) and will therefore force the algorithm to return the correct answer COMPOSITE. Actually, we will see that it would suffice to perform the test in line (8) for just one (specific) d-value. But as this value depends on some properties of the prime numbers p_i the algorithm can't compute this "correct" d-value and has to try all of them.

Let t_i denote the largest integer such that $2^{t_i} \mid (p_i - 1)$. Without loss of generality we may assume that $1 \leq t_1 \leq t_2 \leq \cdots \leq t_r$. Let d denote the largest integer such that $2^{t_1 + d - 1} \mid (n - 1)$. Note that by assumption we have $2^{t_1} \mid (p_1 - 1)$ and $(p_1 - 1) \mid (n - 1)$. Hence, we know that d is at least one. Now we let $m := (n - 1)/2^d$ and observe that the definition of d implies that

$$(p_1 - 1) \mid 2m \qquad \text{but} \qquad (p_1 - 1) \nmid m. \tag{8.5}$$

From now on we will mimic the proof idea of Lemma 8.34. More specifically, let $(*)$ denote the condition

$$(*) \qquad\qquad 1 < \gcd(a^m - 1, n) < n.$$

We will first show that there exists an element a_0 that satisfies $(*)$ and then deduce that this implies that actually at least half of all elements in $\mathbb{Z}_n^*$ satisfy $(*)$.

To find the element a_0 we let $x \in \mathbb{Z}_{p_1}^*$ be a generator of $\mathbb{Z}_p^*$ (which exists according to Lemma 8.33) and observe that this implies that

$$x^s \equiv 1 \pmod{p_1} \qquad \text{if and only if} \qquad (p_1 - 1) \mid s. \qquad (8.6)$$

Now we apply the Chinese Remainder Theorem to find an integer $a_0 \in \mathbb{Z}_n$ such that

$$\begin{aligned} a_0 &\equiv x \pmod{p_1} \quad \text{and} \\ a_0 &\equiv 1 \pmod{\frac{n}{p_1}}. \end{aligned}$$

Then $\gcd(a_0, n) = 1$. Furthermore, $(a_0)^m \equiv x^m \not\equiv 1 \pmod{p_1}$, while $(a_0)^m \equiv 1 \pmod{p_2}$. That is, $p_2 \leq \gcd(a_0^m - 1, n) \leq n/p_1$, showing that a_0 is indeed an element in $\mathbb{Z}_n^*$ that satisfies $(*)$.

Finally, let $u_1, \ldots, u_k$ denote all the elements in $\mathbb{Z}_n^*$ that do not satisfy $(*)$. Consider $w_i := a_0 u_i \bmod n$. Then we easily check (in the same way as in the proof of Lemma 8.34) that the w_i's are pairwise different and belong to $\mathbb{Z}_n^*$. We claim that all the w_i's satisfy $(*)$. Consider w_i for an arbitrary $1 \leq i \leq k$. As u_i does not satisfy $(*)$ we know that either

$$u_i^m \equiv 1 \pmod{p_j} \qquad \text{for all } j = 1, \ldots, r,$$

or that

$$u_i^m \not\equiv 1 \pmod{p_j} \qquad \text{for all } j = 1, \ldots, r.$$

In the first case we deduce that $w_i^m \equiv a_0^m \cdot u_i^m \equiv a_0^m \not\equiv 1 \pmod{p_1}$, while $w_i^m \equiv a_0^m \cdot u_i^m \equiv 1 \pmod{p_2}$, implying that $p_2 \leq \gcd(w_i^m - 1, n) \leq n/p_1$, and hence proving that w_i satisfies $(*)$. To deduce the same fact for the second case we first observe that since x is a generator of $\mathbb{Z}_{p_1}^*$ there exists an $s \in \mathbb{N}$ so that $u_i \equiv x^s \pmod{p_1}$. As $x^{sm} \equiv u_i^m \not\equiv 1 \pmod{p_1}$, (8.5) and (8.6) imply that s is odd. We conclude, again by using (8.5) and (8.6), that $w_i^m \equiv a_0^m \cdot u_i^m \equiv x^m \cdot x^{sm} \equiv 1 \pmod{p_1}$, while $w_i^m \equiv a_0^m \cdot u_i^m \equiv u_i^m \not\equiv 1 \pmod{p_2}$, again implying that w_i satisfies $(*)$. □

Combining Lemmas 8.31, 8.32, 8.34, and 8.35 we deduce:

Theorem 8.36 *Algorithm* `PrimalityTest` *always gives the correct answer if n is prime and gives the correct answer with probability at least $1/2$ if n is composite. The running time of the algorithm is polynomial in $\log n$.* □

Cryptographic Protocols – RSA

As already mentioned in the introduction of this section, primality tests play a major rôle within cryptography. We close this section by outlining the basic principle of one of the fundamental and most well-known cryptographic protocols: RSA, named after its inventors Ron Rivest, Adi Shamir and Leonard Adleman.

RSA is a protocol which allows secure secret communication between two parties. In fact, the use of the word "secure" requires some explanation. Actually, RSA (and basically all other cryptographic protocols) are indeed easily breakable if we provide sufficient computing power. What we want in order to call a protocol *secure* is, that it is not breakable by an efficient, e.g. polynomial time, algorithm. At present, no such algorithm is known for breaking RSA. However, also the non-existence of such an algorithm has not been proven yet. In particular, it is *not* known that breaking RSA is $\mathcal{NP}$-hard. Nevertheless RSA is considered as one of the most secure means of secret communication.

An important property of RSA is that it does *not* need any private setup (like exchange of some secret keys) prior to its use. All the necessary initialization can be done in public over non-secure means of communication. Here is how it works. We have a *sender* **S** and a *receiver* **R**. We assume that **S** wants to send a message m to the receiver **R**. For sake of simplicity we assume that the message m consists just of a (not too large) positive integer. Sender and receiver communicate over an (insecure) channel, where all information can be intercepted by an eavesdropper, who nevertheless should not be able to gain any knowledge about the message m.

Sender **S**	Channel	Receiver **R**
1. Start the protocol by sending an initialization command;	$\xrightarrow{\text{init}}$	
	$\xleftarrow{n,\ k}$	2. Compute two large primes p and q, let $n = pq$ and compute a (random) element $k \in \mathbb{Z}_{\phi(n)}^* \setminus \{1\}$;
3. Compute $s := m^k \bmod n$;	$\xrightarrow{s}$	
		4. Compute the inverse l of k, i.e. an element $l \in \mathbb{Z}_{\phi(n)}^*$ such that $kl \equiv 1 \pmod{\phi(n)}$, and obtain $m := s^l \bmod n$;

Let us look at the single steps in some more detail. For computing each of p and q the receiver chooses randomly an odd number r of appropriate size (current recommendations of RSA Data Security, Inc. are to choose at least 768-bit numbers) and checks whether it is prime. If it is not, $r + 2$ is checked next, etc. The celebrated so-called prime number theorem states that there are approximately $n/\ln n$ prime numbers in the intervall 1 to n. Thus on average every $\ln(2^{768})$th 768-bit number is prime. That is, we may expect to find a prime number within the first 500 tries. Once p and q and hence $n = pq$ are found, $\phi(n)$ is easily determined as $\phi(n) = (p - 1)(q - 1)$, cf. Problem 8.13. The element k is computed as follows: we randomly choose an integer k in $\mathbb{Z}_n \setminus \{0,1\}$ and use Euclid's algorithm to check whether k is relatively prime to $\phi(n)$. If not, we choose a new candidate for k, etc. Observe that according to Corollary 8.28 the output of Euclid's algorithm contains for all k which are relatively prime to $\phi(n)$ also the inverse of k, i.e., the integer l of step 4.

To see that the protocol is correct we just have to verify that $s^l \bmod n$ is indeed equal to m. By the definition of s this is equivalent to $m^{kl} \equiv m \pmod{n}$. Recall that $kl = t\phi(n) + 1 = t(p - 1)(q - 1) + 1$ for an appropriate $t \in \mathbb{N}$. Fermat's Theorem 8.30 therefore implies that $m^{p-1} \equiv 1 \pmod{p}$ and hence $m^{kl} \equiv m \pmod{p}$ for all m that are not divisible by p. For m such that $p \mid m$ the latter fact holds trivially. Similarly, we obtain that for all m also $m^{kl} \equiv m \pmod{q}$ holds. As $n = pq$ we therefore deduce that also $m^{kl} \equiv m \pmod{n}$, as desired.

Finally, what can we say about the secureness of the RSA protocol? There are two ways to break it: either an eavesdropper (who knows k, n, and s) could try to factor n, which would enable him to calculate $\phi(n)$ and hence also l and $m = s^l \bmod n$, or he could try to find a value m such that $m \equiv s^k \pmod{n}$ directly. The latter problem is known as the *discrete logarithm problem*. Like the factoring problem it is believed to be difficult, even though this has not been proven yet.

RSA is a so-called public-key cryptosystem. It should be clear why. Instead of waiting for an initializing request from some (unknown) sender the receiver could also "post" his (public) keys k and n in advance. Then everyone who wants to send him a message only has to look up his public keys, and use them to encode the message. – Actually, posting public keys has also another advantage. It allows to use RSA for *authentificating* (instead of encrypting) messages as well. Here the sender **S** uses his private key (instead of the public key of the receiver) to create a "signed" version of his message. The receiver then uses the public key of the sender to recover the message, verifying at the same time that it is indeed **S**'s message.

Problems

8.1 Show that $\mathcal{NP} \subseteq \mathcal{BPP}$ implies that $\mathcal{RP} = \mathcal{NP}$.

8.2 (Example 8.1, cont.) Let p be a polynomial in n variables and let d be the maximum degree of each variable. Show that p is either identically zero or it has at most $\frac{1}{2}(2d)^n$ zeros in $\{1,2,\ldots,2d\}^n$.

8.3   Prove that a decision problem $\Pi = \langle \mathcal{I}, \text{Sol} \rangle$ is in $\mathcal{ZPP}$ if and only if there exists a randomized polynomial time algorithm $\mathcal{A}$ such that $\mathcal{A}(I) \in \{\text{ACCEPT}, \text{REJECT}, \text{DON'T KNOW}\}$ and such that

$$
\begin{array}{lll}
\Pr\left[\mathcal{A}(I) = \text{REJECT}\right] & = & 0, \quad \text{if } \text{Sol}(I) \neq \emptyset, \\
\Pr\left[\mathcal{A}(I) = \text{ACCEPT}\right] & = & 0, \quad \text{if } \text{Sol}(I) = \emptyset, \text{ and} \\
\Pr\left[\mathcal{A}(I) = \text{DON'T KNOW}\right] & \leq & \frac{1}{2}, \quad \text{for all } I \in \mathcal{I}.
\end{array}
$$

8.4   Deduce from Problem 8.3 that for every decision problem $\Pi = \langle \mathcal{I}, \text{Sol} \rangle$ in $\mathcal{ZPP}$ there exists a randomized algorithm that decides Π correctly and whose *expected* running time is polynomial in $|I|$ for every instance $I \in \mathcal{I}$.

8.5   Fill in the missing details of Example 8.5. In particular, give bounds for the number of necessary repetitions of each step. The overall probability of failure of your algorithm should be at most $1/2$. Determine the running time of the algorithm.

8.6   Given a graph $G = (V, E)$, a *minimum cut* is a partition of the vertex set $V = X \cup Y$ such that the number of edges with one endpoint in X and the other in Y is minimized. Design a randomized algorithm with error probability at most $1/2$ for finding such a minimum cut. (Hint: Proceed as follows. Choose an edge uniformly at random and contract it (removing loops but keeping multiple edges). Continue until the graph consists only of two vertices.)

8.7   Consider $p = \{\{i_1, i_2\}, \{i_3, i_4\}, \dots, \{i_{2j-1}, i_{2j}\}, \dots, \{i_{2n-1}, i_{2n}\}\}$,
      $p' = \{\{i_2, i_1\}, \{i_3, i_4\}, \dots, \{i_{2j-1}, i_{2j}\}, \dots, \{i_{2n-1}, i_{2n}\}\}$ and
      $p'' = \{\{i_{2j-1}, i_{2j}\}, \{i_3, i_4\}, \dots, \{i_1, i_2\}, \dots, \{i_{2n-1}, i_{2n}\}\}$.
      Show that the functions $\sigma(\cdot)$ and $\rho(\cdot)$ from page 141 satisfy
      a) $\sigma(p') = -\sigma(p)$, and $\rho(p') = -\rho(p)$.
      b) $\sigma(p'') = \sigma(p)$, and $\rho(p'') = \rho(p)$.

8.8   Deduce from Problem 8.7 that the pfaffian of a skew-symmetric matrix is well defined.

8.9   Prove Propositions 8.15 and 8.16.

8.10 It is not true that if Algorithm 8.20 outputs a spanning tree that this tree is then necessarily a *minimum* spanning tree. Verify this claim by providing a suitable example.

8.11 Consider the following generalization of Example 8.5 to arbitrary graphs. For a graph $G = (V, E)$ with vertex set $V = \{v_1, \dots, v_m\}$ define an $n \times n$ matrix $M = (m_{ij})$ as follows:

$$
m_{ij} = \begin{cases}
x_{ij}, & \text{if } \{v_i, v_j\} \in E, \text{ and } i < j \\
-x_{ji}, & \text{if } \{v_i, v_j\} \in E, \text{ and } i > j \\
0, & \text{otherwise.}
\end{cases}
$$

One can show that also in this case the determinant $\det(M)$ of the matrix M has the property that $\det(M)$ is identically zero if and only if G does not contain a perfect matching.

a) Use this fact to construct a randomized algorithm for finding a maximum matching in a graph $G = (V,E)$.

b) Combine this property with the techniques of Section 8.2 to develop a randomized pseudopolynomial time algorithm for finding a maximum matching in a network $N = (V,E,\ell)$.

8.12 Recall that in the proof of the upper bound of Theorem 8.22 we considered only very special Steiner trees. Argue that it is nevertheless not possible to further reduce the bound on p in the statement of Theorem 8.22.

8.13 Show that
a) $\phi(p^k) = (p-1)p^{k-1}$ for primes p.
b) $\phi(mn) = \phi(m)\phi(n)$ for integers m,n such that $\gcd(m,n) = 1$.
c) $\phi(p_1^{e_1} \cdots p_k^{e_k}) = \prod_{i=1}^{k}(p_i-1)p_i^{e_i-1}$ for pairwise different primes $p_1,\ldots,p_k$ and integers $e_1,\ldots,e_k \in \mathbb{N}$.

8.14 Complete the proof of Lemma 8.26.

8.15 Show that for the RSA protocol there exists for all primes p, q a "bad choice" for k. Namely, an integer $k \in \mathbb{Z}^*_{\phi(n)} \setminus \{1\}$ such that $m^k \equiv m \pmod{n}$ for all $m \in \mathbb{Z}_n$.

Notes

In fields like numerical analysis and statistical physics, randomized algorithms or *Monte Carlo methods*, as they are usually called in this context, have been used very successfully since the beginning of the computer age. The first rigorous definitions and models in the context of theoretical computer science can be found in an article by de Leeuw, Moore, Shannon, and Shapiro (1955). Their ideas were further pursued by Rabin (1963) and Gill (1977), who laid the foundation of the modern concepts of probabilistic complexity theory. Among the first examples of thoroughly analyzed randomized algorithms for concrete problems were two algorithms by Rabin (1980) and Solovay and Strassen (1977) for primality testing. Our algorithm in Section 8.4 is based on the first paper. For an in-depth treatment of many fascinating aspects of randomized algorithms we refer the reader to the excellent textbook by Motwani and Raghavan (1995). More information and references about probabilistic complexity theory can be found in the books by Bovet and Crescenzi (1994) and Papadimitriou (1994).

Example 8.1 is based on a paper by Schwartz (1980). Due to his work in this area the result from Problem 8.2 is also known as the 'Lemma of Schwartz'. The connection between matchings in graphs and matrix determinants was established by Edmonds (1967) for bipartite graphs and by Tutte (1947) for arbitrary graphs. In fact, one can even relate the size of a maximum matching to the rank of the matrix, see Lovász (1979), Rabin and

Vazirani (1989). In the latter paper one can also find an elegant algorithm for *finding* a maximum matching (which is slightly faster than our simple approach in Example 8.5). Note, however, that even though these randomized algorithms can easily be stated and analyzed, their running times are slower than those of the most efficient deterministic algorithms (which is $\mathcal{O}(n\sqrt{m})$, cf. Hopcroft and Karp (1973) for bipartite graphs and Micali and Vazirani (1980), Vazirani (1994) for arbitrary graphs).

As we have seen in Chapter 7, the problem of finding a minimum spanning tree in a k-uniform hypergraph is known to be $\mathcal{NP}$-hard whenever $k \geq 4$. The status of the case $k = 3$, however, is not yet completely decided. For the unweighted case Lovász (1978) provided a complicated $\mathcal{O}(n^{17})$ algorithm, which was later improved to an $\mathcal{O}(n^4)$ algorithm by Gabow and Stallmann (1986). In his paper connecting the maximum matching problem with matrix algebra, Lovász (1979) observed also the relation of the spanning tree problem in 3-uniform hypergraphs to the pfaffian of appropriately defined matrices (namely the one we presented in Section 8.2). As a consequence he obtained a simple randomized algorithm for the spanning tree problem in unweighted 3-uniform hypergraphs. This was later generalized by Camerini, Galbiati and Maffioli (1992) to a pseudopolynomial randomized algorithm for the weighted case. In fact, neither Lovász (1979) nor Camerini et al. (1992) did study the minimum spanning tree problem directly. They both considered the matroid parity problem over linearly representable matroids, which contains the minimum spanning tree problem in 3-uniform hypergraphs as a special case. The observation that these algorithms can be applied to the Steiner problem and the approximation schemes presented in Section 8.2 are from Prömel and Steger (2000).

Section 8.3 is based on a paper by Kucera, Marchetti-Spaccamela, Protasi, and Talamo (1986). For more information on results about random graphs from an algorithmic point of view we refer the reader to the survey by Frieze and McDiarmid (1997). The book by Habib, McDiarmid, Ramirez-Alfonsin, and Reed (1998) provides an excellent introduction and overview on the use of probabilistic tools for the design and analysis of combinatorial algorithms.

The existence of infinitely many Carmichael numbers was long an open problem. It was only demonstrated some years ago by Alford, Granville and Pomerance (1994). Assuming that the so-called *Generalized Riemann Hypothesis* is true, one can show that for composite odd numbers n there exist not only $\frac{1}{2}n$ many a's for which the primality tester of Section 8.4 returns the correct result "Composite", but that in addition at least one of these a's has size at most $2(\ln n)^2$. This allows to turn this randomized algorithm into a deterministic polynomial time algorithm by just checking all a's less than or equal to $2(\ln n)^2$. This result is due to Miller (1976). The fact that PRIMES is contained in $\mathcal{NP}$ (cf. Problem 3.4) was first shown by Pratt (1975).

The RSA cryptoscheme was invented by Rivest and Shamir and Adleman (1978). These authors also founded RSA Data Security, Inc., which is by now one of the leading companies in the encryption market. For more information on the theoretical background and applications of public-key cryptography we recommend the textbooks by Buchmann (2000) and Stinson (1995).

9

Limits of Approximability

In the previous chapters we went through considerable effort to design approximation algorithms for the Steiner problem with successively better performance ratios. On the other hand, we saw in Section 7.3 that MINIMUM STEINER PROBLEM IN GRAPHS does not admit a *fully* polynomial time approximation scheme. This, however, leaves open the question whether a polynomial time approximation scheme might exist. In this chapter we will see that this is, unless $\mathcal{P} = \mathcal{NP}$, in fact not the case. As MINIMUM STEINER PROBLEM IN GRAPHS belongs to the class $\mathcal{APX}$, this then locates the problem precisely within the hierarchy of optimization problems introduced in Section 7.3. In fact, we will push the limits of approximability even slightly further by showing that the MINIMUM STEINER PROBLEM IN GRAPHS is among the most difficult problems in $\mathcal{APX}$. The tools necessary to prove these results belong to the most celebrated achievements in theoretical computer science. They rely on a characterization of the class $\mathcal{NP}$ in terms of so-called probabilistically checkable proofs. We will explain this connection in Section 9.3.

9.1 Reducing optimization problems

The reference problem throughout this chapter which will lead us to the limits of the approximability of the MINIMUM STEINER PROBLEM IN GRAPHS is the MAX3SAT problem. An easy observation shows that the MAX3SAT problem admits a polynomial time approximation algorithm with constant performance bound. In fact, even MAXSAT has such an approximation algorithm.

Lemma 9.1 *There exists a linear time approximation algorithm for* MAXSAT *with performance ratio 2.*

Proof.   Let F be a Boolean formula in conjunctive normal form with variables $x_1,\dots,x_n$. In the ith step, we assign to x_i the value "true" if by this choice at least half of the clauses of F which contain x_i but none of the variables $x_1,\dots,x_{i-1}$ become true, and "false" otherwise. Obviously, this leads to a linear time approximation algorithm which satisfies at least $1/2$ of the clauses of F. This simple algorithm has thus a performance ratio of 2.   □

With slightly more care one can show that there also exists a polynomial time algorithm which finds for a Boolean formula in conjunctive normal form with *exactly three* variables in each clause a truth assignment which satisfies at least 7/8th of all clauses. The details are left to the reader, cf. Problem 9.1. It is one of the most surprising and spectacular results in theoretical computer science that such a simple approximation algorithm is, unless $\mathcal{P} = \mathcal{NP}$, essentially best possible. Before we state this result we start with a weaker one, which will imply that MAXSAT does not admit a polynomial time approximation algorithm.

Theorem 9.2 *There exists $\varepsilon > 0$ and two functions f and g, computable in polynomial time, such that for every instance $\varphi \in$ SAT the following properties are fulfilled:*

(1) $f(\varphi)$ *is an instance of* 3SAT.

(2) *If φ is satisfiable, then $f(\varphi)$ is satisfiable.*

(3) *If φ is not satisfiable, then every truth assignment satisfies at most $(1-\varepsilon)m$ clauses of $f(\varphi)$, where m denotes the number of clauses in $f(\varphi)$.*

(4)g *Given a truth assignment τ which satisfies more than $(1-\varepsilon)m$ clauses of $f(\varphi)$, $g(\varphi,\tau)$ is a satisfying truth assignment of φ.*

The proof of this highly nontrivial theorem is far beyond the scope of this book. In Section 9.3 we will at least explain some key ideas used in the proof. For now we just state an immediate consequence.

Corollary 9.3 MAX3SAT $\notin \mathcal{PTAS}$, *unless $\mathcal{P} = \mathcal{NP}$.*

Proof.   Assume there exists a polynomial time $(1 + \varepsilon)$-approximation algorithm $\mathcal{A}$ for MAX3SAT. For an arbitrary satisfiability instance φ we first transform it into a 3SAT instance $f(\varphi)$ using the function f from Theorem 9.2. We then apply the approximation algorithm $\mathcal{A}$ to $f(\varphi)$. As $\mathcal{A}$ is by assumption an $(1+\varepsilon)$-approximation algorithm, it has to satisfy the following two properties:

If φ and hence also $f(\varphi)$ is satisfiable, then $\mathcal{A}$ returns a truth assignment that satisfies at least $m/(1 + \varepsilon) > m(1 - \varepsilon)$ clauses.

If φ is not satisfiable, then every truth assignment, and hence in particular the one returned by algorithm $\mathcal{A}$, of $f(\varphi)$ satisfies at most $m(1 - \varepsilon)$ clauses.

That is, the outcome of algorithm $\mathcal{A}$ allows us to decide whether φ is satisfiable or not. As this is not possible unless $\mathcal{P} = \mathcal{NP}$, this concludes the proof of the corollary.   □

As already mentioned above, the result from Corollary 9.3 can be improved significantly.

Theorem 9.4 *Let $\varepsilon > 0$. Then there cannot exist a polynomial time approximation algorithm for* MAX3SAT *with performance ratio $8/7 - \epsilon$, unless $\mathcal{P} = \mathcal{NP}$.*

The proof of this theorem is again far beyond the scope of this book. See the notes for some references.

Let us now return to the Steiner problem. In order to show that it also does not admit a polynomial time approximation scheme, we will show that the MINIMUM STEINER PROBLEM IN GRAPHS is at least as difficult to approximate as MAX3SAT. The crucial point in doing so is to define a proper notion of a reduction between optimization problems. We start with some notation.

Notation 9.5 *Let $\Pi = \langle \mathcal{I}, \mathrm{val}, \mathrm{Sol}, \mathrm{goal} \rangle$ be an optimization problem, $I \in \mathcal{I}$ and $x \in \mathrm{Sol}(I)$. Then*

$$R_\Pi(I,x) := \max \left\{ \frac{\mathrm{opt}(I)}{\mathrm{val}(I,x)}, \frac{\mathrm{val}(I,x)}{\mathrm{opt}(I)} \right\}.$$

Definition 9.6 *An optimization problem $\Pi = \langle \mathcal{I}, \mathrm{Sol}, \mathrm{val}, \mathrm{goal} \rangle$ is AP-reducible to an optimization problem $\Pi^* = \langle \mathcal{I}^*, \mathrm{Sol}^*, \mathrm{val}^*, \mathrm{goal}^* \rangle$ – in symbols $\Pi \leq_{AP} \Pi^*$ – if and only if there exist functions f and g and a constant $\alpha > 0$ such that:*

(1) *For any $\delta > 0$, for any $I \in \mathcal{I}$, $f(I,\delta) \in \mathcal{I}^*$.*

(2) *For any $\delta > 0$, for any $I \in \mathcal{I}$, and $y \in \mathrm{Sol}^*(f(I,\delta))$, $g(I,y,\delta) \in \mathrm{Sol}(I)$.*

(3) *For any fixed $\delta > 0$, the functions f and g are computable in polynomial time.*

(4) *For any $I \in \mathcal{I}$, for any $\delta > 0$, and for any $y \in \mathrm{Sol}^*(f(I,\delta))$:*
    $R_{\Pi^*}(f(I,\delta),y) \le 1 + \delta \quad \Longrightarrow \quad R_\Pi(I,g(I,y,\delta)) \le 1 + \alpha \cdot \delta.$

The triple (f,g,α) is called an AP-reduction from Π to Π^*.

We first observe that the relation $\leq_{AP}$ is transitive.

Proposition 9.7 *Let* Π_0,Π_1 *and* Π_2 *be optimization problems. If* $\Pi_0 \leq_{AP} \Pi_1$ *and* $\Pi_1 \leq_{AP} \Pi_2$ *then* $\Pi_0 \leq_{AP} \Pi_2$.

The proof of this easy but nevertheless very useful observation is left to the reader, cf. Problem 9.2. Secondly, we need that the class $\mathcal{PTAS}$ is an ideal with respect to the relation $\leq_{AP}$. More precisely, the following is true.

Proposition 9.8 *Let* $\Pi = \langle \mathcal{I},\mathrm{Sol},\mathrm{val},\mathrm{goal}\rangle$ *and* $\Pi^* = \langle \mathcal{I}^*,\mathrm{Sol}^*,\mathrm{val}^*,\mathrm{goal}^*\rangle$ *be optimization problems. If* $\Pi^* \in \mathcal{PTAS}$ *and* $\Pi \leq_{AP} \Pi^*$ *then* $\Pi \in \mathcal{PTAS}$.

Proof. We have to show that for every $\varepsilon > 0$ there exists a polynomial time approximation algorithm for Π with performance ratio $1 + \varepsilon$. Fix some $\varepsilon > 0$.

Let (f,g,α) be an AP-reduction from Π to Π^* and choose $\delta = \varepsilon \cdot \alpha^{-1}$. Since $\Pi^* \in \mathcal{PTAS}$, there exists a polynomial time approximation algorithm, say $\mathcal{A}_\delta$, for Π^* with performance ratio $1 + \delta$. Let $I \in \mathcal{I}$ and consider the instance $f(I,\delta) \in \mathcal{I}^*$ which can be computed in polynomial time. $\mathcal{A}_\delta$, then, computes in polynomial time a solution $y \in \mathrm{Sol}^*(f(I,\delta))$ such that

$$R_{\Pi^*}(f(I,\delta),y) \ \leq \ 1+\delta.$$

But then the solution $g(I,y,\delta) \in \mathrm{Sol}(I)$, which can also be computed in polynomial time, has the property that

$$R_\Pi(I,g(I,y,\delta)) \ \leq \ 1+\alpha \cdot \delta \ = \ 1+\varepsilon.$$

$\square$

Combining Proposition 9.8 with Corollary 9.3 shows that proving the relation MAX3SAT $\leq_{AP}$ MINIMUM STEINER PROBLEM IN GRAPHS immediately yields that MINIMUM STEINER PROBLEM IN GRAPHS $\notin \mathcal{PTAS}$ (unless $\mathcal{P} = \mathcal{NP}$), the result which we are aiming for in this section. Hence, it remains to show that MAX3SAT admits an AP-reduction to the MINIMUM STEINER PROBLEM IN GRAPHS. This we will prove in two steps. First we show that MAX3SAT is AP-reducible to a problem called MAX3SAT(6). Subsequently, we define an AP-reduction from MAX3SAT(6) to MINIMUM STEINER PROBLEM IN GRAPHS. The intuitive reason why we have to take this detour is that in the reduction from a satisfiability problem to the Steiner tree problem we can only handle SAT-instances which have the property that the number of clauses is of the same order of magnitude as the number of variables.

MAX3SAT(k):

Given: A Boolean formula in conjunctive normal form such that every clause contains at most 3 variables and every variable occurs in at most k clauses in negated form and in at most k clauses in unnegated form.

Find: A truth assignment which satisfies a maximum number of clauses.

In order to define the reduction from MAX3SAT to MAX3SAT(6) we need some tools which also are of interest on their own.

Definition 9.9 *An (n,d,c)-expander is a d-regular bipartite graph $G = (A \cup B, E)$ with $|A| = |B| = n$ such that for every set $X \subseteq A$ its neighborhood has size at least $|X|(1 + c(1 - |X|/n))$.*

It is relatively easy to prove the *existence* of (n,d,c)-expanders (cf. Problem 9.3). Unfortunately, for our purpose just knowing of the existence of expanders does not suffice. If we want to use expanders in AP-reductions we need an algorithm which *constructs* them in polynomial time. In the following example we present an explicit construction of expanders. It is easy to see that this implies that expanders can indeed be constructed in polynomial time. Note, however, that the proof that this construction indeed defines an expander is far from trivial. In fact it is much too involved to be included here.

Example 9.10 CONSTRUCTION OF EXPANDERS. For every $n \in \mathbb{N}$ let $\mathbb{Z}_n$ denote the group on $\{0, \ldots, n-1\}$ with addition modulo n. Let A and B be two disjoint copies of $\mathbb{Z}_n \times \mathbb{Z}_n$. Connect every vertex (x,y) in A with the following (not necessarily different) five vertices in B: (x,y), $(x,x+y)$, $(x,x+y+1)$, $(x+y,y)$, and $(x+y+1,y)$. Then this graph satisfies the required expansion property with respect to the constant $c = (2 - \sqrt{3})/4$. By adding edges between vertices in A and B until all vertices have degree 5 we then obtain an $(n^2, 5, \frac{2-\sqrt{3}}{4})$-expander.

For a graph $G = (V, E)$ and a subset $X \subseteq V$ of vertices the *cut* induced by X is the set of edges which have exactly one endpoint in X. The definition of expanders implies that the cut of every subset $X \subseteq A$ of size $|X| \leq |A|/2$ contains at least $|X|(1 + c(1 - |X|/n)) \geq |X|(1 + \frac{1}{2}c)$ edges. Our next aim is to prove a similar result for arbitrary subsets X.

Proposition 9.11 *If $G = (A \cup B, E)$ is an expander from Example 9.10 and $X \subseteq A \cup B$ is a set of vertices such that $|X \cap A| \leq |A|/2$, then the cut induced by X contains at least $\frac{1}{24}|X \cap A|$ edges.*

Proof. Let $X_A = X \cap A$ and $X_B = X \cap B$. Furthermore, let ξ be the number of edges of the cut induced by X. As $|X_A| \leq |A|/2$ the construction in Example 9.10 implies that

$$|\Gamma(X_A)| \geq (1 + \tfrac{2-\sqrt{3}}{8})|X_A| \geq (1 + \tfrac{1}{30})|X_A|.$$

We now distinguish two cases. First assume that $|X_B| \leq (1 - \frac{1}{120})|X_A|$. Then

$$\xi \geq |\Gamma(X_A) \setminus X_B| \geq |\Gamma(X_A)| - |X_B| \geq (\tfrac{1}{30} + \tfrac{1}{120})|X_A| = \tfrac{1}{24}|X_A|.$$

Now assume that $|X_B| > (1 - \frac{1}{120})|X_A|$. Let $e(X_A, X_B)$ denote the number of edges with one endpoint in X_A and the other endpoint in X_B. As G is 5-regular, exactly $5 \cdot |X_A|$ edges leave the set X_A. At least $|\Gamma(X_A) \setminus X_B|$ of these have to end outside X_B. Hence,

$$
\begin{aligned}
e(X_A, X_B) &\leq 5|X_A| - |\Gamma(X_A) \setminus X_B| \leq 5|X_A| - |\Gamma(X_A)| + |X_B| \\
&\leq (4 - \tfrac{1}{30})|X_A| + |X_B|
\end{aligned}
$$

and therefore

$$
\begin{aligned}
\xi &= 5|X_A| + 5|X_B| - 2e(X_A, X_B) \geq (-3 + \tfrac{2}{30}) \cdot |X_A| + 3|X_B| \\
&\geq (\tfrac{2}{30} - \tfrac{3}{120}) \cdot |X_A| = \tfrac{1}{24}|X_A|
\end{aligned}
$$

in this case as well.   □

Definition 9.12 *A graph* $G = (V, E)$ *is an* amplifier *for a set* $S \subseteq V$ *if for all* $X \subseteq V$ *such that* $|S \cap X| \leq \frac{1}{2}|S|$ *the cut induced by* X *contains at least* $|S \cap X|$ *edges.*

Example 9.13 CONSTRUCTION OF AMPLIFIERS. Let S be a finite set of cardinality $|S| = n \geq 7$. For every vertex $s \in S$ we construct a complete 5-ary tree $T(s)$ of depth 2 with root s. (That is, $T(s)$ is a tree on 31 vertices with 25 leaves.) Let A denote the union of the leaves of these trees $T(s)$ together with some extra vertices so that A has cardinality $|A| = \lceil\sqrt{25n}\rceil^2$. From Example 9.10 we know how to construct an expander on the set A and an additional set B of the same cardinality. We claim that the graph $G(S)$ which is the union of the trees $T(s)$ and this expander is an amplifier for S. Let X be an arbitrary subset of vertices such that $|X \cap S| \leq \frac{1}{2}|S|$. With respect to X we define

$$
\begin{aligned}
X_1 &:= \{s \in X \cap S \mid T(s) \text{ is completely contained in } X\}, \\
X_2 &:= (X \cap S) \setminus X_1 \quad \text{and} \\
X_A &:= A \cap \bigcup_{s \in X_1} T(s).
\end{aligned}
$$

Then $|X_A| = 25|X_1| \leq \frac{25}{2}|S| = \frac{1}{2}|A|$. Proposition 9.11 therefore implies that at least $\frac{1}{24} \cdot |X_A| = \frac{25}{24}|X_1| > |X_1|$ edges of the expander are in the cut of X. By definition of X_2 we also know that for every vertex $s \in X_2$ at least one edge of the tree $T(s)$ is in the cut of the set X. Hence, at least $|X_1| + |X_2| = |X \cap S|$ edges of $G(S)$ are in the cut of X. This shows that $G(S)$ is indeed an amplifier for S.

Note that $G(S)$ has maximum degree 6, contains $6n + 2\lceil\sqrt{25n}\rceil^2 \leq 64n$ vertices and $30n + 5\lceil\sqrt{25n}\rceil^2 \leq 175n$ edges (the second inequality follows from $\lceil\sqrt{25n}\rceil^2 \leq (\sqrt{25n} + 1)^2 \leq 29n$ for $n \geq 7$), and has the property that every vertex in S has degree exactly 5.

Now we are in the position to prove the reductions we are aiming for.

Theorem 9.14 MAX3SAT $\leq_{AP}$ MAX3SAT(6).

Proof. The idea of the construction we are about to describe is the same as the one in the proof of Theorem 3.10. Namely, we replace each occurrence of a variable x by new variables $x^1, x^2, \ldots$ and add some new clauses in order to ensure that the variables $x^1, x^2, \ldots$ will all have the same truth assignment. The difference lies in these additional clauses. In the proof of Theorem 3.10 we added the clauses corresponding to the implications $(x^1 \Rightarrow x^2), (x^2 \Rightarrow x^k), \ldots, (x^k \Rightarrow x^1)$. This simple construction, however, will not work here. To see why, just observe that by violating a single of these implications, $(x^l \Rightarrow x^{l+1})$ say, we can set $x_1, \ldots, x_l$ to TRUE and $x^{l+1}, \ldots, x_k$ to FALSE, and may thus satisfy up to $k/2$ additional clauses.

Fix some $\delta > 0$ and consider an arbitrary instance I of MAX3SAT. For every variable x in I let k_x denote the number of clauses of I which contain either x or $\overline{x}$. If $k_x \geq 7$ we replace each occurrence of the variable x by k_x new variables $x^1, \ldots, x^{k_x}$, so that each of the variables x^i is used exactly once. To construct the additional clauses we combine for each of the (original) variables x the new variables x^i in a set $S_x = \{x^1, \ldots, x^{k_x}\}$ and construct an amplifier $G(S_x)$ for S_x according to Example 9.13. This amplifier has at most $175 \, k_x$ edges. Now, we define for each edge $\{x^i, x^j\}$ contained in the amplifier two new clauses $(x^i \vee \overline{x}^j)$ and $(x^j \vee \overline{x}^i)$. Let $I^* = f(I, \delta)$ be the resulting instance of MAX3SAT(6). Clearly, I^* is constructible in polynomial time from I.

Let m respectively m^* denote the number of clauses of I respectively I^*. From the construction of I^* we know that $m^* = m + \tilde{m}$, where $\tilde{m} \leq 350 \sum k_x$. Let y be a truth assignment for I^* and consider an arbitrary variable x of the original 3SAT instance I and its copies in the new instance I^*. A priori, there is no reason why the copies of a variable should all have the same truth value in y. We claim, however, that starting from y we can always find a truth assignment that satisfies at least as many clauses as y and that does have the property that all copies of the variable x have the same truth value. In order to do so consider what happens when we change the truth value of all variables in $G(S_x)$ to that of the majority of the variables in S_x (breaking ties arbitrarily). Assume this changes the truth value of $t \leq k_x/2$ variables in S_x. Then we may have "lost" at most t of the old clauses, but according to the definition of an amplifier we also "gained" at least t of the new clauses. Note that this argument implies in particular that $\mathrm{opt}(I^*) = \mathrm{opt}(I) + \tilde{m}$.

Assume now that a truth assignment y of I^* is given that satisfies at least $\mathrm{opt}(I^*)/(1 + \delta)$ clauses of I^*. We have to show that starting from y we can construct a truth assignment for I which satisfies at least $\mathrm{opt}(I)/(1 + \alpha\delta)$ of the clauses in I, for an appropriate constant α. We will see that choosing $\alpha = 4201$ will do. Note that since there is a simple polynomial time algorithm with performance ratio $1/2$ for MAX3SAT, cf. Lemma 9.1, we may hence assume without loss of generality that $\delta \leq 1/4201$.

Recall that we already convinced ourselves that we may assume without loss of generality that y has the property that all variables within the sets S_x have the same truth value. By taking these truth values for the variables x we deduce that y induces a truth assignment $g(I,y,\delta)$ for I that satisfies exactly

$$\mathrm{val}^*(I^*,y) - \tilde{m} \geq \frac{\mathrm{opt}(I^*)}{1+\delta} - \tilde{m} = \frac{\mathrm{opt}(I) + \tilde{m}}{1+\delta} - \tilde{m} = \frac{\mathrm{opt}(I) - \delta\tilde{m}}{1+\delta}$$

clauses. Using that $\tilde{m} \leq 350 \sum k_x$, that $\sum k_x \leq 3m$, and the fact that $\mathrm{opt}(I) \geq \frac{1}{2}m$ (cf. Lemma 9.1), we further deduce that

$$\frac{\mathrm{opt}(I) - \delta\tilde{m}}{1+\delta} \geq \frac{\mathrm{opt}(I) \cdot (1 - 2100\delta)}{1+\delta} \geq \frac{\mathrm{opt}(I)}{1 + 4201\delta}$$

for all $\delta \leq 1/4201$, which is the desired bound. $\square$

Theorem 9.15

$$\text{MAX3SAT}(6) \leq_{AP} \text{MINIMUM STEINER PROBLEM IN GRAPHS.}$$

Proof. We reduce MAX3SAT(6) to MINIMUM STEINER PROBLEM IN GRAPHS using essentially the same reduction as in the proof of the $\mathcal{NP}$-completeness of the Steiner tree problem (cf. Theorem 3.9).

Let $x_1,\ldots,x_n$ be the variables and $C_1,\ldots,C_m$ be the clauses in an arbitrary instance I of MAX3SAT(6) and fix some $\delta > 0$. Without loss of generality we will assume that $n \leq m$. Then $f(I,\delta)$ consists of a pair (G,K) where the graph G has the following structure:

- a *variable line* with one honeycomb per variable; the two sides of each honeycomb consist of 6 vertices each, one side representing the unnegated use and the other the negated use of the variable;
- for each clause a vertex C_j which is connected by paths of length 7 to the appropriate sides of the honeycombs corresponding to the variable of the clause – thereby ensuring that each variable of a honeycomb is connected to at most one clause vertex;
- for each clause a path of length 8 connecting the vertex C_j to the vertex z_0.

Figure 9.1 illustrates this construction. As terminal set K we choose

$$K = \{z_0, z_1, \ldots, z_n\} \cup \{C_1, \ldots, C_m\}.$$

We claim that the length $smt(G,K)$ of a Steiner minimum tree in G for K is exactly $7n + 8m - \mathrm{opt}(I)$.

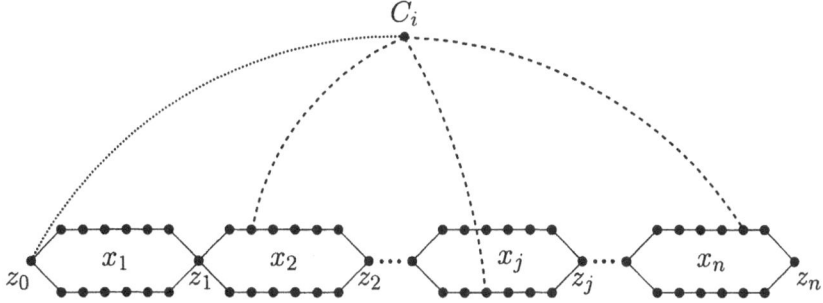

Figure 9.1 Construction of the graph G. Shown are the honeycombs for the variables $x_1,\ldots,x_n$ and the clause $C_i = x_2 \vee \bar{x}_j \vee x_n$. The dashed lines indicate paths of length 7, the dotted line represents a path of length 8.

To see this we first construct a Steiner tree for K of length $7n + 8m - \text{opt}(I)$. We start with a z_0-z_n path P along the variable line corresponding to an optimum assignment for I. This requires $7n$ edges. Next we connect every clause that is satisfied by this assignment to the variable line, more precisely, to a variable that is satisfied. This requires $7 \cdot \text{opt}(I)$ edges. Finally, we connect every unsatisfied clause to z_0. This requires $8 \cdot (m - \text{opt}(I))$ edges.

To see the other direction, let T be a Steiner minimum tree for K. We first claim that we may assume without loss of generality that T contains a z_0-z_n path along the variable line. Assume not. Then consider the components of T restricted to the variable line and add a path between two successive z-vertices in different components to T. This requires at most 7 edges and the path induces a cycle which contains at least two paths between the variable line and clause vertices. As these paths all have length 7, removing any of these paths yields a Steiner tree for K whose restriction to the variable line contains one less component than T. Next we claim that in addition we may also assume without loss of generality that T restricted to the variable line is *exactly* a z_0-z_n path. Assume not. Then the restriction of T to the variable line has to contain at least one leaf v which is not a z-vertex. As v is not a terminal, this implies that v has to be connected to a vertex C_j in T. Remove this connection (that is, the edge connecting v to another vertex on the variable line and the 7 edges of the path from v to C_j) and insert instead the 8 edges of the path from z_0 to C_j. This shows that T has at least $7n + 8m - \text{opt}(I)$ edges.

Obviously, the graph G can be constructed in polynomial time. To see that approximative solutions for the Steiner problem yield approximative solutions for the satisfiability problem, let $\delta > 0$ be an arbitrary constant and T be a Steiner tree in G such that $|T| \le (1 + \delta)smt(G,K)$. From the arguments above it follows that we may assume without loss of generality that the restriction of T to the variable line is exactly a z_0-z_n path. As a truth assignment $g(I,T,\delta)$ for the 3SAT instance I we choose the one corresponding to this path. Let val(I) denote the number of clauses satisfied by this assignment. Then $|T| = 7n + 8m - \text{val}(I)$ and therefore $|T| \leq (1 + \delta)smt(G,K)$ implies

$$\frac{\text{val}(I)}{\text{opt}(I)} = \frac{7n + 8m - |T|}{\text{opt}(I)} \geq \frac{7n + 8m - (1 + \delta)smt(G,K)}{\text{opt}(I)} = 1 - \delta \cdot \frac{smt(G,K)}{\text{opt}(I)}.$$

Trivially, $smt(G,K) \leq 7n + 8m \leq 15m$ and $\text{opt}(I) \geq \frac{1}{2}m$. Thus,

$$\frac{\text{val}(I)}{\text{opt}(I)} \geq 1 - 30\delta \geq \frac{1}{1 + 60\delta},$$

for $\delta \leq \frac{1}{60}$. For $\delta > \frac{1}{60}$ we apply the polynomial time algorithm from Lemma 9.1 to obtain an even better bound. □

Corollary 9.16 *The* MINIMUM STEINER PROBLEM IN GRAPHS *admits no polynomial time approximation scheme, unless* $\mathcal{P} = \mathcal{NP}$. □

This corollary can also be rephrased as follows. There exists a constant $c_0 > 1$ such that the existence of a polynomial time approximation algorithm for the MINIMUM STEINER PROBLEM IN GRAPHS with performance ratio c_0 implies $\mathcal{P} = \mathcal{NP}$. Using Theorem 9.4 and working through the proofs of Theorem 9.14 and Theorem 9.15 shows that e.g. $c_0 = \frac{1}{7} \cdot \frac{1}{4201} \cdot \frac{1}{60} \approx 5.67 \cdot 10^{-7}$ has this property. With more care one can improve this bound considerably, see Problem 9.9 and the notes to this chapter.

9.2 $\mathcal{APX}$–completeness

The last section was devoted to proving that the MINIMUM STEINER PROBLEM IN GRAPHS does not admit a polynomial time approximation scheme, provided $\mathcal{P} \neq \mathcal{NP}$. In this section we will show that the Steiner problem is indeed among the most difficult problems in $\mathcal{APX}$.

Definition 9.17 *An optimization problem* $\Pi^* \in \mathcal{APX}$ *is* $\mathcal{APX}$-*complete if for any other problem* $\Pi \in \mathcal{APX}$ *we have* $\Pi \leq_{AP} \Pi^*$.

Before we proceed, let us recall some facts about $\mathcal{NP}$-completeness. In order to show that a decision problem is $\mathcal{NP}$-complete, we just have to reduce an arbitrary other $\mathcal{NP}$-complete problem to it. Note, however, that this approach relies on the fact, that there do exist $\mathcal{NP}$-complete problems at all. Finding a first $\mathcal{NP}$-complete problem was considerably more work, cf. Theorem 3.4. For $\mathcal{APX}$-completeness we will proceed similarly. First we show that there exists some

$\mathcal{APX}$-complete problem. This then allows us to obtain other $\mathcal{APX}$-complete problems simply by reducing them to this problem. The problem which we will first show to be $\mathcal{APX}$-complete is the optimization variant MAXSAT of the problem SATISFIABILITY which was shown to be $\mathcal{NP}$-complete by Cook, cf. Theorem 3.4.

Theorem 9.18 MAXSAT *is* $\mathcal{APX}$*-complete.*

Since we have already seen that MAXSAT or, more precisely, its special case MAX3SAT, is AP-reducible to MINIMUM STEINER PROBLEM IN GRAPHS (Theorem 9.14 and Theorem 9.15), Theorem 9.18 immediately implies that also the MINIMUM STEINER PROBLEM IN GRAPHS is $\mathcal{APX}$-complete.

Corollary 9.19 MINIMUM STEINER PROBLEM IN GRAPHS *is* $\mathcal{APX}$*-complete.*

The remainder of this section is devoted to the proof of Theorem 9.18. In a first step we will show that we can restrict our consideration to maximization problems.

Lemma 9.20 *For every minimization problem* $\Pi \in \mathcal{APX}$ *there exists a maximization problem* $\Pi^* \in \mathcal{APX}$ *such that* $\Pi \leq_{AP} \Pi^*$.

Proof. Let $\Pi = \langle \mathcal{I}, \text{Sol}, \text{val}, \min \rangle$ be a minimization problem in $\mathcal{APX}$. By definition of $\mathcal{APX}$, there exists some constant $\rho \geq 1$ such that Π admits a polynomial time approximation algorithm $\mathcal{A}$ with performance ratio ρ. Without loss of generality we assume that $\rho \in \mathbb{N}$.

For $I \in \mathcal{I}$, we denote by $\mathcal{A}(I)$ the solution $x \in \text{Sol}(I)$ which is generated by $\mathcal{A}$. We define a maximization problem $\Pi^* = \langle \mathcal{I}^*, \text{Sol}^*, \text{val}^*, \max \rangle$ by $\mathcal{I}^* := \mathcal{I}$, $\text{Sol}^* := \text{Sol}$, and

$$\text{val}^*(I,x) := \max\{1, (1+\rho)\text{val}(I, \mathcal{A}(I)) - \rho \cdot \text{val}(I,x)\}.$$

As val is computable in polynomial time, val^* is also computable in polynomial time, implying that $\Pi^* \in \mathcal{NPO}$. To see that in fact $\Pi^* \in \mathcal{APX}$, we have to exhibit an approximation algorithm $\mathcal{A}^*$ with constant performance ratio. We claim that algorithm $\mathcal{A}$ is such an algorithm. To see this, observe that

$$\text{val}^*(I, \mathcal{A}(I)) = \max\{1, (1+\rho)\text{val}(I, \mathcal{A}(I)) - \rho \cdot \text{val}(I, \mathcal{A}(I))\} = \text{val}(I, \mathcal{A}(I))$$

(recall that $\text{val}(\cdot)$ is positive by definition) and that

$$\text{opt}^*(I) = (1+\rho)\text{val}(I, \mathcal{A}(I)) - \rho \cdot \text{opt}(I) \leq (1+\rho)\text{val}(I, \mathcal{A}(I)) \qquad (9.1)$$

(as $\text{opt}(I)$ is positive). This implies

$$\frac{\text{val}^*(I,\mathcal{A}(I))}{\text{opt}^*(I)} \geq \frac{1}{1+\rho}.$$

That is, $\mathcal{A}$ is an approximation algorithm for Π^* with performance ratio $1+\rho$.

It remains to show that $\Pi \leq_{AP} \Pi^*$. We claim that $(f,g,1+\rho)$, where

$$f(I,\delta) := I, \quad \text{and}$$

$$g(I,y,\delta) := \begin{cases} y, & \text{if } \text{val}(I,y) \leq \text{val}(I,\mathcal{A}(I)) \\ \mathcal{A}(I), & \text{otherwise} \end{cases}$$

is an AP-reduction from Π to Π^*. Conditions (1)-(3) from Definition 9.6 are obvioulsy satisfied. So it remains to verify that (4) holds as well. That is, we have to verify that for every $\delta > 0$:

$$\frac{1}{1+\delta} \leq \frac{\text{val}^*(I,y)}{\text{opt}^*(I)} \implies \frac{\text{val}(I,g(I,y,\delta))}{\text{opt}(I)} \leq 1 + (1+\rho)\delta.$$

To see this observe first that the definitions of val^* and g imply

$$\text{val}^*(I,y) \leq (1+\rho)\text{val}(I,\mathcal{A}(I)) - \rho \cdot \text{val}(I,g(I,y,\delta)).$$

From the assumption $\text{val}^*(I,y)/\text{opt}^*(I) \geq 1/(1+\delta)$ we can therefore deduce that

$$
\begin{aligned}
\frac{\text{val}(I,g(I,y,\delta))}{\text{opt}(I)}
&\leq \frac{1+\delta}{\rho} \cdot \frac{\rho \cdot \text{val}(I,g(I,y,\delta))}{\text{opt}(I)} \\
&\leq \frac{1+\delta}{\rho} \cdot \frac{(1+\rho)\text{val}(I,\mathcal{A}(I)) - \text{val}^*(I,y)}{\text{opt}(I)} \\
&\leq \frac{1+\delta}{\rho} \cdot \left[\frac{(1+\rho)\text{val}(I,\mathcal{A}(I))}{\text{opt}(I)} - \frac{\text{opt}^*(I)}{(1+\delta)\text{opt}(I)} \right] \\
&\overset{(9.1)}{=} 1 + \frac{\delta(1+\rho)}{\rho}\frac{\text{val}(I,\mathcal{A}(I))}{\text{opt}(I)} \leq 1 + \delta(1+\rho),
\end{aligned}
$$

where the last inequality follows from the fact that $\mathcal{A}$ is an approximation algorithm for Π with performance ratio ρ. This proves that $(f,g,1+\rho)$ is indeed an AP-reduction and concludes the proof of Lemma 9.20. □

Using Lemma 9.20 and Theorem 9.2 we are now able to prove Theorem 9.18.

Proof of Theorem 9.18. By Lemma 9.20 and the transitivity of $\leq_{AP}$, it suffices to prove that for any maximization problem $\Pi = \langle \mathcal{I}, \text{Sol}, \text{val}, \max \rangle \in \mathcal{APX}$ we have that $\Pi \leq_{AP} \text{MAX3SAT}$. So consider an arbitrary maximization problem $\Pi = \langle \mathcal{I}, \text{Sol}, \text{val}, \max \rangle$. By definition of $\mathcal{APX}$, we know that there exists an approximation algorithm $\mathcal{A}$ for Π with performance ratio ρ for some constant $\rho \geq 1$. Let $\varepsilon > 0$ be the constant from Theorem 9.2. Based on these two constants we will later define a suitable constant $\alpha = \alpha(\rho,\varepsilon)$. Let $\delta > 0$ be given. We have to construct suitable functions f and g. First assume that δ is large enough so that $\rho \leq 1 + \alpha \cdot \delta$. In this case the approximation algorithm $\mathcal{A}$ already yields the required performance ratio. To be formally correct, we may map every instance I to some trivial MAX3SAT instance, say $f(I,\delta) \equiv x_1$ and define $g(I,y,\delta) := \mathcal{A}(I)$. Then

$$\frac{\text{opt}(I)}{\text{val}(I,g(I,y,\delta))} = \frac{\text{opt}(I)}{\text{val}(I,\mathcal{A}(I))} \le \rho \le 1 + \alpha \cdot \delta,$$

and therefore the AP-condition is satisfied.

In the following we may therefore assume that $\delta > 0$ is given such that $\rho > 1 + \alpha \cdot \delta$. To simplify notation, put $c := 1 + \alpha \cdot \delta$, $k := \lceil \log_c \rho \rceil$ and $\text{val}(\mathcal{A}_I) := \text{val}(I,\mathcal{A}(I))$. We partition the interval $[\text{val}(\mathcal{A}_I),\rho\text{val}(\mathcal{A}_I)]$ into k subintervals as follows:

$$[\text{val}(\mathcal{A}_I),c \cdot \text{val}(\mathcal{A}_I)],[c \cdot \text{val}(\mathcal{A}_I),c^2 \cdot \text{val}(\mathcal{A}_I)],..,[c^{k-1} \cdot \text{val}(\mathcal{A}_I),\rho \cdot \text{val}(\mathcal{A}_I)]$$

Since $\text{val}(\mathcal{A}_I) \le \text{opt}(I) \le \rho \cdot \text{val}(\mathcal{A}_I)$, the optimum value $\text{opt}(I)$ belongs to one of the above subintervals. For $i = 0,\ldots,k-1$ consider the $\mathcal{NP}$-problem Π_i of deciding whether

$$\text{opt}(I) \ge c^i \cdot \text{val}(\mathcal{A}_I).$$

Since $\Pi_i \le_p$ 3SAT, Cook's Theorem 3.4 resp. Corollary 3.5 implies that we can compute for every $I \in \mathcal{I}$ in polynomial time a 3SAT instance $\varphi_i := \varphi_i(I)$ such that given a truth assignment σ_i satisfying φ_i, we can compute in polynomial time a solution $x \in \text{Sol}(I)$ so that $\text{val}(I,x) \ge c^i\text{val}(\mathcal{A}_I)$. Next we use Theorem 9.2 in order to compute for every φ_i another 3SAT formula $\psi_i := f(\varphi_i)$. Finally, we define the 3SAT formula $f(I,\delta)$ as

$$f(I,\delta) := \bigwedge_{i=0}^{k-1} \psi_i.$$

Note that, since k is constant, $\psi := f(I,\delta)$ can be computed in polynomial time. In the following we assume that each ψ_i contains the same number, say m, clauses. We can always achieve this by taking sufficiently many copies of each ψ_i. Furthermore, we denote by i_0 the maximum index i such that ψ_i is satisfiable. Note that by our construction this implies that

$$c^{i_0}\text{val}(\mathcal{A}_I) \le \text{opt}(I) < c^{i_0+1}\text{val}(\mathcal{A}_I).$$

Let τ be any truth assignment to the variables of ψ such that

$$\frac{\text{opt}(\psi)}{\text{val}(\psi,\tau)} \le 1 + \delta. \tag{9.2}$$

Assume that for some index i the restriction τ_i of the assignment τ to the variables in ψ_i satisfies

$$\text{val}(\psi_i,\tau_i) \ge (1 - \varepsilon)m. \tag{9.3}$$

According to Theorem 9.2 this can only happen if ψ_i is satisfiable (i.e., if $i \le i_0$). Moreover, Theorem 9.2 also implies that starting from τ_i we can compute in polynomial time a satisfying assignment σ_i for φ_i. As already mentioned above, Cook's Theorem implies that we can then compute, again in polynomial time, a solution $x \in \text{Sol}(I)$ such that $\text{val}(I,x) \ge c^i\text{val}(\mathcal{A}_I)$.

That is, if we can show that (9.3) holds for $i = i_0$, we can compute in polynomial time a solution $x = g(I,\tau,\delta) \in \mathrm{Sol}(I)$ such that

$$\frac{\mathrm{opt}(I)}{\mathrm{val}(I,x)} \leq \frac{c^{i_0+1}\mathrm{val}(\mathcal{A}_I)}{c^{i_0}\mathrm{val}(\mathcal{A}_I)} = c = 1 + \alpha \cdot \delta,$$

i.e., condition (4) of Definition 9.6 would be satisfied.

Thus, in order to complete the proof, it remains to show that $\mathrm{val}(\psi_{i_0},\tau_{i_0}) \geq (1 - \varepsilon)m$. According to (9.2), τ is a truth assignment for the variables of ψ such that

$$\mathrm{opt}(\psi) - \mathrm{val}(\psi,\tau) \leq \frac{\delta}{1 + \delta} \mathrm{opt}(\psi) \leq \frac{\delta}{1 + \delta} k \cdot m.$$

On the other hand, defining ξ by $\mathrm{val}(\psi_{i_0},\tau_{i_0}) = (1 - \xi)m = (1 - \xi)\mathrm{opt}(\psi_{i_0})$, we get

$$\mathrm{opt}(\psi) - \mathrm{val}(\psi,\tau) = \underbrace{\sum_{i \neq i_0}(\mathrm{opt}(\psi_i) - \mathrm{val}(\psi_i,\tau_i))}_{\geq 0} + \underbrace{\mathrm{opt}(\psi_{i_0}) - \mathrm{val}(\psi_{i_0},\tau_{i_0})}_{= \xi m}.$$

Combining these two inequalities, we obtain

$$\xi \leq \frac{\delta}{1 + \delta} \cdot k.$$

It thus suffices to show that we can define α (and thus $k = \lceil \ln \rho / \ln(1 + \alpha\delta) \rceil \leq 2 \ln \rho / \ln(1 + \alpha\delta))$ in such a way that

$$\frac{\delta}{1 + \delta} \cdot k \leq \frac{\delta}{1 + \delta} \cdot \frac{2 \ln \rho}{\ln(1 + \alpha\delta)} < \varepsilon \qquad \text{for all } \delta > 0.$$

As

$$\frac{2\delta}{1 + \delta} \cdot \frac{\ln \rho}{\ln(1 + \alpha\delta)} < \varepsilon \qquad \Longleftrightarrow \qquad \frac{\rho^{\frac{2\delta}{\varepsilon(1+\delta)}} - 1}{\delta} < \alpha$$

and $\frac{1}{x}(\rho^{\frac{2x}{\varepsilon(1+x)}} - 1)$ is monotone decreasing for all sufficiently large x and converges to a constant for $x \to 0$, such an α obviously exists. This completes the proof of Theorem 9.18.   □

9.3   Excursion: Probabilistically checkable proofs

On April 7, 1992 the following article appeared in **The New York Times**.

New Shortcut Found For Long Math Proofs

In a discovery that overturns centuries of mathematical tradition, a group of graduate students and young researchers has discovered a way to check even the longest and most complicated proof by scrutinizing it in just a few spots. [...]

Using this new result, the researchers have already made a landmark discovery in computer science. They showed that it is impossible to compute even approximate solutions for a large group of practical problems that have long foiled researchers. [...]

The aim of this article was to popularize a characterization of the class $\mathcal{NP}$ which can formally be phrased as $\mathcal{NP} = \mathcal{PCP}(\log n, 1)$. As we will see shortly, this is indeed a quite amazing result with far-reaching consequences. In order to roughly explain it, let us first informally recall the definition of the class $\mathcal{NP}$. A decision problem is in the class $\mathcal{NP}$ if there exists an algorithm which can check in polynomial time whether a string x is a solution for a given instance I. Recall that it is important that the algorithm just has to *check* a given solution x. It does not have to *find* such a solution. It therefore makes sense to call such a solution $x \in \text{Sol}(I)$ a *proof* for the fact that $\text{Sol}(I) \neq \emptyset$. The fact that a decision problem is in the class $\mathcal{NP}$ then simply means that these proofs can be checked (or, as we will henceforth also say, *verified*) in polynomial time.

Let us look at some examples. Typical decision problems in $\mathcal{NP}$ are the satisfiability problem, the knapsack problem, or the Steiner tree problem. Here the "proofs" are simple and straightforward. They just consist of a satisfying assignment, a partition of the items, or a Steiner minimum tree. In what follows, it is important to note that these proofs do not contain much redundant information. Consider for example the satisfiability problem. Surely, to distinguish a satisfying assignment from one that satisfies all but one clause one usually has to read the truth value of every variable, i.e., the whole proof.

This is not the case with *probabilistically checkable proofs*. Probabilistically checkable proofs are inspected by *verifiers* (polynomial time algorithms) which proceed as follows. After reading the instance I they generate some random bits. They use these random bits to decide which bits (positions) of the proof they want to read. Subsequently, they either accept the instance I or reject it — based only on the knowledge of the (few) queried bits! A decision problem is said to have a probabilistically checkable proof if for all YES-instances I there exists a proof π_I which the verifier accepts for *all* possible outcomes of the random bits, while for all NO-instances the verifier rejects *all* proofs with probability at least one half.

At first it may seem impossible to construct probabilistically checkable proofs for problems in $\mathcal{NP}$, which can be checked by reading only a constant number of bits. (Try it!) As we will see later on in this chapter, this is, surprisingly enough, not too difficult, at least not if we allow the proof to be arbitrarily long. Highly non-trivial, however, is the fact that every problem in $\mathcal{NP}$ even has a proof of polynomial length with the same property. More precisely, every YES-instance I of length $n := |I|$ admits a proof of length polynomial in n which can be checked probabilistically by reading only a constant number of bits from it. Note that in order to choose a bit from a proof of length t we just need to specify $\lceil \log_2 t \rceil$ bits. That is, in order to specify constantly many positions of a proof of length polynomial in n it suffices to generate $\mathcal{O}(\log n)$ random bits. This explains, roughly, the essence of the "$\mathcal{NP} = \mathcal{PCP}(\log n, 1)$" result: the verifier generates $\mathcal{O}(\log n)$ random bits and reads $\mathcal{O}(1)$ bits from the proof. We now make these definitions precise, compare Figure 9.2 for a visual illustration.

Definition 9.21 *Let $r(n)$ and $q(n)$ be positive integral functions. An $(r(n), q(n))$-restricted verifier for a decision problem $\langle \mathcal{I}, \text{Sol} \rangle$ is an algorithm $\mathcal{V}$ that has access to an input I, a string τ of random bits, and a proof π such that for every input I of length $n := |I|$ the verifier $\mathcal{V}$ reads only the first $\mathcal{O}(r(n))$ bits from τ and reads at most $\mathcal{O}(q(n))$ positions of the proof π.*

Such a verifier is said to decide *$\langle \mathcal{I}, \text{Sol} \rangle$ if for every input I of length $n = |I|$ the verifier $\mathcal{V}$ returns in time polynomial in n either ACCEPT or REJECT such that*

$$\text{Sol}(I) \neq \emptyset \quad \Longrightarrow \quad \exists \pi_0 : \Pr[\mathcal{V}(I, \tau, \pi_0) = \text{ACCEPT}] = 1,$$

and

$$\text{Sol}(I) = \emptyset \quad \Longrightarrow \quad \forall \pi : \Pr[\mathcal{V}(I, \tau, \pi) = \text{REJECT}] \geq \frac{1}{2}.$$

(Here the probability is with respect to the random string τ, assuming that all such 0-1 strings are equally likely.)

The class $\mathcal{PCP}(r(n), q(n))$ denotes the set of all decision problems $\langle \mathcal{I}, \text{Sol} \rangle$ that can be decided by an $(r(n), q(n))$-restricted verifier.

The functions $r(n)$ and $q(n)$ make the definition of the classes $\mathcal{PCP}(\cdot, \cdot)$ rather general. In particular, it contains well-known classes like $\mathcal{P}$ or $\mathcal{NP}$ as special cases.

Lemma 9.22
$$\mathcal{P} = \mathcal{PCP}(0, 0),$$
$$\mathcal{NP} = \mathcal{PCP}(0, poly(n)) := \bigcup_{k \geq 1} \mathcal{PCP}(0, n^k), \text{ and}$$
$$co\mathcal{RP} = \mathcal{PCP}(poly(n), 0) := \bigcup_{k \geq 1} \mathcal{PCP}(n^k, 0).$$

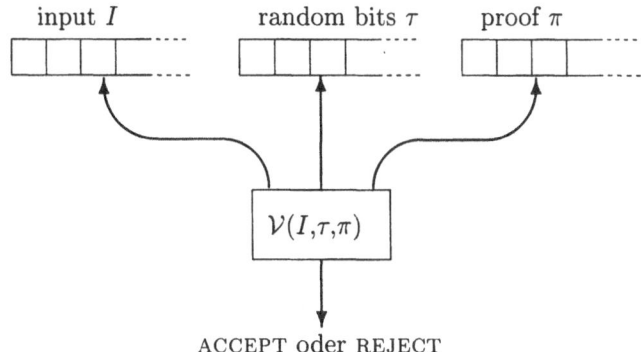

Figure 9.2 Illustration of a verifier for the classes $\mathcal{PCP}(\cdot,\cdot)$.

Proof. Consider first the class $\mathcal{P}$. A decision problem $\langle \mathcal{I}, \text{Sol} \rangle$ belongs to this class if it can be solved by a polynomial time algorithm. It is easy to see that such a polynomial time algorithm is exactly a $(0,0)$-restricted verifier. Indeed, by definition a $(0,0)$-restricted verifier must not read a single random bit nor a single bit from the proof, and it has to return in polynomial time ACCEPT or REJECT. Note also that a $(0,0)$-restricted verifier is a deterministic algorithm. The probability that the verifier returns REJECT is thus for all inputs I either 0 or 1. Definition 9.21 therefore implies that the verifier has to return ACCEPT for all YES-instances and REJECT for NO-instances. Thus, $\mathcal{P} = \mathcal{PCP}(0,0)$. The proof of $\mathcal{NP} = \mathcal{PCP}(0, poly(n))$ is very similar. The important observation here is that a verifier can read at most one bit from the proof in every time step. That is, if it returns an answer ACCEPT or REJECT in polynomial time, then it also can read at most $poly(n)$ bits from the proof. Finally, consider the class $\mathcal{RP}$. It contains those decision problems which can be decided by a randomized polynomial time algorithm. Recall also that in the definition of a randomized algorithm in Section 8.1 we required that it may use only $T_A(|I|)$ random bits. In other words, a randomized polynomial time algorithm is exactly a $(poly(n), 0)$-restricted verifier. Comparing the definition of $co\mathcal{RP}$ and that of the classes $\mathcal{PCP}(\cdot, \cdot)$ then concludes the proof of the lemma.   □

With these definitions at hand we can now formally state the celebrated result referred to in the New York Times.

Theorem 9.23 (PCP-THEOREM) $\mathcal{NP} = \mathcal{PCP}(\log n, 1)$.

The inclusion "$\supseteq$" is fairly easy to prove. We will do that in Lemma 9.24. The inclusion "$\subseteq$" on the other hand is highly nontrivial and beyond the scope of this book. So, instead of proving it, we will only outline the proof of the weaker statement $\mathcal{NP} \subseteq \mathcal{PCP}(poly(n), 1)$. Hopefully, this will give the reader at least some intuition of why Theorem 9.23 might be true at all.

Before we proceed to these proofs let us first show how Theorem 9.23 can be used to prove Theorem 9.2.

Proof of Theorem 9.2. As SATISFIABILITY belongs to $\mathcal{NP}$, Theorem 9.23 implies that there exists a $(\log n,1)$-restricted verifier $\mathcal{V}$ for SATISFIABILITY. Consider an arbitrary instance of SATISFIABILITY, i.e., an arbitrary Boolean formula φ in conjunctive normal form. We will use $\mathcal{V}$ to construct the desired instance $f(\varphi)$ of 3SAT in polynomial time.

The definition of a $(\log n,1)$-restricted verifier implies that there exist constants c and k such that $\mathcal{V}$ will use at most $c \log_2 |\varphi|$ random bits and read at most k bits of a given proof π. Clearly, we may also assume without loss of generality that $\mathcal{V}$ always uses *exactly* $c \log_2 |\varphi|$ random bits and reads *exactly* k bits of the proof π. This implies in particular that $\mathcal{V}$ can access at most $k \cdot 2^{c \log_2 |\varphi|} = k \cdot |\varphi|^c$ different positions of π. We will identify these positions with the variables $x_1, x_2, \ldots$ of the Boolean formula $f(\varphi)$. The clauses of $f(\varphi)$ are constructed as follows. For every string τ_i of length $c \log_2 |\varphi|$ we construct a 3SAT formula F_τ. Assume that $\mathcal{V}$ reads the bits $x_{\tau_1}, \ldots, x_{\tau_k}$ from the proof π. Clearly, there are exactly 2^k different conjunctive(!) clauses which use each of the variables $x_{\tau_1}, \ldots, x_{\tau_k}$ exactly once (negated or unnegated). From these 2^k clauses we keep exactly those which correspond to an assignment for which the verifier accepts (identify TRUE with 1 and FALSE with 0). According to Problem 3.1 this Boolean formula can be transformed into a 3SAT formula F_τ which contains at most $k2^k$ clauses.

Now consider $f(\phi) := \bigwedge_\tau F_\tau$. Obviously, it can be constructed in polynomial time. Furthermore, by construction every satisfying assignment of F corresponds to a proof π for which the verifier $\mathcal{V}$ accepts for *all* random strings τ. Similarly, every proof π which is rejected by the verifier for at least half of all random strings corresponds to an assignment that does not satisfy at least half of all formulas F_τ. In other words, for any such assignment at least $\frac{1}{2} \cdot 2^{c \log_2 |\varphi|}$ clauses of $f(\phi)$ are not satisfied. As $f(\phi)$ contains at most $k2^k \cdot 2^{c \log_2 |\varphi|}$ clauses, this shows that all properties of Theorem 9.2 are satisfied if we let $\varepsilon := \frac{1}{k2^{k+1}}$.   □

Lemma 9.24 $\mathcal{PCP}(\log n,1) \subseteq \mathcal{NP}$.

Proof. Consider an arbitrary decision problem $\langle \mathcal{I}, \text{Sol} \rangle$ in $\mathcal{PCP}(\log n,1)$. Let $\mathcal{V}$ denote a $(\log n,1)$-restricted verifier $\mathcal{V}$ for $\langle \mathcal{I}, \text{Sol} \rangle$. We use it to construct an algorithm $\mathcal{A}$ that shows that $\langle \mathcal{I}, \text{Sol} \rangle \in \mathcal{NP}$. Note that both the verifier $\mathcal{V}$ and the algorithm $\mathcal{A}$ may read (at least parts of) a proof π. The main difference is that $\mathcal{V}$ is randomized, but $\mathcal{A}$ has to be a deterministic algorithm. How can we cope with that? There is a very easy answer to this question: complete enumeration! More precisely, for an input I of length $n := |I|$ and a given proof π the algorithm $\mathcal{A}$ simulates $\mathcal{V}$ for *all* possible strings τ of length $c \log n$ for an appropriate constant c. As there are $2^{c \log n} = n^c$ such strings τ and as $\mathcal{V}$ needs time $poly(n)$ for each such string, the total running time of A is bounded by a polynomial in n. Note that V will accept for all strings τ if and only if $\pi \in \mathrm{Sol}(I)$. That is, algorithm A can indeed decide whether $\pi \in \mathrm{Sol}(I)$.   □

In the remainder of this section we will present some of the main ideas for proving the other inclusion of the PCP-Theorem. We start by studying some basic features of probabilistically checkable proofs.

Basic properties of probabilistically checkable proofs

In order to construct probabilistically checkable proofs, it turns out to be very useful to view a "proof" not simply as string of 0's and 1's, but instead as a vector in $\mathbb{F}_2^n$, where $\mathbb{F}_2 = \{0,1\}$. This allows algebraic manipulations of the proof. For example, we can compute the scalar product $\pi^T x$ of a proof $\pi \in \mathbb{F}_2^n$ with some other vector $x \in \mathbb{F}_2^n$. Note that in this section we will perform all such computations in $\mathbb{F}_2$. That is, the scalar product $\pi^T x$ is computed as $\pi^T x = \sum_{i=1}^{n} \pi_i x_i \bmod 2$.

We explain the construction of probabilistically checkable proofs by considering the following situation. Suppose a person A knows some secret information $a_0 \in \mathbb{F}_2^n$. Another person B claims that he also knows a_0 and wants to convince A of this fact. How can B achieve this?

Clearly, he can give A a vector $a \in \mathbb{F}_2^n$ and ask him to verify that $a_0 \equiv a$. To be sure that a is indeed identical to a_0 A has to compare each of the n bits. In case n is really large this is a lot of work and it is plausible that A would prefer a way which makes it easier for him.

As a more specific example consider the following. Assume $a_0 \in \mathbb{F}_2^n$ are the answers to some exam. Then A is the professor and B is some student who claims that he did solve all problems correctly. Clearly, the professor would be very happy if she could be convinced of this fact without having to mark the whole exam. – But is this really possible? Actually, not if we insist that A makes no mistakes. But what happens, if we do allow A to make some mistakes with a very low probability? Can we then reduce her amount of work? Perhaps. Note however, that B (the student) might be very angry if A (the professor) rejects his solution even though it is correct. We therefore allow A to make mistakes, but only if these are in B's favor. More precisely, if B really knows a_0 (i.e., if $a \equiv a_0$), then A has to accept a. If on the other hand B does not know a_0 (i.e., if $a \not\equiv a_0$), then A may accept a with probability smaller than some given constant, say smaller than $1/100$. Does this help in making A's job easier? — Yes! In fact, in this case A has to look at only a constant(!) number of bits from a, independently of how large n might be! (As we will see, there is however some drawback for B: he has to submit a in an encoded form, which requires considerable more effort from him.) How does this work? The next lemma is a key observation.

Lemma 9.25 *Consider two arbitrary vectors* $a, a_0 \in \mathbb{F}_2^n$. *Then*

$$\Pr_r \left[r^T a = r^T a_0 \right] = \begin{cases} 1, & \text{if } a = a_0 \\ \frac{1}{2}, & \text{if } a \neq a_0 \end{cases}$$

(We use the notation $\Pr_r [\cdots]$ *to indicate that the probability is with respect to the vector* $r \in \mathbb{F}_2^n$ *which is chosen uniformly at random.[1])*

Proof. If $a = a_0$ then there is nothing to show. So assume $a \neq a_0$. Observe that $r^T a = r^T a_0$ if and only if $r^T (a - a_0) = 0$, where the subtraction is computed in $\mathbb{F}_2$. Using that

$$r^T (a - a_0) = \sum_{i=1}^{n} r_i (a - a_0)_i = \sum_{i:\, a_i \neq (a_0)_i} r_i .$$

this implies that $r^T (a - a_0) = 0$ if and only if r contains an even number of 1's at positions i where $a_i \neq (a_0)_i$. As we choose the vector r uniformly at random, this happens with probability exactly $1/2$. To see this, consider the largest index j such that $a_j \neq (a_0)_j$. Choose the entries r_i successively. Clearly, regardless how the r_i's for $i < j$ are chosen, exactly one of the two choices for r_j ensures that $r^T (a - a_0) = 0$. As both choices for r_j are equally likely, this completes the proof of the lemma.   □

Lemma 9.25 suggests the following strategy. B gives A a table T which (supposedly) contains for each $r \in \mathbb{F}_2^n$ the scalar product $r^T a_0$. A chooses a vector $r \in \mathbb{F}_2^n$ uniformly at random and compares $T(r)$ with $r^T a_0$. What we would like to show is that this strategy ensures that A accepts a "false" vector a with probability at most $1/2$, as suggested by Lemma 9.25. Unfortunately, this is not true. The reason is that B can "cheat" not only but choosing a vector $a \neq a_0$, but also by submitting a "false" table T. The following example illustrates this.

Example 9.26 Suppose $a_0 = (11011)$ and assume B submits a table T such that $T(r) = 0$ if and only if $(r_1 = r_4$ and $r_2 = r_5)$ or $(r_1 = r_2$ and $r_4 = r_5)$. Then one easily checks that the probability that $T(r) \neq r^T a_0$ is just $1/8$ and not $1/2$. We leave it to the reader to modify this example in order to see that the probability can in fact be arbitrarily small.

We therefore have to generalize our proof strategy: A also has to check that B does not cheat by handing in a "false" table T. This suggests the following protocol:

---

[1] Note that correctly we should write $\Pr_{r \in \mathbb{F}_2^n} [\cdots]$ instead of $\Pr_r [\cdots]$. We use the short notation for conciseness whenever there is no risk of confusion.

VERIFICATION PROTOCOL

I. B gives A a table T which (supposedly) contains for each $r \in \mathbb{F}_2^n$ the scalar product $r^T a_0$.

II. A repeats each of the following tests 45 times and accepts if and only if all tests are OK:

   (i) Choose vectors $x, y \in \mathbb{F}_2^n$ uniformly at random and verify that $T(x) = T(x+y) + T(y)$.

   (ii) Choose vectors $r, x \in \mathbb{F}_2^n$ uniformly at random and verify that $T(r+x) + T(x) = r^T a_0$.

We say B *knows* the vector a_0 if and only if he submits a table T such that $\Pr_x\left[T(x) = a_0^T x\right] \geq \frac{2}{3}$. This definition makes sense, as it can be shown (Problem 9.10) that this property implies that B is able to reconstruct the vector a_0 from T.

Theorem 9.27 *If B knows a_0 and submits the correct table T, then A will always accept. If B does not know a_0, then A accepts with probability less than $1/100$ regardless of what table T is used.*

For the proof of Theorem 9.27 we need two lemmas whose proofs are left to the reader, cf. Problems 9.11 and 9.12.

Lemma 9.28 *Assume $g : \mathbb{F}_2^n \to \mathbb{F}_2$ satisfies $\Pr_x\left[a^T x = g(x)\right] \geq 1 - \delta$ for some vector $a \in \mathbb{F}_2^n$ and some $0 \leq \delta \leq \frac{1}{2}$. Then*

$$\Pr_x\left[g(x+r) + g(x) \neq a^T r\right] \leq 2\delta \qquad \text{for all } r \in \mathbb{F}_2^n.$$

Lemma 9.29 *Assume $g : \mathbb{F}_2^n \to \mathbb{F}_2$ satisfies $\Pr_r\left[a^T r = g(r)\right] < 1 - \delta$ for some $0 \leq \delta \leq \frac{1}{3}$ and all vectors $a \in \mathbb{F}_2^n$. Then $\Pr_{x,y}\left[g(x) + g(y) \neq g(x+y)\right] > \delta/2$.*

Proof of Theorem 9.27. The case that B knows a_0 is trivial. So assume B does not know a_0. Here we distinguish the two cases that the table T satisfies

$$\Pr_r\left[T(r) = a^T r\right] \geq \tfrac{4}{5}, \qquad \text{for a vector } a \neq a_0, \tag{9.4}$$

and that T does not satisfy this property. Assume first that (9.4) holds. As we do all calculations in $\mathbb{F}_2$, we deduce

$$\Pr_{r,x}\left[T(r+x) + T(x) = a_0^T r\right] = \Pr_{r,x}\left[T(r+x) + T(x) = a_0^T r = a^T r\right]$$
$$+ \Pr_{r,x}\left[T(r+x) + T(x) = a_0^T r \neq a^T r\right]$$
$$\leq \Pr_r\left[a^T r = a_0^T r\right] + \Pr_{r,x}\left[T(r+x) + T(x) \neq a^T r\right]$$
$$\leq \tfrac{1}{2} + \tfrac{2}{5} = \tfrac{9}{10},$$

where last inequality follows from Lemmas 9.25 and 9.28 and our assumption (9.4). The probability that all 45 executions of test (ii) are successful is thus at most $(9/10)^{45} < 1/100$.

Assume now that (9.4) does not hold. Then Lemma 9.29 implies that

$$\mathrm{Pr}_{x,y}\left[T(x) + T(y) \neq T(x+y)\right] \geq \tfrac{1}{10}.$$

The probability that all 45 executions of test (i) are successful is thus at most $(9/10)^{45} < 1/100$. □

Construction of a $(poly(n),1)$-restricted verifier

In the remainder of this section we will show how to use the above ideas to define an $(n^3,1)$-restricted verifier for 3SAT. Again it will turn out to be very useful to work in an algebraic setting. Our first aim is thus to transform an instance of 3SAT into an algebraic problem.

Recall that a *Boolean formula* is an expression built from variables x_i and their negations $\overline{x}_i$ using the operations $\vee$ and $\wedge$. Similarly, an *arithmetic formula* is an expression built from the constants 0, 1 and variables x_i using the operations $+$ and $\cdot$.

The arithmetization of a Boolean formula is achieved by replacing every negated variable $\overline{x}_i$ by $1 - x_i$, every conjunction $\alpha \wedge \beta$ by $\alpha \cdot \beta$, and every disjunction $\alpha \vee \beta$ by $1-(1-\alpha)(1-\beta)$. One easily checks that a Boolean formula F has a satisfying assignment if and only if its arithmetization $A(F)$ is not identically zero. Even more is true. Considered as a polynomial over $\mathbb{F}_2$, the value of $A(F)$ coincides with the value of F, if we identify 0 with *false* and 1 with *true*.

Recall that the input of a 3SAT-problem is a Boolean formula in conjunctive normal form, in which every disjunction contains at most three (potentially negated) variables. In the following we always assume that the input of such a satisfiability problem contains exactly n clauses $C_1,\ldots,C_n$ using m variables $x_1,\ldots,x_m$. Without loss of generality (by adding dummy variables) we may also assume that $n = m$.

The arithmetization $\mathcal{C}(x) = (\hat{C}_1(x),\ldots,\hat{C}_n(x))$ of a satisfiability problem $C_1 \wedge \cdots \wedge C_n$ is obtained by letting $\hat{C}_i$ be the arithmetization of the *complement* of the ith clause. Then the following is immediate.

Lemma 9.30 *A vector $a \in \mathbb{F}_2^n$ corresponds to a satisfying assignment of $C_1 \wedge \cdots \wedge C_n$ if and only if $\mathcal{C}(a) = (\hat{C}_1(a),\ldots,\hat{C}_n(a))$ is identically zero.* □

We are thus back to a situation which is very similar to the one presented in the previous subsection. The only difference being that the "secret" now is a vector $a \in \mathbb{F}_2^n$ such that $\mathcal{C}(a) = (\hat{C}_1(a),\ldots,\hat{C}_n(a))$ is identically zero. We can thus proceed similarly as in the previous subsection. Only the technical realization will become slightly more complicated.

Observe that each of the terms $\hat{C}_i(x)$ is a polynomial of (total) degree at most three. The scalar product of the arithmetization $(\hat{C}_1(x), \dots, \hat{C}_n(x))$ of a satisfiability problem with a vector $r \in \mathbb{F}_2^n$ can thus be written as

$$\sum_{i=1}^n r_i \hat{C}_i(x) \;=\; c(r) + \sum_{i \in S_1(r)} x_i + \sum_{(i,j) \in S_2(r)} x_i x_j + \sum_{(i,j,k) \in S_3(r)} x_i x_j x_k,$$

where the sets $S_1(r)$, $S_2(r)$, $S_3(r)$ and the constant $c(r)$ depend only on the given 3SAT formula and the vector r, but *not* on the assignment x. Observe that, for example, the value $\sum_{i \in S_1} x_i$ corresponds exactly to the scalar product of x with the incidence vector of $S_1(r)$. We thus require that the proof contains all these scalar products (for S_1 as well as for S_2 and S_3).

The rest of this section is devoted to turning these rough ideas into a precise description of an $(n^3,1)$-restricted verifier. For a vector $a \in \mathbb{F}_2^n$ we define three linear functions as follows:

$$A : \mathbb{F}_2^n \to \mathbb{F}_2 \,, \qquad A(x) := \sum_{i=1}^n a_i\, x_i,$$

$$B : \mathbb{F}_2^{n^2} \to \mathbb{F}_2 \,, \qquad B(y) := \sum_{i=1}^n \sum_{j=1}^n a_i\, a_j\, y_{ij}, \qquad \text{and} \qquad (9.5)$$

$$C : \mathbb{F}_2^{n^3} \to \mathbb{F}_2 \,, \qquad C(z) := \sum_{i=1}^n \sum_{j=1}^n \sum_{k=1}^n a_i\, a_j\, a_k\, z_{ijk}.$$

The verifier interprets every proof π as $\pi = \tilde{A}\tilde{B}\tilde{C}$, where $\tilde{A}$ has length 2^n and is considered as a function $\tilde{A} : \mathbb{F}_2^n \to \mathbb{F}_2$. Similarly, $\tilde{B}$ and $\tilde{C}$ have length 2^{n^2} and 2^{n^3}, respectively, and are interpreted as functions $\tilde{B} : \mathbb{F}_2^{n^2} \to \mathbb{F}_2$ and $\tilde{C} : \mathbb{F}_2^{n^3} \to \mathbb{F}_2$. Ideally, $\tilde{A}, \tilde{B}, \tilde{C}$ correspond to the functions A, B and C from (9.5), defined with respect to some vector $a \in \mathbb{F}_2^n$ corresponding to a satisfying assignment.

The verifier needs to achieve two tasks:

(i) Verify that $\tilde{A}$, $\tilde{B}$, $\tilde{C}$ are what they are supposed to be, namely the linear functions from (9.5) defined with respect to the *same* vector $a \in \mathbb{F}_2^n$.

(ii) Verify that this vector a corresponds to a satisfying assignment.

Note that these two tasks correspond to the two tests performed in part II of the verification protocol on page 185. We therefore proceed very similarly. To achieve (i) we first check that the three functions are close to linear functions:

LINEARITY TEST

Choose vectors $x, x' \in \mathbb{F}_2^n$ uniformly at random and verify that
$$\tilde{A}(x) + \tilde{A}(x') = \tilde{A}(x + x').$$
Choose vectors $y, y' \in \mathbb{F}_2^{n^2}$ uniformly at random and verify that
$$\tilde{B}(y) + \tilde{B}(y') = \tilde{B}(y + y').$$
Choose vectors $z, z' \in \mathbb{F}_2^{n^3}$ uniformly at random and verify that
$$\tilde{C}(z) + \tilde{C}(z') = \tilde{C}(z + z').$$

Next we check that these linear functions are consistent, that is, all three of them are defined by the *same* vector $a \in \mathbb{F}_2^n$. Observe that if for $x,x' \in \mathbb{F}_2^n$ we let $x \circ x'$ denote the vector $y \in \mathbb{F}_2^{n^2}$ given by $y_{ij} = x_i \cdot x_j$, then the functions $A(x)$ and $B(y)$ defined in (9.5) satisfy $A(x) \cdot A(x') = B(x \circ x')$ for all $x,x' \in \mathbb{F}_2^n$. Similarly, if for $x \in \mathbb{F}_2^n$ and $y \in \mathbb{F}_2^{n^2}$ we let $x \circ y$ denote the vector $z \in \mathbb{F}_2^{n^3}$ given by $z_{ijk} = x_i \cdot y_{jk}$, then $A(x) = a^T x$, $B(y) = b^T y$, and $C(z) = c^T z$ satisfy $A(x) \cdot B(y) = C(x \circ y)$ for all $x \in \mathbb{F}_2^n$ and $y \in \mathbb{F}_2^{n^2}$. In principle, theses identities are good candidates for a test whether the functions $\tilde{A}$, $\tilde{B}$, and $\tilde{C}$ are consistent. There is just one additional problem. For $x,x' \in \mathbb{F}_2^n$ chosen uniformly at random, the vector $x \circ x'$ is *not* a random element from $\mathbb{F}_2^{n^2}$. Note, however, that if we add to $x \circ x'$ a random element $r \in \mathbb{F}_2^{n^2}$ then $r + x \circ x'$ is (for all fixed $x,x' \in \mathbb{F}_2^n$) also a random element in $\mathbb{F}_2^{n^2}$. Lemma 9.28 thus implies that the following procedure can be used to check the consistency of the three functions $\tilde{A}$, $\tilde{B}$, and $\tilde{C}$:

CONSISTENCY TEST

Choose vectors $x,x' \in \mathbb{F}_2^n$ and $r \in \mathbb{F}_2^{n^2}$ uniformly at random and
   verify that $\tilde{A}(x) \cdot \tilde{A}(x') = \tilde{B}(r) + \tilde{B}(r + x \circ x')$.
Choose vectors $x \in \mathbb{F}_2^n$, $y \in \mathbb{F}_2^{n^2}$, and $r \in \mathbb{F}_2^{n^3}$ uniformly at random and
   verify that $\tilde{A}(x) \cdot \tilde{B}(y) = \tilde{C}(r) + \tilde{C}(r + x \circ y)$.

It remains to design a procedure which enables the verifier to achieve task (ii). This procedure is an immediate consequence of Lemma 9.30:

SATISFIABILITY TEST

Choose a vector $r \in \mathbb{F}_2^n$ uniformly at random and compute
   $c = c(r) \in \mathbb{F}_2$, $S_1 = S_1(r) \in \mathbb{F}_2^n$, $S_2 = S_2(r) \in \mathbb{F}_2^{n^2}$, and $S_3 = S_3(r) \in \mathbb{F}_2^{n^3}$.
Choose vectors $x \in \mathbb{F}_2^n$, $y \in \mathbb{F}_2^{n^2}$, and $z \in \mathbb{F}_2^{n^3}$ uniformly at random and
   verify that $c + \tilde{A}(x) + \tilde{A}(x + S_1) + \tilde{B}(y) + \tilde{B}(y + S_2) + \tilde{C}(z) + \tilde{C}(z + S_3) = 0$.

Lemmas 9.29, 9.28 and 9.30 can now be used to prove the following theorem in a similar way as Theorem 9.27.

Theorem 9.31 *There exists a constant k such that repeating* LINEARITY TEST, CONSISTENCY TEST *and* SATISFIABILITY TEST *k times each, and rejecting whenever any one of these tests fails, is a $(n^3,1)$-restricted verifier for* 3SAT.

We thus have shown that 3SAT is contained in $\mathcal{PCP}(n^3,1)$. As 3SAT is $\mathcal{NP}$-complete this implies $\mathcal{NP} \subseteq \mathcal{PCP}(poly(n),1)$. Reducing the number of random bits from $poly(n)$ to $\log n$ requires considerable more effort. See notes for pointers to the literature.

Problems

9.1   Design an algorithm that constructs for every Boolean formula in conjunctive normal form with *exactly three* variables in each clause in linear time a truth assignment which satisfies at least 7/8th of all clauses.

9.2   Prove Proposition 9.7.

9.3   Let A and B be two disjoint sets of size n. Construct a d-regular bipartite graph as follows. Choose independently and uniformly at random d perfect matchings between A and B. Let G denote the union of these matchings. Show that for d sufficiently large this graph is with probability at least $1/2$ an $(n,d,1/2)$-expander.

9.4   Recall from Problem 3.12 that deciding whether a graph $G = (V,E)$ contains an independent set of size B is $\mathcal{NP}$-complete, if B is part of the input. Show now that the optimization problem BOUNDED INDEPENDENT SET *"Given a graph $G = (V,E)$ with maximum degree 7, find a maximum independent set"* is not contained in the class $\mathcal{PTAS}$, unless $\mathcal{P} = \mathcal{NP}$.

9.5   A *vertex cover* of a graph $G = (V,E)$ is a set $X \subseteq V$ such that $|X \cap e| \geq 1$ for all $e \in E$. Show that optimization problem BOUNDED VERTEX COVER *"Given a graph $G = (V,E)$ with maximum degree 7, find a minimum vertex cover"* is not contained in the class $\mathcal{PTAS}$, unless $\mathcal{P} = \mathcal{NP}$. (Hint: Reduce from BOUNDED INDEPENDENT SET.)

9.6   Show that optimization problem STEINER(1,2) *"Given a complete network $N = (V,E,\ell)$ such that $\ell : E \to \{1,2\}$ and a terminal set $K \subseteq V$, find a Steiner minimum tree T for K"* is not contained in the class $\mathcal{PTAS}$, unless $\mathcal{P} = \mathcal{NP}$. (Hint: Reduce from BOUNDED VERTEX COVER.)

9.7   Consider the following variant of MAX3SAT(k):

MAX3SAT-k:

Given: A Boolean formula in conjunctive normal form such that every clause contains at most 3 variables and every variable occurs in at most k clauses (negated or unnegated).

Find: A truth assignment which satisfies a maximum number of clauses.

a) Use the construction from Theorem 3.10 to define an AP-reduction from MAX3SAT(6) to MAX3SAT-5.
b) Show that MAX3SAT-5 $\leq_{AP}$ MAX3SAT-3.

9.8   Show that MINIMUM METRIC TRAVELLING SALESMAN defined in Problem 7.11 does not belong to $\mathcal{PTAS}$, unless $\mathcal{P} = \mathcal{NP}$.

9.9   Try to improve the constant c_0 defined in the last paragraph of Section 9.1. (Hint: Use that Theorem 9.4 holds also when the input is restricted to 3SAT instances which contain exactly three literals per clause.)

9.10 Consider a function $T : \mathbb{F}_2^n \to \mathbb{F}_2$. Show that there exists at most one vector $a \in \mathbb{F}_2^n$ such that
$$\Pr_x \left[T(x) = a^T x \right] \geq \tfrac{2}{3}.$$
Show furthermore that such a vector a can easily be reconstructed from T.

9.11 Prove Lemma 9.28.

9.12 Assume $g : \mathbb{F}_2^n \to \mathbb{F}_2$ satisfies $\Pr_{x,y} \left[g(x) + g(y) \neq g(x + y) \right] \leq \delta/2$ for some $0 \leq \delta \leq \tfrac{1}{3}$. Define a function h as follows:
$$h(x) := \operatorname{majority}_y \{ g(x + y) + g(y) \},$$
where majority denotes the value occurring most often over all choices of $y \in \mathbb{F}_2^n$. (All computations are in $\mathbb{F}_2$, ties are broken arbitrarily.) Show that h satisfies $\Pr_y \left[h(x) = g(x) \right] \geq 1 - \delta$ and $\Pr_y \left[h(x) \neq g(x + y) + g(y) \right] \leq \delta$ for all $x \in \mathbb{F}_2^n$. Deduce that these facts imply that there exists a vector $a \in \mathbb{F}_2^n$ such that $h(x) = a^T x$ for all $x \in \mathbb{F}_2^n$.

9.13 Prove Theorem 9.31.

Notes

More background information on approximation algorithms for the Maximum Satisfiability Problem can be found in Yannakakis (1994). Recently, Karloff and Zwick (1997) have shown that there exists a polynomial time approximation algorithm for MAX3SAT with performance ratio 8/7. Theorem 9.2 (which is just a reformulation of the PCP-Theorem 9.23 from Section 9.3) and Corollary 9.3 were proven in a seminal paper by Arora, Lund, Motwani, Sudan, and Szegedy (1998). Theorem 9.4 is due to Håstad (1997).

A cornerstone in the development of approximation preserving reductions and the structure of complexity classes for optimization problems was a paper by Papadimitriou and Yannakakis (1991). They also had the idea to use a expanders in order to reduce MAX3SAT to a version in which each variable occurs only a constant number of times. The expander from Example 9.10 is due to Gabber and Galil (1981). For more information on the construction and the applications of expanders we refer the reader to the articles by Alon (1986) and Kahale (1995). A direct probabilistic construction of an amplifier can be found in Bollobás (1988). The non-approximability of the Steiner tree problem up to a constant factor was first shown by Bern and Plassmann (1989); their proof is outlined in Problem 9.4 till 9.6. Currently, the best bound for the non-approximability of the Steiner tree problem is $136/136 \approx 1.0074$. It was proven by Thimm (2001) and is based on the assumption that $co\mathcal{RP} \neq \mathcal{NP}$.

The concept of AP-reductions is due to Crescenzi, Kann, Silvestri, Trevisan (1999). Our proof of the $\mathcal{APX}$-completeness of MAXSAT (Theorem 9.18) adapted from Trevisan (1997). For more information on approximation preserving reductions, non-approximability results, and probabilistically checkable proofs we refer the reader to the survey Steger (2002) and the books by Ausiello, Crescenzi, Gambosi, Kann, Marchetti-Spaccamela, Protasi (1999), Goldreich (1999), and Mayr, Prömel, Steger (1998).

10

Geometric Steiner Problems

Historically, a geometric version of the Steiner problem in graphs was the first Steiner problem to be considered. It dates back to Fermat (1601-1665) who proposed the following problem: "*Given three points in the plane, find a fourth point such that the sum of its distances to the three given points is minimum.*" It is known that Torricelli found a geometric solution to Fermat's problem before 1640. The generalization of Fermat's problem to n instead of 3 points, that is finding a point minimizing the sum of the distances to n given points, was studied by many researchers. One of them was Jacob Steiner (1796-1863), a professor for geometry at the University of Berlin. It was him to whom Courant and Robbins in their book "*What is Mathematics*" attributed in 1941 a problem nowadays known as the *Euclidean Steiner Problem*: Given n points in the plane, find a shortest network which interconnects them. The two mathematicians Jarník and Kössler seem to have been the first who seriously considered this problem around 1930. Whether Jacob Steiner was even aware of it is unknown.

We will study the Euclidean Steiner problem in Section 10.4. As we will see, it is in many respects quite different from the Steiner problem in graphs. For example for the latter it is trivial to argue that there exists some (potentially exponential) algorithm which computes a Steiner minimum tree. To see this recall that there exist only finitely many different subgraphs and one can easily compute their lengths and check whether they form a Steiner tree. For the Euclidean Steiner problem the existence of such an algorithm is on the other hand not so obvious. In fact, constructing one will be our main aim in Section 10.4.

Before we come to that we will first study a geometric variant of the Euclidean Steiner problem which turns out to be computationally much easier. This is the so-called *Manhattan Steiner Problem*. It differs from the Euclidean Steiner problem only in the fact that we measure the distance of two points (x_1, y_1) and (x_2, y_2) not with respect to the Euclidean norm

$$\ell_2((x_1,y_1),(x_2,y_2)) := \sqrt{|x_1 - x_2|^2 + |y_1 - y_2|^2},$$

but with respect to the Manhattan norm

$$\ell_1((x_1,y_1),(x_2,y_2)) := |x_1 - x_2| + |y_1 - y_2|.$$

This version of the Steiner Problem was first considered by Hanan in 1966. As we will see, it is in fact a true special case of the Steiner Problem in Networks, which is still $\mathcal{NP}$-complete, but can be solved much better approximately.

10.1   A characterization of rectilinear Steiner minimum trees

Let $K \subseteq \mathbb{R}^2$ be a finite set of terminals. To solve the Manhattan Steiner problem we have to connect the terminals in K by a Steiner tree T in such a way that all lines are either horizontal (parallel to the x-axis) or vertical (parallel to the y-axis). In this section we will show that these restrictions imply that Steiner minimum trees have a very special shape. In fact, they almost look like real trees.

We start with some notation and definitions. A *segment* is a horizontal or vertical line connecting two points in the plane. A *rectilinear tree* is a connected acyclic collection of line segments, which intersect only at their endpoints. The *degree* of a point is the number of segments incident to it. A *rectilinear Steiner tree* for a given set of terminals is a rectilinear tree such that each terminal is an endpoint of some segment in the tree. Without loss of generality we require that all degree 2 points in a rectilinear tree which are not terminals are adjacent to exactly one vertical and one horizontal segment. Then all points of degree at least 3 which are not terminals are called the *Steiner points* of the tree, while a degree 2 point which is not a terminal is called a *corner point*. The two line segments incident to a corner point will also be called the *legs* of the corner point.

Now let $K \subseteq \mathbb{R}^2$ be a set of terminals and let T be a Steiner minimum tree for K with $m \geq 2$ Steiner points. Then T is called a *fir tree* iff every terminal has degree one in T and one of the following two conditions is satisfied (cf. Figure 10.1):

(i) All Steiner points lie on a vertical line and every Steiner point is adjacent to exactly one horizontal line, and these horizontal lines alternatingly extend to the left and right. In addition, the topmost Steiner point is adjacent to a vertical line ending in a terminal, while the lowest Steiner point is either adjacent to a vertical line or to a corner (with a horizontal leg that extends to the opposite side than the horizontal line of this Steiner point) ending in a terminal.

(ii) All but one Steiner point lie on a vertical line. The extra Steiner point is
connected to the bottommost Steiner point on the vertical line by a corner
from the opposite side than the horizontal line of this Steiner point. Every
Steiner point on the vertical line is adjacent to exactly one horizontal line,
alternating to the left and right. In addition, the topmost and the extra
Steiner point are both adjacent to a vertical line that extends upwards resp.
downwards and ends in a terminal.

The vertical line containing all or all but one of the Steiner points is called the
stem of the fir tree.

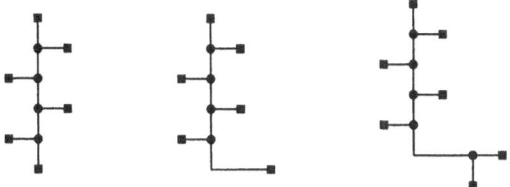

Figure 10.1   The different shapes of a fir tree

With these definitions at hand we are now able to state our main result of
this section. It says that for every terminal set $K \subseteq \mathbb{R}^2$ one of two cases must
hold. Either there exists a rectilinear Steiner minimum tree which is (essentially)
a fir tree or there exists a rectilinear Steiner minimum tree such that at least one
terminal has degree at least 2 (and the tree thus decomposes into at least two
smaller trees).

Theorem 10.1 *Let $K \subseteq \mathbb{R}^2$ be a terminal set such that in every rectilinear
Steiner minimum tree for K all terminals are leaves. Then there exists a Steiner
tree T such that it either is a fir tree or it has one of the following five shapes,
possibly after reflection and/or rotation:*

Before we prove Theorem 10.1 we need some more notation and a lemma.
If we say that two terminals or Steiner points are connected by a line *segment*
this tacitly implies that the line connecting these two points does not contain
any other Steiner point or terminal. For a point x we denote the (maximal) lines
which emanate from x to the top, bottom, left, and right by T_x, B_x, L_x, and
R_x, respectively. Note that these lines may contain other Steiner points and/or
terminals.

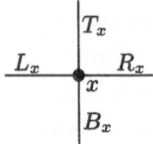

Furthermore, $|T_x|$, $|B_x|$, $|L_x|$, $|R_x|$ denote the lengths of these lines. In the remaining part of this section we will repeatedly use *sliding* and *flipping* of line segments:

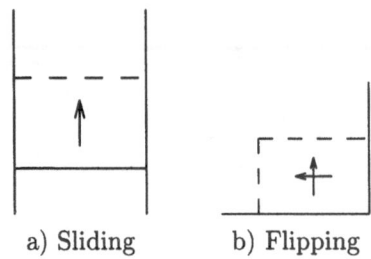

a) Sliding   b) Flipping

It should be immediately obvious that both transformations map Steiner minimum trees onto Steiner minimum trees and that the same is true for rotations of 90° or 180° degrees and/or reflections on vertical or horizontal axes.

Lemma 10.2 *Let $K \subseteq \mathbb{R}^2$ be a terminal set such that in every rectilinear Steiner minimum tree for K all terminals are leaves and let T be one such tree. Then the following properties hold:*

   (i) *Let x be any point in T. If T_x is a leg of a corner turning to the left, then L_x does not exist.*
  (ii) *Let a and b be two Steiner points in T connected by a horizontal line segment. If T_a and T_b exist and $|T_b| \geq |T_a|$ then T_a is the leg of a corner turning away from T_b.*
 (iii) *Let a and b be two Steiner points in T connected by a horizontal line segment. If T_a and T_b exist and T_b contains a terminal or Steiner point, then T_a is the leg of a corner turning away from T_b and $|T_b| \geq |T_a|$.*
  (iv) *No Steiner point in T is adjacent to more than one corner. A Steiner point of degree 4 has to be adjacent to four terminals by four line segments.*
   (v) *Let a and b be two Steiner points in T connected by a horizontal line segment. Then b cannot be connected to a third Steiner point by a vertical line segment.*
  (vi) *Let a and b be two Steiner points in T connected by a horizontal line segment. If T_b exists and is a segment that ends with a terminal then B_a exists and is a segment that ends with a terminal.*
 (vii) *Let a and b be two Steiner points in T connected by a horizontal line segment. If R_b exists and is the leg of a corner turning upwards ending in a Steiner point c, then T_c cannot contain another Steiner point.*

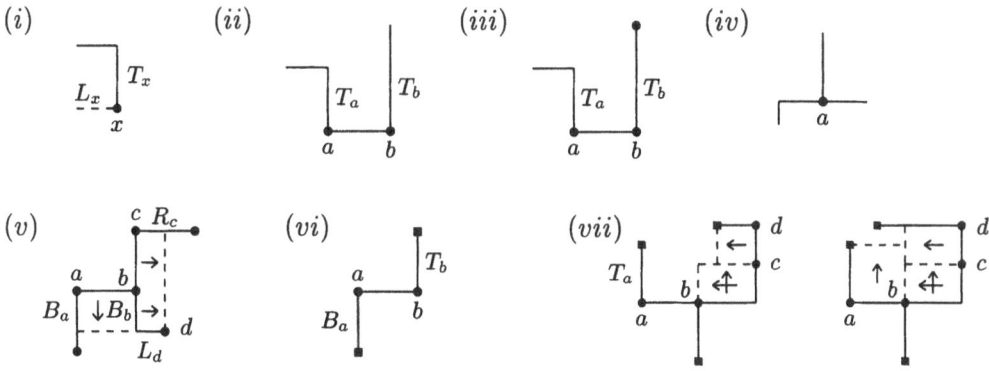

Figure 10.2 Illustrations for the various parts of Lemma 10.2

Proof. (i): Let y be the other endpoint of the corner with leg T_x. Then flipping the corner between x and y would yield a shorter Steiner tree contradicting the minimality of T.

(ii): T_a cannot contain a terminal, as otherwise a horizontal slide of the edge connecting a and b would yield a Steiner tree of the same length in which a terminal has degree 2. Assume T_a contains some Steiner points. Let $a_1, \dots a_r$ denote these Steiner points, ordered by increasing distance from a. The minimality of T implies that for all $1 \leq i \leq r$ the line R_{a_i} cannot exist and that T_{a_r} cannot be the leg of a corner turning towards T_b. This, however, implies that T_{a_r} is the leg of a corner turning away from T_b, which in turn implies by part (i) that L_{a_r} cannot exist, contradicting the assumption that a_r is a Steiner point. So T_a must be the leg of a corner turning away from T_b.

(iii): This follows immediately from part (ii).

(iv): The first part follows immediately from part (i) and the fact that Steiner points have degree at least 3. So assume that a is a Steiner point of degree 4. Then, again by (i), no line can be the leg of a corner. Also, by part (iii) any Steiner point adjacent to a would have to be adjacent to two corners contradicting the first part of (iv), which we already proved.

(v): Assume the contrary, i.e., assume T_b exists and contains a Steiner point c. If T_a exists, part (iii) implies that it is a leg of a corner turning away from T_b. But then, by part (i) and (iv), B_a has to exist and can't be the leg of a corner. Similarly, if T_a does not exist, L_a and B_a have to exist and, again by part (i), B_a can't be the leg of a corner. By symmetry it follows that also R_c has to exist and can't be the leg of a corner. These two facts together in turn imply that, without loss of generality, B_b exists and is a leg of a corner turning away from B_a. Let d be the other endpoint of this corner. From part (iii) it follows that $|B_b| \leq |B_a|$ and $|L_d| \leq |R_c|$. Sliding the horizontal line between a and b downward and the vertical line below c to the right therefore yields a Steiner tree of the same length in which d is either a terminal of degree at least two or a Steiner point of degree

4, contradicting either the assumption or part (iv).

(vi): Similarly as in the proof of part (v), we first deduce from (iii) together with (i) that at most one of T_a and L_a can exist. As a is a Steiner point, this implies that B_a has to exist. If B_a is the leg of corner, then (i) implies that the other leg has to extend to the left. But then, again by (i), it follows that L_a cannot exits. Hence, T_a has to exits and, by (iii), has to be a leg of a corner turning to the left. But this can't be, as sliding both T_a and B_a to the left (and extending R_a) decreases the length of the Steiner tree, contradicting the assumption that T is a Steiner minimum tree. Hence, B_a is not the leg of a corner. But, then (v) implies that B_a can't contain a Steiner point. Hence, B_a is a line segment that ends in a terminal.

(vii): Assume the contrary, i.e., assume d is a Steiner point on T_c. By parts (iii) and (i) T_b can't exist. Therefore B_b has to exist and, by parts (i) and (v), has to end in a terminal. Thus, by part (vi), T_a and, by symmetry, L_d both have to exist and end with terminals. But flipping the corner between b and c, sliding B_d to the left and/or R_a upwards yields a Steiner tree of at most the same the length in which a terminal has degree 2, contradicting the assumption.   □

Proof of Theorem 10.1.   We proceed by induction on $k = |K|$. For $k \leq 2$ the theorem is easily seen to be true. So assume $k > 2$ and let T be any Steiner minimum tree. From Problem 1.2 we know that T contains a Steiner point a which is adjacent to at least two terminals, say x and y. We distinguish several cases. If a has degree 4 then by part (iv) of Lemma 10.2 this can only happen if $|K| = 4$ and T has the fifth shape shown in the theorem.

Next we assume that a has degree 3. Removing a and all incident edges splits T into three components. Let $K_1 = \{x\}$, $K_2 = \{y\}$, and K_3 be the corresponding terminal sets. By the inductive hypothesis there exist Steiner minimum trees T_i for $K_i \cup \{a\}$ which satisfy the requirements of the theorem. Without loss of generality we may assume that for all $i = 1,2,3$ the tree T_i is contained in T. If $|K_3| = 1$ this and part (iv) of Lemma 10.2 implies that T has the third or fourth of the shapes shown in the theorem. Otherwise we know that a is connected to a Steiner point b (contained in T_3) either by a straight line or by a corner. That is, without loss of generality (and the help of parts (i) and (iv) of Lemma 10.2) we may assume that T contains one of the following five configurations.

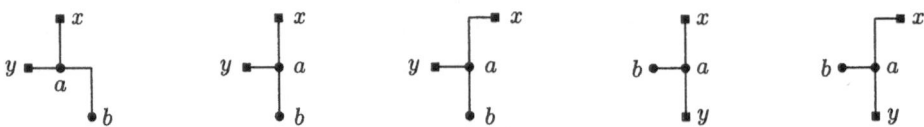

For the first case it is straightforward to check that there are only two possible shapes for T_3: the fourth shape of the theorem and the second version of case (i) of the definition of a fir tree. In both cases T is a fir tree as well.

The fourth case is not possible, as in this case part (vi) of Lemma 10.2 implies that both T_b and B_b are line segments ending in a terminal, contradicting part (iii) of Lemma 10.2. The fifth case is actually the same as the third case, as the corner between a and x can be flipped.

Thus, it remains to consider the second and third case. Here we first deduce from part (vi) of Lemma 10.2 that R_b has to exist and end with a terminal, say z. If b is not adjacent to a corner we easily check that there are only three possible shapes for T_3: the third shape of the theorem and all three versions of a fir tree. In all cases T is already "almost" a fir tree. A "real" fir tree is then obtained by sliding the vertical line through a to the right until we hit x. (By the assumptions on the terminal set this will always be possible.)

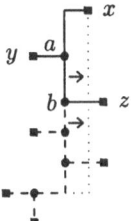

So assume, finally, that b is adjacent to a corner. By part (iii) of Lemma 10.2 there are only two possibilities: either the first leg goes down and the corner bends to the left or the first leg goes to the left and the corner bends downwards. In the first case the tree T_3 either has the fourth shape of the theorem or is a fir tree of type (ii) for which the stem consists of a single Steiner point (cf. part (vii) of Lemma 10.2). For both cases we again easily obtain the desired fir tree by sliding the vertical line through a to the right. In the second case the tree T_3 could again either has the fourth shape of the theorem or it could be a fir tree of type (ii) for which the stem is a vertical line. By sliding the line through a and b to the left as indicated in the following picture

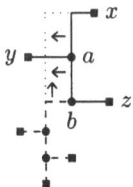

we again obtain a valid fir tree. This concludes the proof of the theorem. □

By definition, the task of a Manhattan Steiner problem is to connect a given set of terminals in the *plane* via horizontal and vertical line segments. In particular, there are therefore no a priori assumptions on the set of possible Steiner points – they may lie anywhere in the plane. Note, however, that at least for Steiner problems where the Steiner minimum tree is a fir tree this freedom is not needed. In a fir tree Steiner points occur only on horizontal and vertical lines that also contain a terminal. Our next result says that this property holds in fact for every Manhattan Steiner problem. To state it precisely we again need some definitions.

A *complete $n \times m$ grid graph*, denoted by $GG(n,m)$, is a graph on the integer points $\{(x,y) \mid 1 \le x \le n, 1 \le y \le m\}$ such that two points (x_1,y_1) and (x_2,y_2) are connected if and only if $|x_1 - x_2| + |y_1 - y_2| = 1$. In case that the spacing between different "rows" and "columns" is not regular we speak of an *irregular $n \times m$ grid graph*. Let $X = \{x_1,\dots,x_n\} \subseteq \mathbb{R}$ and $Y = \{y_1,\dots,y_m\} \subseteq \mathbb{R}$ be two finite sets ordered increasingly. Then the grid graph $GG(X,Y)$ is the graph with vertex set $X \times Y$ such that two vertices (x_{i_1},y_{j_1}) and (x_{i_2},y_{j_2}) are connected if and only if $|i_1 - i_2| + |j_1 - j_2| = 1$. The length of such an edge is then $|x_{i_1} - x_{i_2}| + |y_{j_1} - y_{j_2}|$. (Strictly speaking, $GG(X,Y)$ is of course a grid *network*. However, to be consistent with the existing literature we will continue to use the term grid *graph*.)

For a set $K \subseteq \mathbb{R}^2$ we define the grid graph $G(K)$ induced by K as follows. Let $X = \{x \in \mathbb{R} \mid$ there exists a y such that $(x,y) \in K\}$ be the x-coordinates of K and similarly let Y be the set of y-coordinates of K. Then $G(K)$ is the grid graph induced by X and Y.

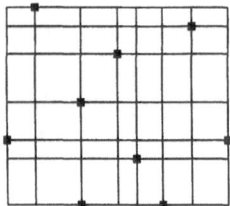

Figure 10.3 The induced grid graph $G(K)$

Theorem 10.3 *For any set $K \subseteq \mathbb{R}^2$ there exists a rectilinear Steiner minimum tree which corresponds to a Steiner minimum tree in $G(K)$.*

Proof.   It suffices to show that for every set $K \subseteq \mathbb{R}^2$ there exists a rectilinear Steiner minimum tree such that all Steiner points are located within the set $X \times Y$. We prove this by induction on k. If $k = 2$ there is nothing to show. So assume the claim holds for some $k \ge 2$ and let K be an arbitrary terminal set of cardinality $k + 1$. We distinguish two cases. If there exists a rectilinear Steiner minimum tree such that at least one terminal has degree at least 2 then the claim follows from the inductive hypothesis. Otherwise we know from Theorem 10.1 that every rectilinear Steiner tree is either a fir tree or one of five special cases. As all these trees satisfy the claimed restriction on the location of the Steiner points, the theorem follows.   □

With Theorem 10.3 we have reduced the Manhattan Steiner Problem to the Steiner Problem in Networks. In particular, we can apply the algorithms of Chapter 5 to solve the Manhattan Steiner Problem exactly or the algorithms of Chapter 6 and 7 to solve it approximately. To justify an approximate solution we would, however, like to know that the problem is indeed difficult. This does not follow from the results in Chapter 3 directly, but was shown by Garey and Johnson. We state their result without proof.

Theorem 10.4 *The Manhattan Steiner Problem is $\mathcal{NP}$-complete.*

10.2   The Steiner ratios

For the Steiner problem in networks the results of Chapter 6 guarantee a performance ratio of 2. Recall that it was the Steiner ratio ρ_2 which was responsible for this factor, where ρ_2 is defined to be the smallest upper bound for the ratio between the length of a minimum spanning tree and that of a Steiner minimum tree. While there exist simple examples which show that this bound is two for general graphs, perhaps we can do better in the Manhattan Steiner Problem? This is indeed the case.

Theorem 10.5 *In the rectilinear metric, $\rho_2 = 3/2$.*

Proof.   One easily checks that the four points $(0, -1)$, $(1,0)$, $(0,1)$ and $(-1,0)$ imply that ρ_2 is at least $3/2$. It therefore suffices to show that it is also at most $3/2$. To see this we proceed by induction on $k = |K|$. For $k \leq 2$ the claim is easily seen to be true. Indeed, in this case the length of a minimum spanning tree equals that of a rectilinear Steiner minimum tree.

So let $K \subseteq \mathbb{R}^2$ be a terminal set of size $k \geq 3$. If there exists a rectilinear Steiner tree for K such that at least one terminal has degree greater or equal to 2, the claim follows immediately from the inductive hypothesis. If no such Steiner tree exists then Theorem 10.1 applies. We distinguish two cases. If $k \leq 4$ we deduce from Theorem 10.1 that there exists a rectangle R of height h and width w that contains all terminals and satisfies $smt(K) \geq h + w$:

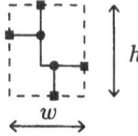

We obtain a spanning tree for K by omitting the longest section of the rectangle. That is,

$$mst(K) \leq \frac{k-1}{k} \cdot 2(h+w) \leq \frac{3}{2}(h+w) = \frac{3}{2}smt(K).$$

So assume $k > 4$. Then Theorem 10.1 tells us that there exists a Steiner minimum tree which is a fir tree. Let T be such a tree. We denote the lengths of the horizontal and vertical edges of T with h_i and v_i as indicated in the following picture (in the picture on the right at least one of h_1 and h_n will always be zero, see below):

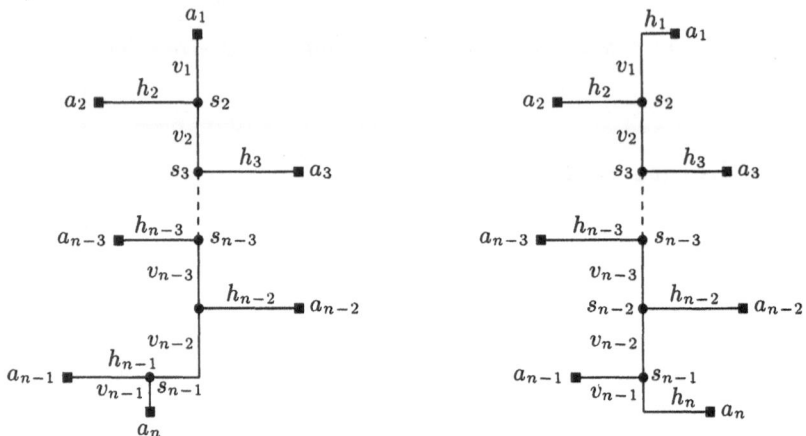

We claim that we may assume without loss of generality that at least one of the following two inequalities holds:

$$h_2 > h_4 \qquad \text{or} \qquad h_{n-3} > h_{n-1}. \tag{10.1}$$

Indeed, this follows easily by considering T as well as its rotation by 180 degrees. (This is the reason for the shape of the second tree: by definition, at least one of h_1 and h_n is zero, but it depends on h_2, h_4, h_{n-3}, and h_{n-1} which one it is.)

In the following we will show that there exists a Steiner point s_q, $2 \leq q \leq n-2$, and a path P through $a_1, \ldots, a_q$ (not necessarily in that order) such that

$$\ell(P) \leq \frac{3}{2}(\ell(T_{q-1}) + v_{q-1}), \tag{10.2}$$

where T_{q-1} is the subtree of T induced by $a_1, \ldots, a_{q-1}, s_2, \ldots s_{q-1}$. From (10.2) the claim follows easily by considering the terminal set $K' = \{a_q, \ldots, a_n\}$. If M' is a minimum spanning tree for K' then $M' \cup P$ is a spanning tree for K. Therefore, by induction,

$$
\begin{aligned}
mst(K) \quad &\leq \quad mst(K') + \ell(P) \\
&\leq \quad \frac{3}{2}smt(K') + \frac{3}{2}(\ell(T_{q-1}) + v_{q-1}) \\
&\leq \quad \frac{3}{2}(\ell(T) - \ell(T_{q-1}) - v_{q-1})) + \frac{3}{2}(\ell(T_q) + v_{q-1}) \\
&= \quad \frac{3}{2}smt(K).
\end{aligned}
$$

To find such a path P we distinguish two cases. Assume first that there exists a $2 \leq q \leq n-1$ so that

$$h_1 + 2\sum_{i=2}^{q-1} h_i + h_q + \sum_{i=1}^{q-1} v_i \leq \frac{3}{2}\left(\sum_{i=1}^{q-1} h_i + \sum_{i=1}^{q-1} v_i\right). \tag{10.3}$$

Then the path P through $a_1, a_2, \ldots, a_q$, in this order, satisfies inequality (10.2). If (10.3) holds for no q between 2 and $n-1$ then in particular we have that

$$\sum_{i=1}^{q-1} v_i \leq \sum_{i=1}^{q} h_i + h_q \qquad \text{for all } 2 \leq q \leq n-1. \tag{10.4}$$

By assumption (10.1) we know that there exists a $4 \leq q \leq n-1$ such that

$$h_q < h_{q-2}, \qquad \text{but } h_i \geq h_{i-2} \text{ for all } 3 \leq i \leq q-1.$$

Choose this q. Then there exists a (closed) tour P_T through $a_1, \ldots, a_q$ whose length corresponds to the perimeter of the circumscribing rectangle (cf. Figure 10.4):

$$\ell(P_T) = 2\left(h_{q-2} + h_{q-1} + \sum_{i=1}^{q-1} v_i\right).$$

Let P' be a path connecting $a_{q-4}, a_{q-2}, a_q, a_{q-1}, a_{q-3}$ in this order. Removing the

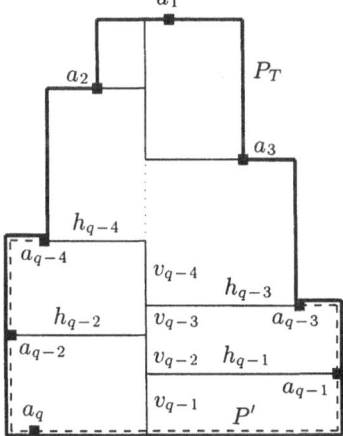

Figure 10.4   The Steiner ratio for $r = 2$.

maximum edge in P' from P_T we obtain a path P containing $a_1, \ldots, a_q$. From (10.4) we deduce that the length of this path satisfies

$$
\begin{aligned}
\ell(P) \;\leq\;& \ell(P_T) - \frac{1}{4}\ell(P') \\[2mm]
=\;& \ell(P_T) - \frac{1}{4}\Big(\ell(P_T) - h_{q-3} - h_{q-4} - \sum_{i=1}^{q-4} v_i - \sum_{i=1}^{q-5} v_i\Big) \\[2mm]
\leq\;& \frac{3}{4}\ell(P_T) + \frac{1}{4}\Big(h_{q-3} + h_{q-4} + \sum_{i=1}^{q-3} h_i + h_{q-3} + \sum_{i=1}^{q-4} h_i + h_{q-4}\Big) \\[2mm]
\leq\;& \frac{3}{4}\ell(P_T) + \sum_{i=1}^{q-3} h_i \\[2mm]
=\;& \frac{3}{2}\Big(h_{q-1} + h_{q-2} + \sum_{i=1}^{q-1} v_i\Big) + \sum_{i=1}^{q-3} h_i \\[2mm]
\leq\;& \frac{3}{2}\big(\ell(T_{q-1}) + v_{q-1}\big).
\end{aligned}
$$

□

Corollary 10.6 *Let $K \subseteq \mathbb{R}^2$ and let T_{opt} be a rectilinear Steiner minimum tree for K. Then Mehlhorn's algorithm computes in $\mathcal{O}(k^2 \log k)$ steps a Steiner tree T_M for K such that*

$$
\ell(T_K) \;\leq\; \frac{3}{2} \cdot \ell(T_{\mathrm{opt}}).
$$

Proof.   The grid graph $G(K)$ induced by K contains at most k^2 vertices and at most $2k(k-1)$ edges. The complexity bound follows.   □

We note, however, that the bound on the complexity in Corollary 10.6 is not best possible. In fact, computing first the grid graph $G(K)$ (which has k^2 vertices) and then a Voronoi partition of these vertices seems rather wasteful. It is much better to compute the Voronoi regions directly for the points in the plane. It can be shown that this can be done in $\mathcal{O}(k \log k)$ time, which we also saw to be the best known bound for computing the minimum spanning tree for k points in the plane.

To also improve the result of Corollary 8.10 we need a better bound for ρ_3.

Theorem 10.7 *In the rectilinear metric, $\rho_3 = 5/4$.*

Sketch of the Proof.   The same four points as in the proof of Theorem 10.5 show that $\rho_3 \geq 5/4$. To see that ρ_3 is also at most $5/4$ we proceed by induction on k. In the inductive step we may assume without loss of generality that T is a fir tree. Our aim is to construct four spanning trees $T_1, \ldots, T_4$ in $H_3(N)$ such that $\sum w_3(T_i) \leq 5|T|$. (Recall that the vertex set of $H_3(N)$ is K, while the edge set consists of all 2- and 3-element subsets of K, and the weight $w_3(e)$ of an edge is the length of Steiner minimum tree for e in the original network, i.e., in our case here, the length of a rectilinear Steiner minimum tree for e.) We exhibit the proof only for case (i) of the definition of a fir tree, that is, for trees T of the form

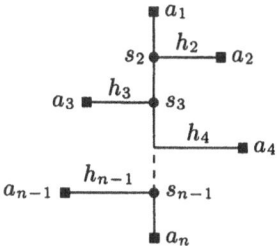

To simplify the following construction we will assume that all the h_i are different. The necessary modifications for the general case are left to the reader.

For the construction of T_1 and T_2 we double all horizontal line segments which are "local maxima", i.e., all line segments $s_i a_i$ such that $h_i > h_{i-2}$ and $h_i > h_{i+2}$. For T_1 we double all those segments to left, for T_2 we double those to the right of the stem of the tree. All full components in T_1 and T_2 look then essentially as the one shown on the left hand side of the following picture:

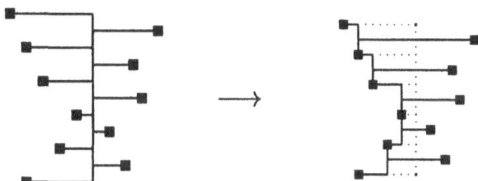

This picture also indicates how such a tree can be modified (without increasing its length) so that all full components contain at most three terminals, implying that one can find a spanning tree in $H_3(N)$ of the same length.

Now we come to the construction of T_3 and T_4. Here we double all remaining horizontal line segments (the left ones for T_3 and the right ones for T_4) and selected vertical line segments (indicated below). After doubling of the horizontal line segments all full components either contain at most three terminals or are centered around a local maximum:

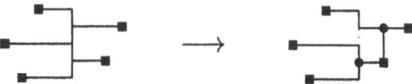

Again, a spanning tree in $H_3(K)$ of the same length is easily found. (Note that for T_3 and T_4 we will never have to double the same vertical segment.)   □

Corollary 10.8 *For all $\epsilon > 0$ there exists a randomized polynomial time approximation algorithm with performance ratio $5/4 + \epsilon$ for the Manhattan Steiner Problem.*

Proof.   From Corollary 8.8 we know that we can find in randomized polynomial time in the hypergraph $H_3(N)$ a spanning tree T of weight at most $w(T) \leq (1 + \epsilon) \cdot mst(H_3(N))$. By definition of $H_3(N$ and ρ_3, we have $mst(H_3(N)) = mst_3(N) \leq \rho_3 \cdot smt(N)$. The corollary is thus an immediate consequence of Theorem 10.7.   □

10.3   An almost linear time approximation scheme

In Chapter 9 we saw that the Steiner tree problem in graphs and networks is $\mathcal{APX}$-complete. Therefore, unless $\mathcal{P} = \mathcal{NP}$, there exists a constant $\epsilon_0 > 0$ such that no polynomial time algorithm can achieve a performance ratio of $1 + \epsilon_0$ or less. In the previous section we have seen that the approximation algorithms from Chapter 6 and 7 yield better performance ratios if we restrict the input to grid graphs $G(K)$ for terminal sets $K \subseteq \mathbb{R}^2$. In this section we will show that even something much stronger holds: there exists an approximation algorithm with performance ratio $1 + \epsilon$ for every $\epsilon > 0$. In other words, the Manhattan Steiner Problem belongs to the class $\mathcal{PTAS}$!

Theorem 10.9 *For every $c \in \mathbb{N}$ there exists a polynomial time approximation algorithm for the Manhattan Steiner Problem with performance ratio $1 + 1/c$.*

Theorem 10.9 is a celebrated result of Arora and Mitchell, who independently proved it in 1996. A year later Arora modified his original algorithm to dramatically reduced the running time. In this section we present this improved version of Arora's algorithm.

We proceed as follows. First we will show (Lemma 10.10) that we may restrict our considerations to terminal sets K which have some special properties. Then we develop the main ideas for the algorithm and analyze its time complexity (Lemma 10.12). The most difficult part (Lemma 10.14) will be the last step, where we will show that the Steiner tree constructed by the algorithm does have the desired performance ratio.

Lemma 10.10 *It suffices to prove Theorem 10.9 for terminal sets K such that $K \subseteq \{1,3,5,\ldots,4ck - 1\}^2$.*

Proof. First observe that we may assume without loss of generality that $K \subseteq [0,M]^2$ for an appropriate constant M and that there exists y_1 and y_2 such that $(0,y_1) \in K$ and $(M,y_2) \in K$. Clearly, if we now multiply all coordinates with $12ck/M$ we just scale the problem, but do not change the structure of a rectilinear Steiner minimum tree. That is, we may in fact assume without loss of generality that $K \subseteq [0, 4 \cdot (3c) \cdot k]^2$ and that the length of a Steiner minimum tree is at least $12ck$.

Now observe what happens if we move every point in K to the closest point with odd integer coordinates. Let K' denote this rounded terminal set, and let T_{opt} (T'_{opt}) denote a rectilinear Steiner minimum tree with respect to the terminal set K (K'). As the (Manhattan) distance between a point $(x,y) \in K$ and its copy in K' is at most two, we know that

$$\ell_1(T'_{\text{opt}}) \le \ell_1(T_{\text{opt}}) + 2k.$$

Similarly, if T' is a rectilinear Steiner tree for K' we can easily transform it into a rectilinear Steiner tree T for K such that

$$\ell_1(T) \le \ell_1(T') + 2k.$$

Observe that this implies that if T' satisfies

$$\ell_1(T') \le (1 + \frac{1}{3c})\ell_1(T'_{\text{opt}}),$$

then

$$\ell_1(T) \le (1 + \frac{1}{3c})(\ell_1(T_{\text{opt}}) + 2k) + 2k \le (1 + \frac{1}{3c})\ell_1(T_{\text{opt}}) + 6k \le (1 + \frac{1}{c})\ell_1(T_{\text{opt}}),$$

where the last inequality follows from the assumption that $\ell_1(T_{\text{opt}}) \ge 12ck$. $\square$

In the following paragraphs we develop the main ingredients of Arora's algorithm. In order to do so we first fix integers t and L such that $L = 2^{t+1} > 4ck \ge 2^t$ and let a and b denote arbitrary *even* integers in $\{0, 2, \dots, L-2\}$. (We will later show how to choose these integers or, more precisely, show that choosing them uniformly at random will be a good strategy.)

Within the algorithm we will only consider horizontal and vertical lines through *even* integers. We will denote such lines by H resp. V. Note that the use of the letter H or V will always tacitly imply that the line runs through *even* integers. A vertical line through points with x-coordinate m is said to be at level i with respect to a if $m = a + s \cdot \frac{L}{2^i}$ for some *odd* (potentially negative) integer s. Similarly, we say a horizontal line through points with y-coordinate n is said to be at level i with respect to b if $n = a + s \cdot \frac{L}{2^i}$ for some *odd* integer s. Observe that the square $[0, 4ck]^2$ (which contains all terminals) is intersected by exactly one horizontal and one vertical line at level 0 (the one with y-coordinate b, resp. x-coordinate a) and that these two lines split the original square into at most four rectangles of smaller size. We will say that these rectangles are at level 0. Now consider the vertical and horizontal lines at level 1. They split every rectangle at level 0 into at most four rectangles at level 1. Continuing in this way we finally end up with rectangles at level t (which have size at most 2×2). Figure 10.5 illustrates this procedure for $k = 3$, $c = 1$, $t = 3$ and $a = 6$, $b = 12$.

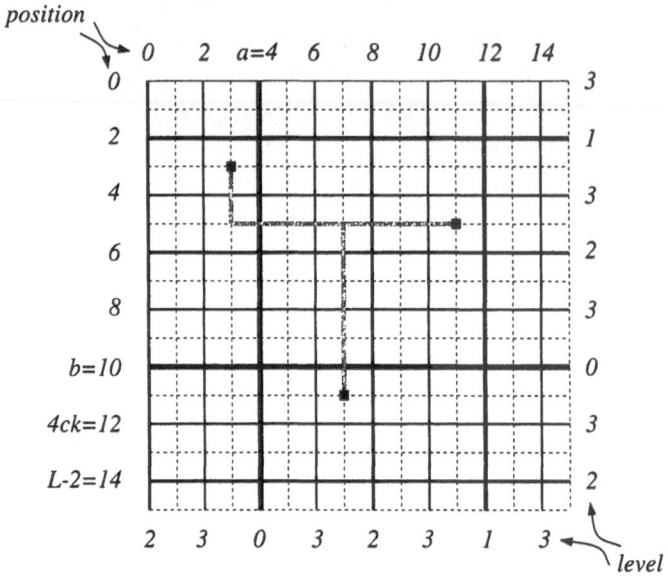

Figure 10.5 Subdivision of the square $[0,4ck]^2$ into rectangles at levels 0 up to t.

Observe that all terminal sets are contained *within* some rectangle at level t. This follows from our assumption that all coordinates of terminals in K are odd integers, while a and b are even integers. Recall also that by Theorem 10.3 there exists a rectilinear Steiner minimum tree T_{opt} that is contained in the grid induced by the terminal set. Note that this implies that all Steiner points of T_{opt} are contained *within* some rectangle at level t and that no horizontal or vertical line H or V can contain a line segment from T_{opt}, cf. also Figure 10.5.

These facts suggest a bottom-up strategy. Namely, first compute all possible solutions for the rectangles at level t and then use these to compute all solutions for rectangles at level $t-1$. Then use these to compute all solutions for rectangles at level $t-2$. And so on, up to the rectangles at level 0, from which we can easily compute a rectilinear Steiner tree for the whole terminal set K.

There are a couple of points we have to specify more precisely. First let us clarify what we mean by a "solution". For a rectangle that contains only a (proper) subset of the terminal set K, a solution is a collection of rectilinear trees such that every tree is connected to at least one point on the boundary of the rectangle and that every terminal is contained in exactly one of the trees. Note, however, that we do allow trees that contain no terminal. For rectangles that contain *all* terminals from K, a solution is a rectilinear Steiner tree in the usual sense. (Note that this implies that a solution for the original square $[0,4ck]^2$ is exactly a rectilinear Steiner tree, as desired.)

It should be obvious that this approach leads to a valid algorithm (i.e., one that really computes a rectilinear Steiner minimum tree). There is, however, one drawback: it is not a polynomial time algorithm! The reason for this is that we may have to store exponentially many different solutions for the rectangles at level $i \geq 1$.

In what follows we will therefore modify this algorithm in order to achieve polynomial running time. The main difference will be that for a rectangle at level $i \geq 1$ we compute and store not *all* possible solutions, but only those which have a very special structure. We will call these special solutions (m,r)-standardized. An (m,r)-*standardized solution* is a solution such that:

(i) It shares no line segment with the boundary of the rectangle.

(ii) It contains at most r points on each of the four boundary lines of the rectangle and all these points are contained in a restricted subset of points from the boundary, which are called *portals*. Portals on a horizontal (vertical) boundary of a rectangle at level i are points that have an x-coordinate (y-coordinate) of the form $a + s\frac{L}{m2^i}$ (respectively $b + s\frac{L}{m2^i}$) for some integer s. (Note that a rectangle at level i has side length at most $L/2^i$ and that hence every such rectangle contains at most $m - 1$ portals on each of its four sides. Note also that we do *not* require that the portals have integer coordinates.)

We need one more observation. Namely, that in order to combine the solutions of rectangles at level i to solutions for rectangles at level $i - 1$ it is *not* necessary to know the *precise shapes* of the trees contained in the solutions for the rectangles at level i. In fact, it suffices to know which points of the boundary belong to the same tree.

As every (m,r)-standardized solution contains at most $4r$ points on the boundary, these points can be partitioned into at most $4r$ nonempty subsets. As there are less than $y^x/y!$ ways to partition x elements in y subclasses, we thus deduce that it suffices to store at most

$$\left(\sum_{j=0}^{r} \binom{m-1}{j} \right)^4 \cdot \left(\sum_{i=1}^{4r} \frac{i^{4r}}{i!} \right) \leq (4mr)^{4r}$$

different (m,r)-standardized solutions for every rectangle. (Note that this number is independent of the level of the rectangle.)

Algorithm 10.11 (ARORA'S APPROXIMATION SCHEME)
Input: An integer $c \in \mathbb{N}$ and a terminal set K such that $K \subseteq \{1,3,\ldots,4ck\text{-}1\}^2$.
Output: A rectilinear Steiner tree T.

{ Initialization }
Let $t := \lceil \log_2(4ck + 1) \rceil$, $L := 2^t$, $m := 4ct$, and $r := 8c$.
Choose integers $a,b \in \{0,2,\ldots,L-2\}$ uniformly at random.
Compute for all horizontal and vertical lines its level, and determine for all $0 \le i \le t$ the set of rectangles at level i. Compute for all rectangles at level t the set of all (m,r)-standardized solutions.
{ Recursion }
for $i := t - 1$ **down to** 0 **do**
    for all rectangles R at level i **do**
        Consider all combinations of solutions for the four rectangles at level $i + 1$ which are contained in R. Store all (m,r)-standardized solutions. If there is more than one solution for a given set of portals and a given partition of the point set, store only a solution of minimum length.
{ Computation of the output }
Consider all combinations of solutions for the at most four rectangles at level 0 that have the property that their union is a rectilinear Steiner tree for K. Return the one which has minimum length.

Lemma 10.12 *The complexity of Algorithm 10.11 is $\mathcal{O}(mk^2 \cdot (4mr)^{16r})$.*

Proof.   Observe that there are at most 4^{i+1} rectangles at level i. Now let us bound the time required to compute the solution set for some rectangle at level i. For each of the four rectangles at level $i + 1$ contained in it, we have stored at most $(4mr)^{4r}$ solutions. Checking the type and the length of a solution which arises by combining four of these solutions can certainly be done in time linear in m and the number of terminals contained in the rectangle. That is, the overall complexity of the algorithm can be bounded by

$$\sum_{i=0}^{t} \left((4mr)^{4r}\right)^4 \cdot \mathcal{O}(m \cdot 4^{i+1} + k) \; = \; (4mr)^{16r} \cdot \mathcal{O}(m \cdot \sum_{i=0}^{t} 4^{i+1} + \sum_{i=0}^{t} k)$$

$$= \; \mathcal{O}(mk^2 \cdot (4mr)^{16r}).$$

$\square$

Remark 10.13 An alert reader might be wondering whether Algorithm 10.11 is really best possible. Indeed it is not. The reason is that we spend a lot of time combining solutions for rectangles containing *no* terminal at all. This is not necessary. In fact it suffices to consider rectangles which contain at least one terminal. In other words, we do not subdivide a rectangle at level i if it contains only one terminal. For the terminal set in Figure 10.5, for example, we could stop with the four rectangles at level 0. It is not difficult to check that for

a) Reducing the number of crossings.   **b)** Moving crossings to portals.

Figure 10.6 Modifying an arbitrary Steiner tree into an (m,r)-standardized tree.

a rectangle which contains only one terminal, the set of all (m,r)-standardized solutions can easily be computed from scratch. One can show that the complexity of the algorithm modified in this way is reduced to $\mathcal{O}(|R| \cdot m(4mr)^{16r})$, where $|R|$ denotes the total number of rectangles containing at least one terminal. It is also not difficult to verify that $|R| = \mathcal{O}(k \log(4ck))$. A detailed proof of these facts is left to the reader, cf. Problem 10.5.

Lemma 10.14 *Algorithm 10.11 computes with probability at least 1/2 a rectilinear Steiner tree T that satisfies*

$$\ell_1(T) \leq (1 + \frac{1}{c}) \ell_1(T_{\text{opt}}).$$

Proof. Recall that our assumption that all terminals have odd integer coordinates together with Theorem 10.3 ensures that there exists a rectilinear Steiner minimum tree T_{opt} such that all Steiner points of T_{opt} are contained in the interior of some rectangle at level t and such that none of the horizontal and vertical lines H and V intersects T_{opt} in more than one point.

We now modify T_{opt} step by step so that in the end we have a rectilinear Steiner tree T for which the intersection with *every* rectangle at level $1 \leq i \leq t$ is an (m,r)-standardized solution. We again proceed in a bottom-up fashion:

$T := T_{\text{opt}};$
for $i := t$ **down to** 0 **do**
    for all rectangles R at level i **do**
        If T intersects some boundary of R in more than r points,
        modify T according to Figure 10.6 a).
        Modify T by moving every crossing with the boundary of R
        to the closest portal according to Figure 10.6 b).
  If T contains cycles, remove line segments until T is a tree.

(Note that the fact that we proceed "bottom-up" implies that all changes which occurred within a rectangle at level i will never increase the number of crossings on the boundary of rectangles at level less than i. We will use this fact heavily in the further analysis of the algorithm.)

We now bound the difference $\ell_1(T) - \ell_1(T_{\mathrm{opt}})$. We start with some notation. For a horizontal line H we let $lev(H,b)$ denote the level of the line H with respect to b and we let $\pi(H)$ denote the number of crossings of T_{opt} with H. For a vertical line V we define $lev(V,a)$ and $\pi(V)$ similarly. Before we continue with our analysis we note that

$$\sum_H \pi(H) + \sum_V \pi(V) \; = \; \frac{1}{2}\ell_1(T_{\mathrm{opt}}). \qquad (10.5)$$

This immediately follows if we consider the line segments in T_{opt} separately. Consider for example a horizontal line segment of length ξ. As this line segment starts and ends in points with odd integer coordinates, it crosses exactly $\xi/2$ vertical lines V.

We first bound the total increase due to modifications according to Figure 10.6 a). For a horizontal line H let $c(H,a,b,i)$ denote the number of times a modification according to Figure 10.6 a) occurred along H due to a rectangle at level i. Note that, clearly, $c(H,a,b,i) = 0$ for all $i < lev(H,b)$. Note also that if b_1 and b_2 are two values such that $lev(H,b_1) \geq lev(H,b_2)$, then $c(H,a,b_1,i) = c(H,a,b_2,i)$ for all $i \geq lev(H,b_2)$ (as the rectangles at level $\geq lev(H,b_2)$ are in fact identical in both cases). That is, we know that there exist values $\tilde{c}(H,a,i)$ (that do not depend on b) such that

$$c(H,a,b,i) = \begin{cases} \tilde{c}(H,a,i), & \text{if } i \geq lev(H,b), \\ 0, & \text{otherwise.} \end{cases}$$

For a vertical line V we define $\tilde{c}(V,b,i)$ similarly. Observe that every application of the construction in Figure 10.6 a) reduces the number of crossings by at least r. That is, we have

$$\sum_{i=1}^{t} \tilde{c}(H,a,i) \leq \frac{\pi(H)}{r} \qquad \text{and} \qquad \sum_{i=1}^{t} \tilde{c}(V,b,i) \leq \frac{\pi(V)}{r}, \qquad (10.6)$$

for all horizontal and vertical lines H and V. With the above notation at hand, we can bound the total increase due to modifications according to Figure 10.6 a) by

$$\sum_H \sum_{i=lev(H,b)}^{t} \tilde{c}(H,a,i) \cdot \frac{2L}{2^i} + \sum_V \sum_{i=lev(V,a)}^{t} \tilde{c}(V,b,i) \cdot \frac{2L}{2^i}. \qquad (10.7)$$

Next we aim at bounding the total increase due to modifications according to Figure 10.6 b). Observe what happens to crossings which are modified according to Figure 10.6 b). If such a modification takes place for a rectangle at level i, the increase is at most $2 \cdot \frac{L}{2m2^i} = \frac{L}{m2^i}$. We can therefore bound the total increase due to modifications according to Figure 10.6 b) by

$$\sum_{H} \sum_{i=lev(H,b)}^{t} \pi(H) \cdot \frac{L}{m2^i} + \sum_{V} \cdot \sum_{i=lev(V,a)}^{t} \pi(V) \cdot \frac{L}{m2^i}. \qquad (10.8)$$

With the help of (10.7) and (10.8) we are now able to compute the expectation of $\ell_1(T) - \ell_1(T_{\text{opt}})$, where the expectation is with respect to the random choice of a and b in $\{0,2,4,\ldots,L-2\}$.

$$\text{Ex}\left[\ell_1(T) - \ell_1(T_{\text{opt}})\right] \leq \frac{1}{L^2/4} \sum_{a} \sum_{b} \left[\sum_{H} \sum_{i=lev(H,b)}^{t} \left(\frac{2L\tilde{c}(H,a,i)}{2^i} + \frac{L\pi(H)}{m2^i} \right) \right.$$

$$\left. + \sum_{V} \sum_{i=lev(V,a)}^{t} \left(\frac{2L\tilde{c}(V,b,i)}{2^i} + \frac{L\pi(V)}{m2^i} \right) \right].$$

In order to simplify this expression we first consider only the horizontal lines.

$$\sum_{a} \sum_{b} \sum_{H} \sum_{i=lev(H,b)}^{t} \left(\frac{2L\tilde{c}(H,a,i)}{2^i} + \frac{L\pi(H)}{m2^i} \right)$$

$$= \sum_{a} \sum_{H} \sum_{i=1}^{t} \underbrace{\#\{b : lev(H,b) = i\}}_{\leq 2^i} \left(\frac{2L\tilde{c}(H,a,i)}{2^i} + \frac{L\pi(H)}{m2^i} \right)$$

$$\leq \sum_{a} \sum_{H} \sum_{i=1}^{t} \left(2L\tilde{c}(H,a,i) + \frac{L\pi(H)}{m} \right)$$

$$\overset{(10.6)}{\leq} \sum_{a} \sum_{H} \left(\frac{2L\pi(H)}{r} + \frac{tL\pi(H)}{m} \right)$$

$$= \sum_{H} \left(\frac{L^2\pi(H)}{r} + \frac{tL^2\pi(H)}{2m} \right) = \left(\frac{L^2}{r} + \frac{tL^2}{2m} \right) \cdot \sum_{H} \pi(H).$$

Proceeding similarly for the vertical lines we obtain

$$\text{Ex}\left[\ell_1(T) - \ell_1(T_{\text{opt}})\right] = \frac{4}{L^2} \left(\frac{L^2}{r} + \frac{tL^2}{2m} \right) \cdot \left(\sum_{H} \pi(H) + \sum_{V} \pi(V) \right)$$

$$\overset{(10.5)}{=} \left(\frac{4}{r} + \frac{2t}{m} \right) \cdot \frac{1}{2} \ell_1(T_{\text{opt}})$$

$$= \frac{1}{2c} \ell_1(T_{\text{opt}}),$$

by choice of r and m.

For the final step recall Markov's inequality (Lemma 1.14). It implies in particular that the probability that a random variable is twice as large as its expectation is at most $1/2$. Using this fact for the random variable $\ell_1(T) - \ell_1(T_{\mathrm{opt}})$ we deduce that

$$\Pr[\ell_1(T) - \ell_1(T_{\mathrm{opt}}) \geq \frac{1}{c}\ell_1(T_{\mathrm{opt}})] \leq \frac{1}{2},$$

which concludes the proof of the lemma.   □

Combining Lemma 10.10, Lemma 10.12 (and its improvement according to Remark 10.13, cf. Problem 10.5) and Lemma 10.14 we are finally in the position to state Arora's result in its strongest version.

Theorem 10.15 *There exists a randomized algorithm $\mathcal{R}$ that computes for every $c \in \mathbb{N}$ and terminal set K in time $\mathcal{O}(k \cdot (c \log k)^{\mathcal{O}(c)})$ with probability at least $1/2$ a rectilinear Steiner tree T for K such that $\ell_1(T) \leq (1 + \frac{1}{c})\ell_1(T_{\mathrm{opt}})$, where T_{opt} denotes a rectilinear Steiner minimum tree for K.*   □

Note that there are only $L^2/4 = \mathcal{O}(c^2 k^2)$ many different choices for the pair (a,b). Clearly, if we try *all* these pairs we will deterministically find a rectilinear Steiner tree whose length is close to that of a Steiner minimum tree. Thus we obtain the following corollary of Theorem 10.15.

Corollary 10.16 *There exists a deterministic algorithm $\mathcal{A}$ that computes for every $c \in \mathbb{N}$ and terminal set K in time $\mathcal{O}(k^3 \cdot (c \log k)^{\mathcal{O}(c)})$ a rectilinear Steiner tree T for K such that $\ell_1(T) \leq (1 + \frac{1}{c})\ell_1(T_{\mathrm{opt}})$, where T_{opt} denotes a rectilinear Steiner minimum tree for K.*   □

10.4   Excursion: The Euclidean Steiner problem

We now return to Fermat's problem which we stated in the introduction of this chapter: given three points a, b and c in the plane, find a fourth point that minimizes the sum of its distances to the three given points. Torricelli showed that the three circles circumscribing the equilateral triangles constructed on the sides of and outside the given triangle $\triangle abc$ intersect in the desired point. In his honor this point is nowadays called the *Torricelli point*. In 1647 Cavalieri provided another characterization of the Torricelli point: it is exactly that point which induces angles of 120° with a, b and c.

Yet another characterization is due to Simpson. In 1750 he showed that the three lines joining the outside vertices of the three equilateral triangles to the opposite vertices of the given triangle intersect in the Torricelli point. Moreover, he proved that these three lines, now known as the *Simpson lines*, all have the same length and that this length is exactly the sum of the distances of the Torricelli point to a, b and c.

In this section we first develop an algorithm for finding the Torricelli point in a given triangle. While doing this, in passing we will provide proofs for the three characterizations mentioned above. In a second step we will then generalize this algorithm for computing a Torricelli point to one which computes a Euclidean Steiner minimum tree for a given set of points in the plane.

We start with a few easy lemmas from Euclidean geometry.

Lemma 10.17 *Let $\triangle abc$ be an equilateral triangle with circumscribing circle C. Then all points p on the smaller segment of C between a and b satisfy $\angle apb = 120°$, while all points q on the larger segment satisfy $\angle aqb = 60°$.*

Proof. Let p be any point on the smaller segment of C between a and b and assume without loss of generality that p is at most as close to b as to a:

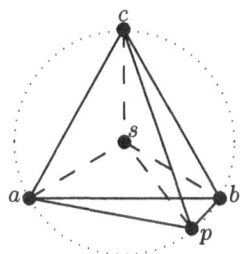

By symmetry, it suffices to show that $\angle apb = 120°$ and $\angle apc = 60°$. Let s be the center of the circle C. Then $\triangle asp$ and $\triangle bsp$ are isosceles triangles. Therefore

$$\angle apb = \tfrac{1}{2}(180° - \angle asp) + \tfrac{1}{2}(180° - \angle bsp) = 180° - \tfrac{1}{2}\angle asb = 120°,$$

as the triangle $\triangle abc$ is equilateral. Similarly, now considering the triangles $\triangle asp$ and $\triangle csp$, we obtain

$$\angle apc = \tfrac{1}{2}(180° - \angle asp) + \tfrac{1}{2}(180° - \angle csp) = 180° - \tfrac{1}{2}(\angle asb + \angle bsc) = 60°.$$

$\square$

Lemma 10.18 *Let p be any point in an equilateral triangle $\triangle abc$ with height h, and let h_a, h_b and h_c be the distances of p to the sides of the triangle opposite from a, b and c, respectively. Then $h = h_a + h_b + h_c$.*

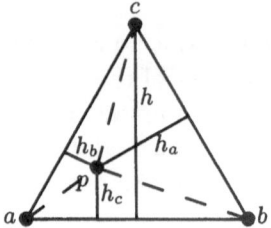

Proof.   Let $|\triangle xyz|$ denote the area of a triangle $\triangle xyz$. Then, obviously

$$|\triangle abc| = |\triangle apb| + |\triangle bpc| + |\triangle cpa|.$$

As the area of any triangle is $\frac{1}{2} \cdot h \cdot |xy|$ if $|xy|$ denotes the length of any one of its sides and h is the height of the triangle with respect to this side, this implies

$$\tfrac{1}{2} \cdot h \cdot |ab| = \tfrac{1}{2} \cdot h_c \cdot |ab| + \tfrac{1}{2} \cdot h_a \cdot |bc| + \tfrac{1}{2} \cdot h_b \cdot |ac|.$$

As $\triangle abc$ is by assumption an equilateral triangle, this obviously concludes the proof of the lemma.   □

For triangles in which all angles are less than $120°$, we can now easily deduce Cavalieri's characterization of the Torricelli point.

Corollary 10.19 *Let a,b,c be three points such that all internal angles of the triangle $\triangle abc$ are less than $120°$, and let p be a point in the interior of $\triangle abc$ such that the angle between any two of the line segments ap, bp and cp is $120°$. Then p is the uniquely defined Torricelli point for the triangle $\triangle abc$.*

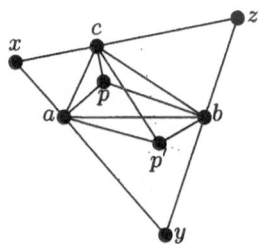

Proof.   Construct a triangle xyz such that xy is perpendicular to ap, xz is perpendicular to cp, and yz is perpendicular to bp. Then $\angle yxz = \angle xyz = \angle xzy = 60°$. That is, the triangle $\triangle xyz$ is equilateral. From Lemma 10.18 we therefore deduce that $|ap| + |bp| + |cp|$ is equal to the height of $\triangle xyz$. Now consider an arbitrary point p' different from p. The sum of the distances from p' to the three points a,b and c is certainly larger than the sum of the distances to the three lines xy, yz, and xz. As the latter is again equal to the height of $\triangle xyz$, this implies that p is the uniquely determined Torricelli point of the triangle $\triangle abc$.   □

Torricelli's and Simpson's characterizations follow also easily from Lemma 10.18. We leave these two proofs to the reader, cf. Problem 10.6 and 10.7. For triangles where one angle is at least 120°, the Torricelli point coincides with one of the three points defining the triangle. In order to prove this we need one more lemma.

Lemma 10.20 *Let a,b and c be any three points. If the Torricelli point p is different from a, b and c then any pair of line segments ap, bp and cp meet in an angle of exactly 120° in p.*

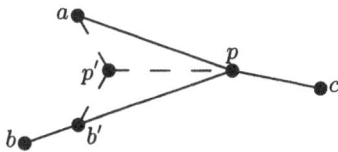

Proof. Assume the contrary, i.e., assume ap and bp meet in an angle $\angle apb < 120°$. Without loss of generality assume that $|ap| \leq |bp|$ and let b' be the point on bp such that $|b'p| = |ap|$. All angles in the isosceles triangle $\triangle ab'p$ are then less than 120°. Corollary 10.19 therefore implies that the Torricelli point p' of the triangle $\triangle ab'p$ is uniquely defined. In particular we therefore have $|ap'| + |b'p'| + |p'p| < |ap| + |b'p|$. But this implies

$$|ap'| + |bp'| + |cp'| \leq |ap'| + |bb'| + |b'p'| + |p'p| + |cp| < |bb'| + |ap| + |b'p| + |cp|,$$

contradicting the fact that p is the Torricelli point of the triangle $\triangle abc$. □

Corollary 10.21 *Let a,b,c be three points such that $\angle abc$ is at least 120°. Then b is the Torricelli point for the triangle $\triangle abc$.*

Proof. It suffices to observe that on the one hand the Torricelli point can't lie outside of the triangle $\triangle abc$ and that on the other hand the angle of any point p in the interior of $\triangle abc$ with a and c is larger than 120°. □

For an algorithmic construction of the Torricelli point we can thus restrict our considerations to triangles in which all internal angles are at most 120°.

Algorithm 10.22 (CONSTRUCTION OF THE TORRICELLI POINT)
Input: $a,b,c \in \mathbb{R}^2$ such that all internal angles of the triangle $\triangle abc$ are less than 120°.
Output: The Torricelli point p for a,b,c.
(1) Construct an equilateral triangle $\triangle abd$ such that d is on one side of the line ab and c on the other.
(2) Construct the circle C circumscribing the triangle $\triangle abd$.
(3) The Torricelli point p is given by the intersection of the line segment cd with the circle C.

Lemma 10.23 *Algorithm 10.22 is correct and the point p satisfies $|ap| + |bp| = |dp|$.*

Proof.   To see that the intersection p of the line segment cd with the circle C is indeed the Torricelli point of $\triangle abc$ by Corollary 10.19 it suffices to check that the angle between any two of the line segments ap, bp and cp is 120°. This, however, follows immediately from Lemma 10.17, as this lemma implies that $\angle apb = 120°$ and $\angle apd = \angle dpb = 60°$.

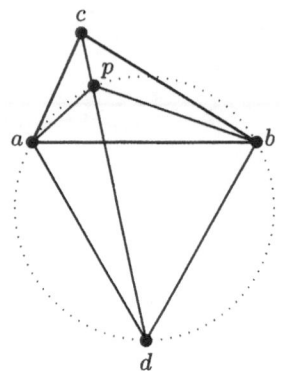

To show the claimed equality of the lengths of the line segments ap, bp and dp we denote by α, β and γ the angles as shown in the following figure and denote by s the center of the circle:

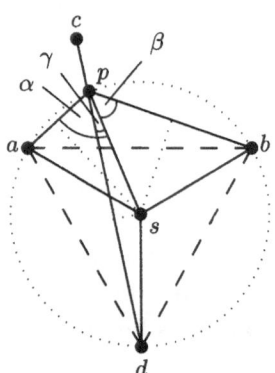

Recalling that in a right-angled triangle the length of the two smaller sides are given by the length of the hypotenuse times the cosines of the included angle we deduce

$$\tfrac{1}{2}|ap| = |ps| \cdot \cos \alpha, \qquad \tfrac{1}{2}|bp| = |ps| \cdot \cos \beta, \qquad \text{and} \qquad \tfrac{1}{2}|dp| = |ps| \cdot \cos \gamma.$$

That is, it suffices to show that $\cos \alpha + \cos \beta = \cos \gamma$. To see this, observe first that $\beta = 120° - \alpha$ and $\gamma = \alpha - 60°$. It is then a routine exercise (which we leave to the reader, use $\cos(x \pm y) = \cos x \cos y \mp \sin x \sin y$) to check that indeed $\cos \alpha + \cos(120° - \alpha) = \cos(\alpha - 60°)$.   □

A *Euclidean Steiner tree* is a connected acyclic collection of line segments in the plane which intersect only in their endpoints. The *degree* of a point is the number of segments incident to it. A Euclidean Steiner tree for a given set of terminals $K \subseteq \mathbb{R}^2$ is a tree such that each terminal is an endpoint of some segment in the tree. Without loss of generality we can obviously require that a Steiner tree contains no degree 2 points which are not terminals. All points of degree at least 3 which are not terminals are called the *Steiner points* of the tree. Note that the arguments in the proof of Lemma 10.20 immediately imply that in a Steiner minimum tree all line segments must meet at angles of exactly 120° degree. In particular, we therefore deduce that all Steiner points must have degree exactly three.

Corollary 10.24 *Let T be a Euclidean Steiner minimum tree for a terminal set K. Then all Steiner points in T have degree 3 and any two line segments incident to a Steiner point meet at an angle of exactly 120°.*   □

Melzak was the first to propose a finite algorithm for the Euclidean Steiner problem. The key ingredient of his algorithm is the algorithmic construction of the Torricelli point together with Lemma 10.23. To see how it is used, assume for the moment that by some means we already know the *topology* of a Steiner minimum tree T for a terminal set K. That is, we know the number of Steiner points contained in T and all pairs of terminals and/or Steiner points which are connected by line segments — but not the precise position of the Steiner points. In Problem 1.2 we have seen that implies that every Steiner minimum tree T which contains at least one Steiner point also contains at least two terminals, a and b say, which are connected to the same Steiner point, p say. Let c be the third neighbor of p. Then Lemma 10.23 tells us that p is the Torricelli point of the triangle $\triangle abc$. And, even more importantly, it tells us that the tree $\tilde{T}$ obtained from T by deleting the two line segments ap and bp and adding the one from p to d has the same length as T. That is, in order to find the Steiner minimum tree T for K it suffices to find a Steiner minimum tree for $(K \setminus \{a,b\}) \cup \{d\}$ in which d is connected to c.

This is almost a complete algorithm for finding a minimum Steiner tree with a given topology. Almost, because we cheated at one point: Lemma 10.23 tells us that p is determined by the equilateral triangle constructed on the side ab *outside* of the triangle $\triangle abc$ — but c might be another Steiner point and we thus don't know its location. That is, to ensure that we end up with the correct Steiner tree we have to consider *both* possibilities. Algorithm 10.25 is an informal sketch of Melzak's algorithm.

Algorithm 10.25 (Melzak's Algorithm)

Input: A set of terminals and a topology.

Output: A shortest Steiner tree with respect to the given topology, if one exists.

{ *Generation of the subproblems* }

while there exists at least one Steiner point **do**

  Determine two terminals a and b which are connected to a Steiner point p.
  Let c be the third neighbor of p. Compute the two points d_1 and d_2 which
  form an equilateral triangle with a and b. Remove a and b from the terminal
  set and the topology.
  If c is a terminal then from d_1 or d_2, that point that lies outside of the
  triangle $\triangle abc$, as a new terminal connected to c.
  Otherwise generate two new problems, one for d_1 and one for d_2. In both
  of them replace the Steiner point p by a terminal with coordinates d_i.
  Handle the generated problems recursively.

{ *Expansion phase* }

while T does not contain all original terminals **do**

  if there exists no Steiner point **then**

    Connect the terminals by line segments according to the current topology.

  else

    {*Let a and b be the two terminals removed at this step, d the inserted
    terminal, and c the neighbor of d in the current Steiner tree T.*}
    If the 120°-segment of the circle circumscribing $\triangle abd$ intersects the line
    segment dc then connect a and b with the intersection point p and delete
    the line segment dp.
    Otherwise **stop**: the given topology admits no valid Steiner minimum
    tree.

Figure 10.7 illustrates the algorithm for a small example. Here we assume
that the topology is $\frac{1}{2})$—$\binom{3}{4}$. In the first step we choose $a = 1$ and $b = 2$, resulting
in two subproblems, both with topology 5—$\binom{3}{4}$. In the second step we choose in
both cases $a = 5$ and $b = 4$, resulting in four subproblems, all with topology
6—3. From here we proceed to the expansion phase. In all four cases we start
with a single line segment connecting the two remaining terminals and a circle
circumscribing the points 4, 5 and 6. In the second and fourth configuration the
stop criterion is satisfied. In the other two we replace the line segment between 3
and 6 by three line segments as indicated in the picture. In the next iteration the
algorithm generates the correct Steiner minimum tree for the first configuration,
while for the third it stops again with failure.

The complexity of Melzak's algorithm is bounded by two to the number of
Steiner points in the given topology. As the latter is at most $k - 2$, we conclude
that $\mathcal{O}(2^k)$ is an upper bound.

To compute a Euclidean Steiner minimum tree for a given terminal set, we
simply enumerate all possible topologies (at most $\mathcal{O}(c^k k!)$, cf. Problem 10.9),
apply Melzak's algorithm for each of them, and return the shortest of the Steiner
tree which are computed.

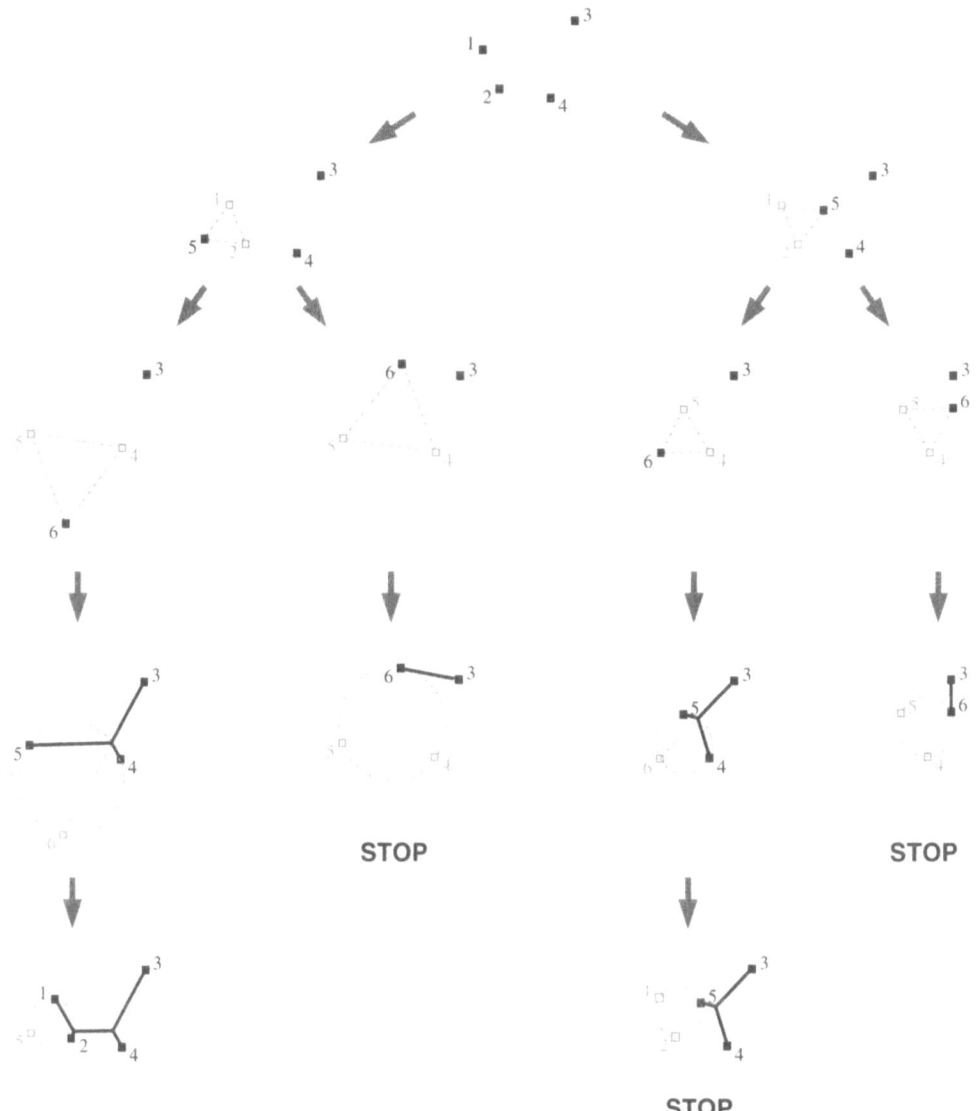

Figure 10.7 Illustration of Melzak's algorithm for four terminals 1, 2, 3 and 4 and topology $\frac{1}{2}\rangle\!-\!\langle\frac{3}{4}$

There is, however, one difficulty which we have neglected so far: how are we going to compare the lengths of two trees? While at first sight this question may sound rather silly, it is in fact highly nontrivial. The reason is that even if all coordinates of the terminals are integers, the lengths of the edges may be irrational. So, when comparing the lengths of two trees we in fact have to compare two sums of square roots – and, till today, the complexity of this problem is open. In particular, no polynomial time algorithm is known.

For practical purposes there is an easy way around that problem: We just approximate the square roots. The most elegant way to do this is to take the integral part of the root. By previously multiplying the coordinates with a suitable integer, M say, we can achieve any desired precision. Indeed, let x and y be any two points, let $\ell_2(\cdot,\cdot)$ denote the usual Euclidean norm, and let $\phi_M(x)$ denote the function which multiplies every coordinate of x by M. Then

$$\ell_2(x,y) - \frac{1}{M}\lfloor \ell_2(\phi_M(x),\phi_M(y))\rfloor \leq \frac{1}{M}.$$

As any Steiner tree on k points contains at most $2k - 2$ edges, this implies that the total deviation of the lengths of the two trees is bounded by $(2k - 2)/M$. That is, for $M = (2k - 2)/\epsilon$ the (absolute) error is bounded by ϵ.

The above argument indicates that the "real" Euclidean Steiner problem and the discretized version are tightly connected. Indeed, while the former is not even known to be in $\mathcal{NP}$, it was shown by Garey, Graham, and Johnson that it is at least as hard as the discretized version, and that the latter is $\mathcal{NP}$-complete. We omit the technically quite involved proof.

Theorem 10.26 *The Euclidean Steiner Problem is $\mathcal{NP}$-hard.*

By suitably refining Arora's approximation scheme (Algorithm 10.11) for the Manhattan Steiner problem one can, however, show that Theorem 10.15 holds for the Euclidean Steiner problem as well, cf. Problem 10.11.

Theorem 10.27 *There exists a randomized algorithm that computes for every $c \in \mathbb{N}$ and terminal set $K \subseteq \mathbb{R}^2$ in time $\mathcal{O}(k \cdot (c\log k)^{\mathcal{O}(c)})$ with probability at least $1/2$ a Euclidean Steiner tree T for K such that $\ell_2(T) \leq (1 + \frac{1}{c})\ell_2(T_{\mathrm{opt}})$, where T_{opt} denotes a Euclidean Steiner minimum tree for K.* $\qquad\square$

Problems

10.1 Show that there exists a polynomial time algorithm for the Manhattan Steiner problem if the terminals lie on the perimeter of a rectangle.

10.2 Fill in the details of the proof of Theorem 10.7.

10.3 Show that in the rectilinear metric $\rho_r \geq 1 + \frac{1}{2r-1}$ for all r ≥ 4.

10.4 One can show that in the rectilinear metric $\rho_r = 1 + \frac{1}{2r-1}$ for all r ≥ 4. Give a proof of the slightly weaker result $\rho_r \leq 1 + \frac{1}{2r-2}$.

10.5 Fill in the details in order to achieve the result outlined in Remark 10.13.

10.6 Let $\triangle abc$ be a triangle such that all internal angles are less than $120°$. Show that the circles circumscribing the equilateral triangles constructed on the sides of and outside the triangle intersect in the Torricelli point.

10.7 Show that the line segments joining the outside vertices of the equilateral triangles defined in Problem 10.6 to the opposite vertices intersect in the Torricelli point.

10.8 A Steiner topology is called *full* if all terminals have degree one. Let $f(k)$ denote the number of full topologies for the Euclidean Steiner problem with k terminals in which all Steiner points have degree 3. Show that $f(k) = 2^{-k+2}(2k-4)!/(k-2)!$.

10.9 Let $t(k)$ denote the number of Steiner topologies for k terminals. Show that there exists a constant c such that $t(k) \leq c^k \cdot k!$ for all $k \geq 2$.

10.10 Prove that $\rho_2 = 2/\sqrt{3}$ for the Euclidean Steiner problem with at most 3 terminals.

10.11 Show how to modify Arora's approximation scheme (Algorithm 10.11) in order to prove Theorem 10.27.

Notes

Theorem 10.1 is due to Hwang (1976) who used it in order to prove Theorem 10.5. Our proof of Theorem 10.1 also uses some ideas from Richards, Salowe (1991). Theorem 10.3 is due to Hanan (1966), who gave a direct proof for it that does not need Theorem 10.1. An $\mathcal{O}(k \log k)$ algorithm for computing the Voronoi regions of k points in the plane can be found in Shamos and Hoey (1975).

Theorem 10.7 is a result of Berman, Fößmeier, Karpinski, Kaufmann, and Zelikovsky (1994). For $r \geq 4$ the Steiner ratios ρ_r for the Manhattan Steiner problem were determined to be $\rho_r = 1 + \frac{1}{2r-1}$ by Borchers, Du, Gao, and Wan (1998). Much less is known about the Steiner ratios for the Euclidean Steiner problem. For $r = 2$ Gilbert, Pollak (1966) conjectured that here $\rho_2 = 2/\sqrt{3}$. They proved this fact for terminal sets containing at most three terminals (cf. Problem 10.10). The proof for general terminal sets, however, turned out to be very difficult and numerous papers were written which contain upper bounds and/or results for restricted terminal sets. The prove of $\rho_2 = 2/\sqrt{3}$ in its full generality is a celebrated result of Du and Hwang (1992). For $r \geq 3$ the values ρ_r are still unknown. A conjecture for the value of ρ_3 can be found in Du, Zhang, and Feng (1991).

The results of Chapter 8.3 imply that in random graphs the lengths of a Steiner minimum tree and that of a minimum spanning tree in the complete distance graph induced by the terminal set are very close to each other. A result of Bern (1988) shows that a similar fact is not true for the Manhattan Steiner problem. More precisely, Bern showed that if k terminals are uniformly distributed on $[0,1]^2$ then the length of a minimum spanning tree divided by $\sqrt{k}$ is almost surely close to a constant β_M. On the other hand, the length of a rectilinear Steiner minimum tree divided by $\sqrt{k}$ is almost surely close to a constant β_S, where $\beta_S \leq \beta_M - 0.0014$.

The result of Problem 10.1 can be strengthened considerably. Cohoon, Richards, Salowe (1990) present a linear time algorithm for this problem. Richards and Salowe (1992) generalized it to an $\mathcal{O}(r^4 k)$ time algorithm if the k terminals lie on the boundary of an r-sided rectilinear convex hull.

The $\mathcal{NP}$-completeness/-hardness results from Theorem 10.4 and Theorem 10.26 are due to Garey and Johnson (1977) and Garey, Graham and Johnson (1977), respectively. As mentioned in the introduction of Section 10.3 polynomial time approximation schemes for the Manhattan (and simultaneously also for the Euclidean) Steiner problem were developed independently by Arora (1998) and Mitchell (1999). Our presentation is based on a Arora's work. Small modifications of these results not only yield a polynomial time approximation scheme for the Euclidean Steiner problem (10.27), but also polynomial time approximation schemes for other geometric optimization problems, like e.g. the Euclidean travelling salesman problem.

Algorithm 10.25 is due to Melzak (1961). Hwang (1986) modified it in order to reduce its running time. During the last years a lot of effort has gone into the development of fast exact algorithms for both the Manhattan and the Euclidean Steiner problem. At the time of writing problems with up to 2000 terminals can be solved in reasonable time. See Warme, Winter, Zachariasen (2000) for details and more references on this topic.

More information about the two geometric Steiner problems considered in this Chapter can be found in the survey by Bern and Graham (1989) and the book by Hwang, Richards, Winter (1992).

Bibliography

Aho, A., Hopcroft, J., and Ullman, J. (1974), *The Design and Analysis of Computer Algorithms*, Addison-Wesley Publishing Company, Reading (MA).

Ahuja, R., Magnanti, T., and Orlin, J. (1993), *Network Flows: Theory, Algorithms, and Applications*, Prentice-Hall, Englewood Cliffs, NJ.

Alford, W., Granville, A., and Pomerance, C. (1994), There are infinitely many Carmichael numbers, *Ann. Math., II. Ser.* **139**, 703–722.

Alon, N. (1986), Eigenvalues and expanders, *Combinatorica* **6**, 83–96.

Appel, K., and Haken, W. (1977), Every planar map is four colorable. Part I. Discharging, *Illinois J. Math.* **21**, 429–490.

Appel, K., Haken, W., and Koch, J. (1977), Every planar map is four colorable. Part II. Reducibility, *Illinois J. Math.* **21**, 491–567.

Arora, S. (1998), Polynomial time approximation schemes for Euclidean traveling salesman and other geometric problems, *J. ACM* **45**, 753–782, Preliminary versions in FOCS'96 and FOCS'97.

Arora, S., Lund, C., Motwani, R., Sudan, M., and Szegedy, M. (1998), Proof verification and the hardness of approximation problems, *J. ACM* **45**, 501–555, Preliminary version in FOCS'92.

Ausiello, G., Crescenzi, P., Gambosi, G., Kann, V., Marchetti-Spaccamela, A., and Protasi, M. (1999), *Complexity and Approximation*, Springer-Verlag, Berlin.

Baker, B. (1983), A new proof for the first fit decreasing algorithm, Bell Laboratories, Murray Hill.

Bellman, R. (1957), *Dynamic Programming*, Princeton University Press, Princeton.

Bellman, R. (1958), On a routing problem, *Quarterly of Applied Mathematics* **16**, 87–90.

Bellman, R., and Dreyfus, S. (1962), *Applied Dynamic Programming*, Princeton University Press, Princeton.

Berman, P., and Karpinski, M. (1999), On some tighter inapproximability results, *26th International Colloquium on Automata, Languages and Programming (ICALP'99)*, Lecture Notes in Computer Science 1644, pp. 200–209.

Berman, P., Fößmeier, U., Karpinski, M., Kaufmann, M., and Zelikovsky, A. (1994), Approaching the 5/4-approximation for rectilinear Steiner trees, *Second Annual European Symposium on Algorithms (ESA '94)*, Lecture Notes in Computer Science 855, pp. 60–71.

Bern, M., and Plassmann, P. (1989), The Steiner problem with edge lengths 1 and 2, *Inform. Process. Lett.* **32**, 171–176.

Bern, M. W., and Graham, R. L. (1989), The shortest-network problem, *Scientific American* **260**, 84–89.

Bern, M. (1988), Two probabilistic results on rectilinear Steiner trees, *Algorithmica* **3**, 191–204.

Bern, M. (1990), Faster exact algorithms for Steiner trees in planar networks, *Networks* **20**, 109–120.

Błazewicz, J., Ecker, K. H., Pesch, E., Schmidt, G., and Węglarz, J. (1996), *Scheduling in Computer and Manufacturing Systems*, Springer-Verlag, Berlin.

Bollobás, B. (1985), *Random Graphs*, Academic Press, London.

Bollobás, B. (1998), *Modern Graph Theory*, Springer-Verlag, Berlin.

Bollobás, B. (1988), The isoperimetric number of random regular graphs, *Europ. J. Combin.* **9**, 241–244.

Borchers, A., and Du, D.-Z. (1997), The k-Steiner ratio in graphs, *SIAM J. Computing* **26**, 857–869.

Borchers, A., Du, D.-Z., Gao, B., and Wan, P. (1998), The k-Steiner ratio in the rectilinear plane, *J. Algorithms* **29**, 1–17.

Borůvka, O. (1926), O jistém problému minimálním, *Práca Moravské Přírodovědecké Společnosti* **3**, 37–58.

Bovet, D., and Crescenzi, P. (1994), *Introduction to the Theory of Complexity*, Prentice Hall, New York.

Buchmann, J. (2000), *Introduction to Cryptography*, Springer-Verlag, Berlin.

Camerini, P. M., Galbiati, G., and Maffioli, F. (1992), Random pseudo-polynomial algorithms for exact matroid problems, *J. Algorithms* **13**, 258–273.

Choukhmane, E.-A. (1978), Une heuristique pour le probleme de l'arbre de Steiner, *RAIRO Rech. Opér.* **12**, 207–212.

Christofides, N. (1976), Worst-case analysis of a new heuristic for the travelling salesman problem, Report 388, Graduate School of Industrial Administration, Carnegie-Mellon University, Pittsburgh.

Coffman Jr., E., Garey, M., and Johnson, D. (1997), Approximation algorithms for bin packing problems: a survey, in: Hochbaum, D. S. (Ed.), *Approximation Algorithms for NP-hard Problems*, PWS Publishing Company, Boston, pp. 46–93.

Cohoon, J. P., Richards, D. S., and Salowe, J. S. (1990), An optimal Steiner tree algorithm for a net whose terminals lie on the perimeter of a rectangle, *IEEE Transactions on Computer-Aided Design* **9**, 398–407.

Cook, S. A. (1971), The complexity of theorem-proving procedure, *3rd Annual Symposium on Foundations of Computer Science*, pp. 431–439.

Cormen, T., Leiserson, C., and Rivest, R. (2001), *Introduction to Algorithms* (2nd edition), The MIT Press, Cambridge, MA.

Courant, R., and Robbins, H. (1941), *What is Mathematics?*, Oxford University Press, London.

Crescenzi, P., Kann, V., Silvestri, R., and Trevisan, L. (1999), Structure in approximation classes, *SIAM J. Comput.* **28**, 1759–1782.

Dantzig, G. (1951), Maximization of a linear function of variables subject to linear inequalities, in: Koopmans, T. (Ed.), *Activity Analysis of Production and Allocation*, Wiley, New York, pp. 339–347.

De Leeuw, K., Moore, E., Shannon, C., and Shapiro, N. (1955), Computability by probabilistic machines, in: Shannon, C., and McCarthy, J. (Eds.), *Automata Studies*, Princeton University Press, Princeton, N.J.

Diestel, R. (1997), *Graph Theory*, Springer-Verlag, Berlin.

Dijkstra, E. (1959), A note on two problems in connexion with graphs, *Numerische Mathematik* **1**, 269–271.

Dreyfus, S. (1969), An appraisal of some shortest-paths algorithms, *Oper. Res.* **17**, 395–412.

Dreyfus, S., and Wagner, R. (1972), The Steiner problem in graphs, *Networks* **1**, 195–207.

Du, D., Zhang, Y., and Feng, Q. (1991), On better heuristic for Euclidean Steiner minimum trees (Extended Abstract), *32nd Annual Symposium on Foundations of Computer Science*, pp. 431–439.

Du, D. (1995), On component-size bounded Steinerg trees, *Discrete Applied Mathematics* **60**, 131–140.

Du, D., and Hwang, F. (1992), A proof of the Gilbert-Pollak conjecture on the Steiner ratio, *Algorithmica* **7**, 121–135.

Edmonds, J. (1967), Systems of distinct representatives and linear algebra, *J. Res. Natl. Bur. Stand.* **71B**, 241–245.

Erdős, P., and Rényi, A. (1960), On the evolution of random graphs, *Publ. Math. Inst. Int. Hungar. Acad. Sci.* **5**, 17–61.

Erickson, R., Monma, C., and Veinott, Jr., A. (1987), Send-and-split method for minimum-concave-cost network flows, *Math. Oper. Res.* **12**, 634–664.

Floren, R. (1991), A note on "A faster approximation algorithm for the Steiner problem in graphs", *Inform. Process. Lett.* **38**, 177–178.

Floyd, R. (1962), Algorithm 97 (SHORTEST PATH), *Communications of the Association for Computing Machinery* **5**, 345.

Ford, L. (1956), Network flow theory, Technical report, Rand Corp., Santa Monica, CA.

Fredman, M., and Tarjan, R. (1987), Fibonacci heaps and their uses in improved network optimization algorithms, *J. ACM* **34**, 596–615.

Fredman, M., and Willard, D. (1993), Surpassing the information theoretic bound with fusion trees, *J. Comput. System Sci.* **47**, 424–436.

Fredman, M., and Willard, D. (1994), Trans-dichotomous algorithms for minimum spanning trees and shortest paths, *J. Comput. System Sci.* **48**, 533–551.

Frieze, A., and McDiarmid, C. (1997), Algorithmic theory of random graphs, *Random Structures & Algorithms* **10**, 5–42.

Gabber, O., and Galil, Z. (1981), Explicit constructions of linear-sized superconcentrators, *J. Comput. System Sci.* **22**, 407–420.

Gabow, H., and Stallmann, M. (1986), An augmenting path algorithm for linear matroid parity, *Combinatorica* **6**, 123–150.

Garey, M., Graham, R., and Johnson, D. (1977), The complexity of computing Steiner minimal trees, *SIAM J. Appl. Math.* **32**, 835–859.

Garey, M., and Johnson, D. (1976), Approximation algorithms for combinatorial problems: an annotated bibliography, in: Traub, J. (Ed.), *Algorithms and Complexity: New Directions and Recent Results*, Academic Press, New York, pp. 41–52.

Garey, M., and Johnson, D. (1977), The rectilinear Steiner tree problem is NP-complete, *SIAM J. Appl. Math.* **32**, 826–834.

Garey, M., and Johnson, D. (1978), Strong NP-completeness results: motivation, examples, and implications, *J. ACM* **25**, 499–508.

Garey, M., and Johnson, D. (1979), *Computers and Intractibility. A Guide to the Theory of NP-Completeness*, W.H. Freeman and Co., San Francisco.

Garey, M., and Johnson, D. (1981), Approximation algorithms for bin packing problems: a survey, in: Ausiello, G., and Lucertini, M. (Eds.), *Analysis and Design of Algorithms in Combinatorial Optimization*, Springer-Verlag, Berlin, pp. 147–172.

Gilbert, E., and Pollak, H. (1966), Steiner minimal trees, *SIAM J. Appl. Math.* **16**, 1–29.

Gill, J. (1977), Computational complexity of probabilistic Turing machines, *SIAM J. Comput.* **6**, 675–695.

Goldreich, O. (1999), *Modern cryptography, probabilistic proofs and pseudo-randomness*, Springer-Verlag, Berlin.

Graham, R. L., and Hell, P. (1985), On the history of the minimum spanning tree problem, *Annals of the History of Computing* **7**, 43–57.

Graham, R. (1966), Bounds for certain multiprocessing anomalies, *Bell System Technical Journal* **45**, 1563–1581.

Graham, R. (1969), Bounds on multiprocessing timing anomalies, *SIAM J. Appl. Math.* **17**, 416–429.

Graham, R., Knuth, D., and Patashnik, O. (1989), *Concrete Mathematics*, Addison-Wesley Publishing Company, Reading (MA).

Gröpl, C., Hougardy, S., Nierhoff, T., and Prömel, H. J. (2001), Approximation algorithms for the Steiner tree problem in graphs, in: Cheng, X., and Du, D.-Z. (Eds.), *Steiner Trees in Industries*, Kluwer Academic Publishers, Norvell, Massachusetts, pp. 235–279.

Grötschel, M., Lovász, L., and Schrijver, A. (1988), *Geometric Algorithms and Combinatorial Optimization*, Springer-Verlag, Berlin.

Habib, M., McDiarmid, C., Ramirez-Alfonsin, J., and Reed, B. (1998), *Probabilistic Methods for Algorithmic Discrete Mathematics*, Springer-Verlag, Berlin.

Hagerup, T. (1998), Sorting and searching on the word ram, *15th Annual Symposium on Theoretical Aspects of Computer Science*, Lecture Notes in Computer Science 1373, pp. 366–398.

Hakimi, S. (1971), Steiner's problem in graphs and its applications, *Networks* **1**, 113–133.

Hall, L. A. (1997), Approximation algorithms for scheduling, in: Hochbaum, D. S. (Ed.), *Approximation Algorithms for NP-hard Porblems*, PWS Publishing Company, Boston, pp. 1–45.

Hanan, M. (1966), On Steiner's problem with rectilinear distance, *SIAM J. Appl. Math.* **14**, 255–265.

Håstad, J. (1997), Some optimal inapproximability results, *Proceedings of the Twenty-Ninth Annual ACM Symposium on Theory of Computing*, pp. 1–10.

Hochbaum, D. S. (1997), Various notions of approximations: Good, better, best, and more, in: Hochbaum, D. S. (Ed.), *Approximation Algorithms for NP-hard Problems*, PWS Publishing Company, Boston.

Hochbaum, D. S., and Shmoys, D. B. (1987), Using dual approximation algorithms for scheduling problems: Theoretical and practical results, *J. ACM* **34**, 144–162.

Hopcroft, J. E., and Karp, R. M. (1973), An $n^{5/2}$ algorithm for maximum matchings in bipartite graphs, *SIAM J. Comput.* **2**, 225–231.

Horowitz, E., and Sahni, S. K. (1978), *Fundamentals of Computer Algorithms*, Pitman, Potomac.

Hwang, F. K. (1976), On Steiner minimal trees with rectilinear distance, *SIAM J. Appl. Math.* **30**, 104–114.

Hwang, F. K. (1986), A linear time algorithm for full Steiner trees, *Oper. Res. Lett.* **4**, 235–237.

Hwang, F. K., Richards, D. S., and Winter, P. (1992), *The Steiner Tree Problem*, Vol. 53 of *Annals of Discrete Mathematics*, North-Holland, Amsterdam.

Ibarra, O. H., and Kim, C. E. (1975), Fast approximation algorithms for the knapsack and sum of subset problems, *J. ACM* **22**, 463–468.

Iwainsky, A., Canuto, E., Taraszow, O., and Villa, A. (1986), Network decomposition for the optimization of connection structures, *Networks* **16**, 205–235.

Janson, S., Łuczak, T., and Ruciński, A. (2000), *Random Graphs*, John Wiley & Sons, Inc., New York.

Jarník, V. (1930), O jistém problému minimálním, *Práca Moravské Přírodovědecké Společnosti* **6**, 57–63.

Johnson, D. (1973), Near-optimal bin packing algorithms, PhD thesis, MIT, Cambridge.

Kahale, N. (1995), Eigenvalues and expansion of regular graphs, *J. ACM* **42**, 1091–1106.

Karger, D., Klein, P., and Tarjan, R. (1995), A randomized linear-time algorithm to find minimum spanning trees, *J. ACM* **42**, 321–328.

Karloff, H., and Zwick, U. (1997), A 7/8-approximation algorithm for MAX 3SAT?, *38th Annual Symposium on Foundations of Computer Science*, pp. 406–415, Remark: According to the authors conjectures 4.3 and 4.5 are now proven.

Karmarkar, N., and Karp, R. (1982), An efficient approximation scheme for the one-dimensional bin-packing problem, *23rd Annual Symposium on Foundations of Computer Science*, pp. 312–320.

Karp, R. (1972), Reducibility among combinatorial problems, in: Thatcher, J., and Miller, R. (Eds.), *Complexity of Computer Computations*, Plenum Press, New York, pp. 85–103.

Khachiyan, L. (1979), A polynomial algorithm in linear programming (in Russian), *Doklady Akademii Nauk SSSR* **244**, 1093–1096, English translation: *Soviet Mathematics Doklady* **20** (1979), pp. 191–194.

Knuth, D. (1973), *The Art of Computer Programming, Vol. 1, Fundamental Algorithms* (2nd edition), Addison-Wesley Publishing Company, Reading (MA).

Korte, B., and Schrader, R. (1981), On the existence of fast approximation schemes, in: Magasarian, O., Meyer, R., and Robinson, S. (Eds.), *Nonlinear Programming 4*, Academic Press, New York, pp. 415–437.

Kou, L., Markowsky, G., and Berman, L. (1981), A fast algorithm for Steiner trees, *Acta Inform.* **15**, 141–145.

Kruskal, J. (1956), On the shortest spanning subtree of a graph and the traveling salesman problem, *Proceedings of the American Mathematical Society* **7**, 48–50.

Kučera, L., Marchetti-Spaccamela, A., Protasi, M., and Talamo, M. (1986), Near optimal algorithms for finding minimum Steiner trees on random graphs, *12th Syposium on Mathematical Foundations of Computer Science*, Lecture Notes in Computer Science 233, pp. 501–511.

L.A. Wolsey (1982), An analysis of the greedy algorithm for the submodular set covering problem, *Combinatorica* **2**, 385–393.

Lawler, E. L., Lenstra, J. K., Rinooy Kan, A. H. G., and Shmoys, D. B. (1993), Sequencing and Scheduling: Algorithms and Complexity, in: Graves, S. C., Rinnooy Kan, A. H. G., and Zipkin, P. (Eds.), *Handbooks in Operations Research and Management Science, Volume 4: Logistics of Production and Inventory*, North-Holland, Amsterdam.

Lawler, E. (1976), *Combinatorial Optimization: Networks and Matroids*, Holt, Rinehart and Winston, New York.

Lenstra, H. (1983), Integer programming with a fixed number of variables., *Math. Oper. Res.* **8**, 538–548.

Levin, A. (1971), Algorithm for the shortest connection of a group of graph vertices, *Soviet Math. Dokl.* **12**, 1477–1481.

Lichtenstein, D. (1982), Planar formulae and their uses, *SIAM J. Comput.* **11**, 329–393.

Lovász, L. (1978), The matroid matching problem, *Algebraic Methods in Graph Theory*.

Lovász, L. (1979), On determinants, matchings and random algorithms, in: Budach, L. (Ed.), *Fundamentals of Computing Theory*, Vol. 9, pp. 565–574.

Lovász, L., and Plummer, M. (1993), *Matching Theory* (2nd edition), North-Holland Publications.

Mayr, E. W., Prömel, H. J., and Steger, A. (Eds.) (1998), *Lectures on Proof Verification and Approximation Algorithms*, Lecture Notes in Computer Science 1367, Springer-Verlag, Berlin.

Mehlhorn, K. (1988), A faster approximation algorithm for the Steiner problem in graphs, *Inform. Process. Lett.* **27**, 125–128.

Melzak, Z. (1961), On the problem of Steiner, *Canad. Math. Bull.* **4**, 143–148.

Micali, S., and V.V.Vazirani (1980), An $O(V^{1/2}E)$ algorithm for finding maximum matching in general graphs, *21st Annual Symposium on Foundations of Computer Science*, pp. 17–27.

Miller, G. L. (1976), Riemann's hypothesis and tests for primality, *J. Comput. System Sci.* **13**, 300–317.

Mitchell, J. S. B. (1999), Guillotine subdivisions approximate polygonal subdivisions: A simple polynomial-time approximation scheme for geometric TSP, k-MST, and related problems, *SIAM J. Comput.* **28**, 1298–1309.

Motwani, R., and Raghavan, P. (1995), *Randomized Algorithms*, Cambridge University Press.

Nemhauser, G. (1966), *Introduction to Dynamic Programming*, Wiley, New York.

Oxley, J. (1992), *Matroid Theory*, Oxford University Press.

Papadimitriou, C. (1994), *Computational Complexity*, Addison-Wesley, New York.

Papadimitriou, C., and Yannakakis, M. (1991), Optimization, approximation, and complexity classes, *J. Comput. System Sci.* **43**, 425–440.

Plesnik, J. (1981), A bound for the Steiner problem in graphs, *Math. Slovaca* **31**, 155–163.

Pratt, V. R. (1975), Every prime has a succinct certificate, *SIAM J. Comput.* **4**, 214–220.

Preparata, F. P., and Shamos, M. I. (1985), *Computational Geometry: An Introduction*, Springer-Verlag, Berlin.

Prim, R. (1957), Shortest connection networks and some generalizations, *Bell Systems Technical Journal* **36**, 1389–1401.

Prömel, H. J., and Steger, A. (2000), A new approximation algorithm for the Steiner problem, *J. Algorithms* **36**, 89–101.

Rabin, M. O. (1963), Probabilistic automata, *Information and Control* **6**, 230–245.

Rabin, M. O. (1980), Probabilistic algorithms for testing primality, *Journal of Number Theory* **12**, 128–138.

Rabin, M. O., and Vazirani, V. (1989), Maximum matchings in general graphs through randomization, *J. Algorithms* **10**, 557–567.

Rado, R. (1957), Note on independence functions, *Proceedings of the London Mathematical Society* **7**, 300–320.

Richards, D. S., and Salowe, J. S. (1991), A simple proof of Hwang's theorem for rectilinear Steiner minimal trees, *Ann. Oper. Res.* **33**, 549–556.

Richards, D. S., and Salowe, J. S. (1992), A linear-time algorithm to construct a rectilinear Steiner minimal tree for k-extremal point sets, *Algorithmica* **7**, 247–276.

Rivest, R. L., Shamir, A., and Adleman, L. (1978), A method for obtaining digital signatures and public key cryptosystems, *Communications of the Association for Computing Machinery* **21**, 120–126.

Robertson, N., Sanders, D. P., Seymour, P. D., and Thomas, R. (1997), The four colour theorem, *J. Combin. Theory*, Ser. B **70**, 2–44.

Robins, G., and Zelikovsky, A. (2000), Improved Steiner tree application in graphs, *Proceedings of the 11th Annual ACM-SIAM Symposium on Discrete Algorithms*, pp. 770–779.

Rothkopf, J. (1966), Scheduling independent tasks on parallel processors, *Management Science* **12**, 347–447.

Sahni, S. K., and Gonzales, T. F. (1976), P-complete approximation algorithms, *J. ACM* **23**, 555–565.

Sahni, S. K. (1976), Algorithms for scheduling independent tasks, *J. ACM* **23**, 116–127.

Schrijver, A. (1986), *Theory of Linear and Integer Programming*, John Wiley & Sons Ltd., Chichester.

Schwartz, J. (1980), Fast probabilistic algorithms for verification of polynomial identities, *J. ACM* **27**, 701–717.

Sedgewick, R. (1998), *Algorithms* (2nd edition), Addison-Wesley, Reading (MA).

Shamos, M. I., and Hoey, D. (1975), Closest-point problems, *16th Annual Symposium on Foundations of Computer Science*, pp. 151–162.

Simchi-Levi, D. (1994), New worst-case results for the bin-packing problem, *Naval Research Logistics* **41**, 579–585.

Solovay, R., and Strassen, V. (1977), A fast Monte-Carlo test for primality, *SIAM J. Comput.* **6**, 84–85, Erratum: *SIAM J. Comput.* **7**, 1978, 118.

Steger, A. (2002), Approximability of NP-Optimization Problems, in: Reed, B., and Sales, C. L. (Eds.), *Algorithms and Combinatorics*, Springer-Verlag, to appear.

Stinson, D. (1995), *Cryptography: Theory and Practice*, CRC Press, Boca Raton.

Takahashi, H., and Matsuyama, A. (1980), An approximate solution for the Steiner problem in graphs, *Math. Jap.* **24**, 573–577.

Tarjan, R. E. (1975), Efficiency of a good but not linear set union algorithm, *J. ACM* **22**, 215–225.

Tarjan, R. E. (1985), Amortized computational complexity, *SIAM J. Algebraic Discrete Methods* **6**, 306–318.

Tarjan, R. (1983), *Data structures and network algorithms*, CBMS-NSF Regional Conference Series in Applied Mathematics **44**, SIAM, Philadelphia.

Thimm, M. (2001), On the approximability of the Steiner tree problem, *26th Symposium on Mathematical Foundations of Computer Science*, Lecture Notes in Computer Science 2136, pp. 678–689.

Thorup, M. (1997), Undirected single source shortest paths in linear time, *37th Annual Symposium on Foundations of Computer Science*, pp. 12–21.

Trevisan, L. (1997), *Reductions and (Non)-Approximability*, Ph.D. thesis, Computer Science Department, University of Rome "La Sapienza".

Tutte, W. T. (1947), The factorization of linear graphs., *J. London Math. Soc. (2)* **22**, 107–111.

van der Waerden, B. (1931), *Modern Algebra*, Springer-Verlag, Berlin.

van Leeuwen, J. (1990), Graph Algorithms, in: van Leeuwen, J. (Ed.), *Handbook of Theoretical Computer Science, Volume A*, Elsevier, Amsterdam, pp. 525–631.

Vazirani, V. (1994), A theory of alternating paths and blossoms for proving correctness of the $O(\sqrt{V}E)$ general graph maximum matching algorithm, *Combinatorica* **14**, 71–109.

Warme, D., Winter, P., and Zachariasen, M. (2000), Exact algorithms for plane Steiner tree problems: a computational study, in: Du, D.-Z., Smith, J., and Rubinstein, J. (Eds.), *Advances in Steiner Trees*, Kluwer Academic Publishers, Norvell, Massachusetts, pp. 81–116.

Warshall, S. (1962), A theorem on boolean matrices, *J. ACM* **9**, 11–12.

Welsh, D. J. A. (1976), *Matroid Theory*, Academic Press, London.

Whitney, H. (1935), On the abstract properties of linear dependence, *American Journal of Mathematics* **57**, 509–533.

Widmayer, P. (1986a), Fast approximation algorithms for Steiner's problem in graphs, Habilitationsschrift, Institut für Reine und Angewandte Informatik und Formale Beschreibungsverfahren, Universität Karlsruhe.

Widmayer, P. (1986b), On approximation algorithms for Steiner's problem in graphs, in: Tinhofer, G., and Schmidt, G. (Eds.), *Proceedings of the International Workshop on Graph-Theoretic Concepts in Computer Science*, Lecture Notes in Computer Science 246, pp. 17–28.

Woeginger, G. J. (1999), When does a dynamic programming formulation guarantee the existence of an FPTAS?, *Proceedings of the Tenth Annual ACM-SIAM Symposium on Discrete Algorithms*, pp. 820–829.

Yannakakis, M. (1994), On the approximation of maximum satisfiability, *J. Algorithms*, 475–502.

Zelikovsky, A. (1993), An 11/6- approximation algorithm for the network Steiner problem, *Algorithmica* **9**, 463–470.

Zelikovsky, A. (1996), Better approximation bounds for the network and Euclidean Steiner tree problems, Technical Report CS-96-06, Department of Computer Science, University of Virginia.

Index

Symbol Index

Milnor's Textbook on Dynamics

John Milnor
Dynamics in One Complex Variable
Introductory Lectures

2. ed. 2000. viii, 257 pp. Softc. € 26,00   ISBN 3-528-13130-6

Contents: Chronological Table - Riemann Surfaces - Iterated Holomorphic Maps - Local Fixed Point Theory - Periodic Points: Global Theory - Structure of the Fatou Set - Using the Fatou Set to study the Julia Set - Appendices

This text studies the dynamics of iterated holomorphic mappings from a Riemann surface to itself, concentrating on the classical case of rational maps of the Riemann sphere. It is based on introductory lectures given by the author at Stony Brook, NY, in the past ten years. The subject is large and rapidly growing. These notes are intended to introduce the reader to some key ideas in the field, and to form a basis for further study. The reader is assumed to be familiar with the rudiments of complex variable theory and of two-dimensional differential geometry, as well as some basic topics from topology. The exposition is clear and enriched by many beautiful illustrations.

Abraham-Lincoln-Straße 46
65189 Wiesbaden
Fax 0611.7878-400   Stand 1.10.2001. Änderungen vorbehalten.
www.vieweg.de   Erhältlich im Buchhandel oder im Verlag.
vieweg